Edoardo Favari · Franca Cantoni
Editors

Complexity and Sustainability in Megaprojects

MeRIT Workshop 2022

Springer

Editors
Edoardo Favari
DIG
Politecnico di Milano
Milan, Italy

Franca Cantoni
Dipartimento di Scienze economiche e sociali
Università Cattolica del Sacro Cuore
Piacenza, Italy

ISSN 2366-2557 ISSN 2366-2565 (electronic)
Lecture Notes in Civil Engineering
ISBN 978-3-031-30878-9 ISBN 978-3-031-30879-6 (eBook)
https://doi.org/10.1007/978-3-031-30879-6

This Springer imprint is published by the registered company Springer Nature Switzerland AG
The registered company address is: Gewerbestrasse 11, 6330 Cham, Switzerland

Preface

Megaprojects are an area of study that lends itself to multidisciplinarity. A megaproject must be studied from the perspectives of economics, sociology, human resource management, law, and engineering. All these disciplines are of equal value and equal importance to the success of a megaproject.

The experience of the authors of this volume prompts us to ask some questions: is on-time and on-budget so important if value is not generated for stakeholders? what role do stakeholders play in the success of a megaproject? what is the recognized role of sustainability in the implementation of a megaproject?

This volume encompasses the reflections of several authors who work in universities, are engaged in consultancy, and follow the different phases of implementation megaprojects. Its ultimate goal is to take a tiny step beyond what has traditionally been known and recognized.

The first part of the volume aims to collect the contributions gathered under the macro theme "The edge of megaprojects research" while section two illustrates some case studies.

If we want to find a common thread among the various contributions, we can do so by searching for the most recurring terms. These are:

- Management for stakeholder
- Stakeholder engagement
- Project success factors
- New iron triangle
- Project transformation
- From a linear approach to chaos
- From global to local
- New metrics for performance management
- Change of paradigm
- Change of approach
- New competencies
- Value creation

In a nutshell, the contents of the chapters can be summarized as follows.

Chapter 1 "How Circular Economy Can Contribute to the Sustainability of Megaprojects: An Overview of Research Hotspots and Developing a Research Agenda" written by *Shams Esfandabadi, Cottafava, Corazza, and Scagnelli* aims to present a general overview of the main research themes and hotspots within the literature on megaprojects and provide the potential of the circular economy in supporting the megaprojects.

Chapter 2 "Mega-projects and Social Impact Evaluation: The Difficult (Un)Balanced Inclusion of Social Needs in Current Practices" written by *Caccialanza, De Nito, Canonico, and Favari* highlights that the better understanding of the interactions between megaproject and social impact assessment disciplines is crucial, and an effective implementation of a multi-dimensional perspective has been positively correlated with project success and failures avoidance.

Chapter 3 "Megaproject Cost Growth: A Stakeholder Perspective" *Gil*—using the case of High-Speed 2, a new railway under planning and construction in the UK—argues that the root cause of cost growth is the limitations of project appraisal methods to formally recognize the social and environmental benefits that society demands from capital investments, which are likely to go above and beyond the law and regulations.

Chapter 4 "Understanding Project Stakeholders Management as a Key Driver for Managing Complexity Within Mega-projects" authored by *Arcuri, Armenia, Di Nauta Iandolo, and Simone* aims at discussing the impacts of projects' structure and dynamics on their performances, with a specific focus on the influence of stakeholder management.

In Chapter 5 "The Concept of Sustainable Infrastructure: A Content Analysis of Construction Companies Reports," the authors *Gordano, Torchia, and Corazza* argue that the variety of definitions and assessment frameworks used to categorize sustainability components by academics and professional bodies call for the need to further investigate and reflect on the concept.

Chapter 6 "Integrating Risk and Stakeholder Management in Complex Mega-projects: A Multilayer Network Analysis Approach" written by *Castelblanco, Fenoaltea, De Marco, Demagistris, Petruzzi, and Zeppegno* aims to constitute the first effort to develop multilayer analysis to unravel behaviors that are hidden in unidimensional analysis.

Chapter 7 "Stakeholder Involvement in the Development of Megaprojects: A Dedicated Framework for Transport Infrastructure" authored by *Cellerino and Mancini* remarks the development of an applicable model for stakeholder inclusion in transport infrastructure.

Chapter 8 "Overtaking the Traditional Leadership Style in Megaprojects" written by *Virtuani, Barabaschi and Cantoni* makes us reflect on the fact that shift the focus from a traditional concept of leadership acknowledged to a specifical individual within a project to an organization-wide, aware, and sustainable (WAS) leadership exerted by key people.

Chapter 9 "Communications in Megaprojects: Two Projects, Two Approaches" by *Clemente* illustrates the difficulties that may be caused by communications in a megaproject in the Middle East and compares the communication management choices for two megaprojects in the Arabic peninsula.

Chapter 10 "Knowledge Discovery Framework for Decision Support Systems in Tendering, Cost Analysis and Construction Phases" from *Mirarchi, Simeone, and Pavan* presents a DSS knowledge framework highlighting the key data sources and the techniques that may be applied to extract knowledge contained in the different data sources (drawings, reports, etc.) that constitutes the basis of design and construction activities.

Chapter 11 "Practice Enterprise and MOOCs in the Design and Implementation of Megaprojects. Some Lessons from European Projects" by *Bianchi* describes the use of didactics based on Practice Enterprise (PE) and Massive Open Online Courses (MOOCs) experimented in some Tempus and Erasmus projects.

Chapter 12 "Extended Reality (XR) as an Interaction Tool for Digital Twins: Application to Safety Management in Construction Megaprojects" written by *Muñoz La Rivera, Atencio, and Mora-Serrano* integrates safety management processes and actions through XR interaction environments linked to the digital twins of construction.

Chapter 13 "Towards the Integration Between Construction Projects and the Organization: The Connections Between BIM and Enterprise Architecture" written by *Atencio, Muñoz-La Rivera, and Mancini* seeks to identify the connections between BIM and EA through a literature review showing the advantages and synergies of their combined application.

Chapter 14 "Pillar Two and Tax Competition: What is the Future of Tax Incentives?" by *Arginelli and Reboli* analyzes, from a theoretical perspective, the evolution of the use of tax incentive and considers the foreseeable implications that the Pillar Two Proposal could have in their respect.

Chapter 15 "Megaproject, Taxation and the (New) Relationship Between the Italian Revenue Agency and Taxpayers" by *Allena* proposes to verify how the Italian legal system allows for a preventive dialogue between tax authorities and taxpayers in order to prevent situations of tax uncertainty that may characterize a given economic transaction.

Chapter 16 "Managing Inflation Risk in Megaprojects: Contract Law and International Best Practices" written by *Zecchin* deals with the problem of the current inflation's level in megaprojects, in order to avoid that it leads to the failure of megaprojects.

Chapter 17 "Megaproject, ESG and Taxation" written by *Purpura* investigates the relationship between megaprojects, environmental sustainability, sustainable development, ESG factors, and taxation.

Chapter 18 "Communication Screen Shot in Projects of Multicultural Diversity" written by *Nassar* collects the pros and cons of working in a multicultural environment and provides a literature review of the hottest managerial topics related to multiculturalism and diversity through the teamwork.

Chapter 19 "A New Path for Green Hydrogen" written by *Gobbi, Bongiorni, Bricchi, and Cantoni* describes that the idea is to build up a compact system to be placed in containers able to produce and store green hydrogen.

Chapter 20 "Infrastructure and Value Creation for Local Systems: The Case of Brescia-Padua A4" authored by *Rizzi, Turci, and Favari* presents the development of the Brescia-Padua A4 highway in the last 70 years, in parallel with the socio-economic evolution of the provinces where the highway was built ("Alta Padana").

Chapter 21 "Implementation of a Project Management Office at the Service of a Mega Mobility Project: The Case of A4 Highway/HSL Corridor Brescia-Padua" written by *Motter Caregnatto and Masoli* describes the fundamental role of an external Project Management Office (PMO) focused on the smooth delivery of the project, a pioneering decision for the Italian infrastructure market.

Chapter 22 "Risk Analysis in Private Building Projects: A Pilot Study in Chile" written by *Castelblanco, Mesa, and Serra* aims for the identification, assessment, and allocation of the risks in private building projects in Chile.

Chapter 23 "Case Study: The Olona Valley Masterplan. Soft Mobility Infrastructures as Innovative Strategies for the Sustainable and Adaptive Regeneration of a Territory" written by *Fregni and Sermasi* highlights a masterplan that brings together economic subjects, representatives of the public administration, researchers and innovative companies and local communities toward a vision of the future based on resilience, adaptability, and the rediscovery of the intrinsic potential of the territories.

Chapter 24 "Sustainability Challenges of High-Speed Railway Megaprojects from a Systems Thinking Lens" by *Shams Esfandabadi, Cottafava, Corazza, and Scagnelli* sheds light on the significance of considering both short-term and long-term effects of the activities directly or indirectly linked with megaprojects and confirms systems thinking as an appropriate approach to analyze the sustainability issues of these projects.

Contents

The Edge of Megaproject Research

How Circular Economy Can Contribute to the Sustainability of Megaprojects: An Overview of Research Hotspots and Developing a Research Agenda

Zahra Shams Esfandabadi, Dario Cottafava, Laura Corazza, and Simone Domenico Scagnelli

Abstract Megaprojects with high levels of complexity and innovation face many sustainability challenges, including environmental concerns, economic resources, and social impacts over decades. Due to the large scale of these projects and their significant consequences for the communities, scientific production focusing on different aspects of megaprojects has been increasingly developed. This research aims to first, present a general overview of the main research themes and hotspots within the literature on megaprojects, and second, provide the potential of the circular economy (CE) in supporting the megaprojects. To do so, a systematic review using keyword co-occurrence analysis was conducted on the megaproject scientific production. The results revealed five major research themes in megaproject-related research, including (i) construction and built environment, (ii) mega events and urban environment, (iii) risk management, (iv) project management, and (v) sustainability and governance. Moreover, challenges and prospects of incorporating the CE in megaprojects were provided and discussed. The findings shed light on the body of knowledge of megaproject research and can support policymakers, officials, and practitioners involved in the megaprojects toward the CE transition.

Keywords Mega-project · Built environment · Urban environment · Change of approach · Change of paradigm · Stakeholder engagement · End of life

Z. Shams Esfandabadi (✉) · D. Cottafava · L. Corazza · S. D. Scagnelli
Department of Management, Corso Unione Sovietica, University of Turin, 218 Bis, 10134 Torino, Italy
e-mail: zahra.shamsesfandabadi@unito.it

D. Cottafava
e-mail: dario.cottafava@unito.it

L. Corazza
e-mail: laura.corazza@unito.it

S. D. Scagnelli
e-mail: s.scagnelli@ecu.edu.au

S. D. Scagnelli
School of Business and Law, Edith Cowan University, 270 Joondalup Dr, Joondalup 6027, Australia

E. Favari and F. Cantoni (eds.), *Complexity and Sustainability in Megaprojects*, Lecture Notes in Civil Engineering 342,
https://doi.org/10.1007/978-3-031-30879-6_1

1 Introduction

The sustainability of megaprojects due to their significant impacts on the natural environment (e.g., biodiversity and ecological processes), and economic and social developments (e.g., public health, poverty eradication, and immigrant settlement) has gained momentum [1, 2]. Megaprojects, which are generally defined as large-scale and complex projects typically costing over US$1 billion, have a key role in the development of societies as they provide the crucial infrastructure required by a country for its economic development [3, 4]. During past decades, several pieces of research have addressed specific aspects of megaprojects, such as project management [5] and its critical success factors [6], megaprojects' complexity [7–9], and social responsibility within the context of megaprojects [10, 11]. Furthermore, the concern of sustainability in megaprojects, highlighting their significant impacts on social and economic developments as well as the natural environment, has been investigated and discussed in the literature [1]. More remain to be done in the context of the social and environmental impacts of megaprojects. One could think of the World Cup Qatar 2022, or the infrastructural legacy of several Olympic games with their white elephants.

A unique and comprehensive classification for megaprojects does not exist, yet, and various classifications have been presented by researchers and organizations in this regard (for instance, Thacker et al. [12] and German Cooperation and UNEP [13]). Even though the majority of megaprojects consist of physical infrastructure, any other collaborative or capital project that requires knowledge, skills, or resources exceeding what is readily or conventionally available to the key participants is also a megaproject [14]. In this vein, redevelopment of urban settings [15], Universal EXPO [16], the Olympic Games [17], huge aerospace programs such as the Apollo Programs of the United States [18], and international infrastructural long-term policy projects, such as the Belt and Road Initiative [19] are some other megaprojects to be named as examples. Regardless of the type of megaprojects, environmental concerns, as a part of sustainability challenges, are associated with the activities linked with these projects [20, 21].

The circular economy (CE), with a particular focus on the waste hierarchy, intends to close the supply chain loops as much as possible toward making a sustainable and zero-waste environment [22]. The environment-economy interaction requires a focus on waste resulting from the processing of resources, and therefore, the interlinkages between the economy and environment would be circular rather than linear [23]. Despite the potential role the CE can play in making megaprojects more sustainable in their various lifecycle stages from the planning or design phase to the decommissioning phase, this role has been overlooked in the literature. Therefore, this research aims at highlighting the footprint of the CE in the published scientific literature regarding megaprojects and analyzing further potential contributions that CE can make in turning megaprojects more sustainable.

The remainder of this article is organized as follows. Section 2 describes the search protocol for data collection and analysis. The main research hotspots within the literature on megaprojects are disclosed in Sect. 3. The potentials of the CE to incorporate in megaprojects are discussed as a research agenda for megaprojects governance towards the CE transition in Sect. 4. Finally, Sect. 5 concludes the remarks of the research along with presenting the research limitations.

2 Methodology

The research began with an attempt to scrutinize the literature on the CE footprint in megaprojects to identify the CE-related activities reflected in the scientific literature on megaprojects. In this regard, first, the "circular economy" and "megaproject", as the main two keywords of this research, were employed to catch the relevant articles from the target literature. On this basis, different writing forms and synonyms of these two keywords were used to design the following search string: ("megaproject*" OR "mega project*" OR "mega-project*" OR "sustainable infrastructure*" OR "mega infrastructure*" OR "mega-infrastructure*" OR "major project*" OR "major-project*" OR "critical infrastructure*" OR "critical-infrastructure*") AND ("circular economy" OR "circular-economy" OR "closed-loop suppl*" OR "closed loop suppl*" OR "circular business*" OR "circular transaction*"OR "end of life" OR "end-of-life" OR "re-us*" Or "reus*" OR "refurbish*" OR "repair*" OR "recycl*"). Although the focus of this review is on megaprojects, the terms 'sustainable infrastructure' and 'critical infrastructure' was also used in the designed search string to ensure capturing the articles that have focused on a megaproject but used sustainable infrastructure or critical infrastructure to point at it in their topic. Besides, concentrating on the 5Rs strategies in the CE (i.e., reuse, refurbish, repair, recycle, and reduce) all these strategies are used in the search string except 'reduce', which returns a huge number of irrelevant records due to the generality of its application in different fields. Although running the search string in both the Web of Science (WoS) and Scopus databases returned 316 peer-reviewed research and review articles, screening these articles and carefully scrutinizing the potentially most relevant articles, we found that the literature still lacks sufficient and effective research on the link between the CE and megaprojects.

To overcome this issue, in the next step, a two-stage process was considered. First, a science mapping approach by using the keyword-based analysis adopted from Ranjbari et al. [24] was used to explore the main research hotspots reflected in the megaprojects literature. And second, based on the available academic and grey literature, the potential contribution of CE strategies to the CE transition in megaprojects was highlighted.

To this end, a systematic literature review using a structured search string was conducted on the megaprojects research in the Scopus database. In this vein, the following search string was designed to search in the title of the articles: ("megaproject*" OR "mega project*" OR "mega-project*" OR "mega infrastructure*" OR "mega-infrastructure*" OR "large project*" OR "major project*" OR "big project*") on October 7, 2022. The search was limited to articles, conference papers, book chapters, and reviews in the English language. As a result, a total of 1701 articles remained as the final sample for conducting the keyword-based analysis.

A keyword co-occurrence analysis was conducted on the authors' keywords (3103 keywords) of the articles (1701 articles). In order to increase the reliability of the analysis, a data cleaning was performed on the keywords in a reasonable manner through (i) removing some general keywords that do not convey insightful meaning based on the main focus of the present study, such as case study, country names, systematic literature review, etc., (ii) unifying different styles of spelling as well as plural and singular forms of the keywords, such as mega-project, mega-projects, megaprojects, mega project, mega projects, and (iii) removing parentheses and merging full forms and their abbreviations. Consequently, a total of 2977 keywords were used for keyword co-occurrence analysis employing the VOSviewer software version 1.6.16 developed by van Eck and Waltman [25]. The keyword co-occurrence analysis has been widely used by scholars in the literature to uncover the main research themes and hotspots of the body of knowledge in different contexts and domains related to megaprojects, such as sustainability of megaprojects [1], and project success research [26].

3 Results

The keyword co-occurrence analysis was conducted on the authors' keywords of the articles within the dataset. On this basis, having cleaned the keywords as described in the methodology section, a total of 2977 keywords were used as input for keyword co-occurrence analysis. Then, the co-occurrence network of the keywords within the megaproject literature was constructed as visualized in Fig. 1. Due to the large number of keywords, in order to increase the readability of the network map, only keywords with more than 3 occurrences were included in the network. As can be seen in Fig. 1, five major research hotspots of megaproject-related research include (i) construction and built environment, (ii) mega events and urban environment, (iii) risk management, (iv) project management, and (v) sustainability and governance. These are the main subjects of research in the megaproject domain, which have considered different aspects of megaprojects through a wide range from industry-related issues to environmental and sustainability concerns.

The research in the construction and built environment cluster has been mainly focused on barriers, success factors, and analysis frameworks of integrating building information modeling and lean construction practices [27–29], and minimizing the environmental impacts of megaprojects [30]. Scholars within the mega events and urban environment have been paying special attention to megaprojects such as the Olympic games in terms of public perceptions, feelings, and approval regarding these megaprojects [31], impacts on urban policies [32], tourism and regional mobility [33], and restructuring of urban governance [34]. The sustainability and governance of megaprojects has appeared as another major research area in the megaproject

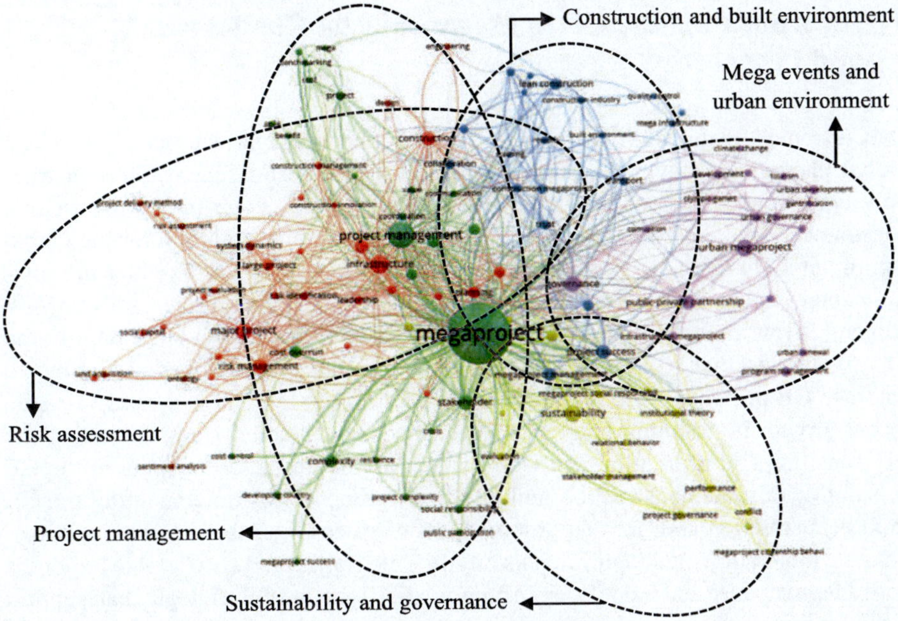

Fig. 1 Keyword co-occurrence network in the megaproject literature

literature. In this regard, the literature has covered some areas of research, such as the involvement of stakeholder networks in megaprojects [35–37], the efficacy of trust for the governance of uncertainty [38], inter-organizational conflict management [39] and the role of contractual flexibility in conflict and project success [40], contracts from the legal perspective [41], and sustainability of megaprojects [42, 43].

Project management as a critical area in megaprojects has been investigated from different points of view, including inter alia, developing decision-making supporting systems [44], competitiveness, cost minimization, and techno-economic analysis of megaprojects [45], benchmarking studies [46], and alignment between the project management process and business strategies [47]. Moreover, risk assessment as a critical factor in megaproject success has been widely examined in the literature focusing on objectives, namely developing risk assessment frameworks of public–private partnerships in megaprojects [48], identifying megaprojects supply chains infectious risks [49], reconciling complexity [50], and social risk perception and assessment [51].

4 Incorporating the CE in Megaprojects: Challenges and Prospective

Due to rapid urbanization worldwide, construction megaprojects have become a global phenomenon over the last two decades [52]. In this regard, the construction industry and its associated activities are significantly contributing to resource consumption and the generation of waste, and air and noise pollution for the global community. Therefore, there is an urgent need to embrace the CE to reduce the environmental impacts of the existing linear production and consumption system [53]. Çimen [54] in an inclusive study on the construction and built environment in the CE highlighted that (i) although the construction industry has been struggling to embrace CE principles, the research in this area still is in its infancy stage despite recent growth in academic interest, (ii) the two most studied areas, including the CE transition and promotion and waste valorization, are major hotspots of the CE research within the construction and built environment, (iii) implementing the CE in the construction and built environment needs to consider the diversity of stakeholders, motivation, and their impacts on lifecycle stages, and (iv) flexible buildings with adaptive reuse and modularity in combination with the CE strategies can support increasing efficiency and human well-being.

Delivering a sustainable megaproject is a challenging complex issue with various difficulties in effectiveness and interaction mechanisms. Besides, managing all participants in megaprojects in terms of their contributions and interests requires an inclusive participative planning approach [55]. Environmental concerns have become one of the main factors influencing the capacity, performance, and impact of megaprojects in urban settings, highlighting the importance of fostering sustainability in cities and regions in megaprojects even though urban leaders face pressures from investors [56]. The contribution of different stakeholders is required to create a circular ecosystem; however, differences and conflicts of these stakeholders can potentially lead to hampering the long-term sustainability of the ecosystem [57]. Thus, this complexity calls for more effective governance to ensure economic, social, and environmental sustainability. Arfaoui et al. [58] in a study to understand how the governance arrangement matters in the CE identified three governance arrangement factors that are crucial for successfully implementing the CE as follows: first, facilitative leadership and sufficient technological expertise along with the participation of all relevant stakeholders in the governance process; second, self-organizing activities and information sharing to support effective collaboration; and finally, applying constitutional rules by actors while elaborating collective choice rules that allow a bottom-up and collaborative governance. Nevertheless, the contribution of the CE to the sustainability of megaprojects lacks sufficient studies. As a potential research direction for further developments in the future, the megaproject research community needs to investigate how the CE can support the governance and sustainability of megaprojects.

Urban megaevents, such as the Olympic games have massive effects on the surrounding environment and can change cities at different scales through a wide range of changes from clear and calculable to extensive and unpredictable changes over a long period of time [59]. Clearly, the huge and costly effects of such projects in long term could pose a burden on the national economy and the social community. Megaprojects such as the Olympic games bring many challenges for governments which need significant efforts and capabilities to properly manage the post-Olympic era in the host countries. For instance, productively utilizing the extensive infrastructure provided for the Olympic games, the lack of governmental long-term and inclusive planning, and bureaucracy are just a few challenges for host countries [60]. Moreover, environmental damages and their restoration processes, and the sustainability of social infrastructure built for such mega events are also serious issues to be taken into account [61]. The CE strategies, such as reuse, recover, repurpose, remanufacture, repair, reduce, and rethink as a sustainable solution to deal with challenges posed by megaprojects such as the Olympic games and their long-term effects deserve to be more investigated and incorporated in the agenda of these megaprojects. Hence, the governments, decision-makers, and stakeholders of such projects are still struggling with effectively managing the long-term effects of these megaprojects. The potential benefits of implementing the CE strategies to increase efficiency, maximize retention of the economic value of materials and products, and close production loops [62] call for more action, research, and support by all stakeholders of megaprojects in the future.

Studies on various megaprojects worldwide have shown that within the domain of project management, several challenges in economic, social, and political aspects exist, which require proper risk management to overcome [46]. For instance, in the case of the construction of Panama Canal in South America, elements such as lack of risk management, inappropriate design, inadequate planning, non-availability of finances, and health and safety of the staff resulted in the failure of a French team in the completion of the planned project within time and budget [46]. This is while successful megaprojects have three elements in common, namely total alignment, clear strategic vision, and adapting to complexity [63]. The complexity within the megaprojects includes both technical and social aspects [64], however, rebound dynamics should be strictly considered in project risk management planning [65]. Furthermore, due to the increasing attention towards sustainability and sustainable development during the past decade [66], the sustainability of megaprojects throughout their lifecycle has also achieved more importance [42]. Depending on the project type and the project lifecycle stage, the CE may have the potential to contribute to the sustainability of a megaproject. Nevertheless, when it comes to project management and risk management, the potential to use the CE approaches might arise that are not or are less considered in the available documents. For instance, sharing economy, as an enabler for circularity [67] can be used to reduce the need for investing in resources to be used or consumed by the staff and workers (e.g. through sharing vehicles, office consumable products, etc.) and avoid waste generation; if applicable and cost-effective, renewable sources of energy can be consumed during the construction phase of the project to save non-renewable resources and avoid pollution; and circular designs and reused or recycled materials can be used to build temporary cabins and accommodation places for the staff and workers.

5 Conclusion

This research explored the main research themes and hotspots of megaprojects within the academic literature. Having conducted a keyword co-occurrence analysis on a total of 2977 authors' keywords from the target literature, five main research areas of megaproject were identified and presented. On this basis, megaproject research has been mainly focused on construction and built environment, mega events and urban environment, risk management, project management, and sustainability and governance issues. Moreover, the potential of the CE to support the sustainability and governance of megaprojects was discussed.

This research has a limitation. In order to analyze the research on megaproject, only articles published in the WoS and Scopus databases were considered. Hence, the useful materials coming from grey literature, such as technical reports, industry-based documents, websites, and media, have been neglected which may provide additional insights for further research.

References

1. Wang G, Wu P, Wu X, Zhang H, Guo Q, Cai Y (2020) Mapping global research on sustainability of megaproject management: a scientometric review. J Clean Prod 259:120831. https://doi.org/10.1016/j.jclepro.2020.120831
2. Corazza L, Cottafava D, Torchia D (2022) Toward sustainable infrastructural megaprojects. In: W. Leal Filho et al (ed) SDGs in the European Region, Implementing the UN Sustainable Development Goals – Regional Perspectives. Springer Nature, Switzerland, 2022 ISBN 9783030912611.
3. Corazza L, Torchia D, Cottafava D, Tipaldo G (2022) Considering the social and economic implications of infrastructure megaprojects: theoretical contributions, practical challenges and managerial implications. In: The Impact of Corporate Social Responsibility, Routledge, pp. 134–160 ISBN 9781003182276
4. Flyvbjerg B (2014) What you should know about megaprojects and why: an overview. Proj Manag J 45:6–19. https://doi.org/10.1002/pmj.21409
5. Rolstadås A, Tommelein I, MortenSchiefloe P, Ballard G (2014) Understanding project success through analysis of project management approach. Int J Manag Proj Bus 7:638–660. https://doi.org/10.1108/IJMPB-09-2013-0048
6. Cepeda DM, Sohail M, Ogunlowo OO (2018) Understanding the critical success factors for delivery of megaprojects in Colombia. Proc Inst Civil Eng. Manage Procure Law 171(2):45–57. https://doi.org/10.1680/jmapl.17.00005
7. Nyarirangwe M, Babatunde OK (2021) Megaproject complexity attributes and competences: lessons from IT and construction projects. Int J Inf Syst Proj Manag 7:77–99. https://doi.org/10.12821/ijispm070404
8. Qiu Y, Chen H, Sheng Z, Cheng S (2019) Governance of institutional complexity in megaproject organizations. Int J Proj Manag 37:425–443. https://doi.org/10.1016/j.ijproman.2019.02.001
9. Chapman RJ (2016) A framework for examining the dimensions and characteristics of complexity inherent within rail megaprojects. Int J Proj Manag 34:937–956. https://doi.org/10.1016/j.ijproman.2016.05.001
10. Zhou Z, Mi C (2017) Social responsibility research within the context of megaproject management: trends, gaps and opportunities. Int J Proj Manag 35:1378–1390. https://doi.org/10.1016/j.ijproman.2017.02.017

11. Corazza L, Cottafava D, Torchia D (2022) Stakeholder identification and mapping in highly contested megaprojects: interpreting stakeholders ecosystems through relational stakeholder theory. SSRN Electron J 1913:0–2. https://doi.org/10.2139/ssrn.4165936
12. Thacker S, Adshead D, Fay M, Hallegatte S, Harvey M, Meller H, O'Regan N, Rozenberg J, Watkins G, Hall JW (2019) Infrastructure for sustainable development. Nat Sustain 2:324–331. https://doi.org/10.1038/s41893-019-0256-8
13. German Cooperation and UNEP Sustainable Infrastructure Tool Navigator.
14. Sykes A (1990) Macro projects: status, prospects, and the need for international cooperation. Technol Soc 12:157–172. https://doi.org/10.1016/0160-791X(90)90006-X
15. Dogan E, Stupar A (2017) The limits of growth: a case study of three mega-projects in Istanbul. Cities 60:281–288. https://doi.org/10.1016/j.cities.2016.09.013
16. Shakirova S (2015) Country image of Kazakhstan: from stereotype and critique to positive national branding. J Eastern Eur Central Asian Res 2(1):78. https://doi.org/10.15549/jeecar.v2i1.78
17. Randeree K (2014) Reputation and mega-project management: lessons from host cities of the olympic games. Change Manage Int J 13(2):1–7. https://doi.org/10.18848/2327-798X/CGP/v13i02/50750
18. Horwitch M (1990) From unitary to distributed objectives. Technol. Soc. 12(2):173–195. https://doi.org/10.1016/0160-791X(90)90007-Y
19. Daye M, Charman K, Wang Y, Suzhikova B (2020) Exploring local stakeholders' views on the prospects of China's belt & road initiative on tourism development in Kazakhstan. Curr. Issues Tour. 23:1948–1962. https://doi.org/10.1080/13683500.2019.1700941
20. Li K, Zhu C, Wu L, Huang L (2013) Problems caused by the three gorges dam construction in the yangtze river basin: a review. Environ Rev 21:127–135. https://doi.org/10.1139/er-2012-0051
21. Norese MF, Corazza L, Cottafava D (2021) Environmental monitoring of the socio-economic components of the impact of a mega project. In: Doumpos M, Ferreira FAF, Zopounidis C (eds) Multiple Criteria Decision Making for Sustainable Development, pp 131–149, ISBN 9783030892777
22. Ranjbari M, Saidani M, ShamsEsfandabadi Z, Peng W, Lam SS, Aghbashlo M, Quatraro F, Tabatabaei M (2021) Two decades of research on waste management in the circular economy: Insights from bibliometric, text mining, and content analyses. J Clean Prod 314:128009. https://doi.org/10.1016/j.jclepro.2021.128009
23. Pearce DW, Turner RK, Turner RK (1990) Economics of natural resources and the environment. Johns Hopkins University Press
24. Ranjbari M, ShamsEsfandabadi Z, Ferraris A, Quatraro F, Rehan M, Nizami A-S, Gupta VK, Lam SS, Aghbashlo M, Tabatabaei M (2022) Biofuel supply chain management in the circular economy transition: an inclusive knowledge map of the field. Chemosphere 296:133968. https://doi.org/10.1016/j.chemosphere.2022.133968
25. van Eck NJ, Waltman L (2010) Software survey: VOSviewer, a computer program for bibliometric mapping. Scientometrics 84:523–538. https://doi.org/10.1007/s11192-009-0146-3
26. He Q, Wang T, Chan APC, Li H, Chen Y (2019) Identifying the gaps in project success research. Eng Constr Archit Manag 26:1553–1573. https://doi.org/10.1108/ECAM-04-2018-0181
27. Evans M, Farrell P (2021) Barriers to integrating building information modelling (BIM) and lean construction practices on construction mega-projects: a Delphi study. Benchmarking 28:652–669. https://doi.org/10.1108/BIJ-04-2020-0169
28. Evans M, Farrell P, Mashali A, Zewein W (2021) Critical success factors for adopting building information modelling (BIM) and lean construction practices on construction mega-projects: a Delphi survey. J. Eng. Des. Technol. 19:537–556. https://doi.org/10.1108/JEDT-04-2020-0146
29. Evans M, Farrell P, Zewein W, Mashali A (2021) Analysis framework for the interactions between building information modelling (BIM) and lean construction on construction mega-projects. J. Eng. Des. Technol. 19:1451–1471. https://doi.org/10.1108/JEDT-08-2020-0328
30. Gharehbaghi K, McManus K, Robson K (2019) Minimizing the environmental impacts of mega infrastructure projects: Australian public transport perspective. J. Eng. Des. Technol. 17:736–746. https://doi.org/10.1108/JEDT-12-2018-0223

31. Streicher T, Schmidt SL, Schreyer D, Torgler B (2020) Anticipated feelings and support for public mega projects: hosting the olympic games. Technol. Forecast. Soc. Change 158:120158. https://doi.org/10.1016/j.techfore.2020.120158
32. Souliotis N, Sayas J, Maloutas T (2014) Megaprojects, neoliberalization, and state capacities: assessing the medium-term impact of the 2004 Olympic games on Athenian urban policies. Environ. Plan. C Gov. Policy 32:731–745. https://doi.org/10.1068/c1136r
33. Ni MC (2017) Mega infrastructure developments in Macao in light of increasing tourism and regional mobility: a review. Worldw. Hosp. Tour. Themes 9:289–299. https://doi.org/10.1108/WHATT-02-2017-0010
34. Carrión A (2016) Megaprojects and the restructuring of urban governance. Lat Am Perspect 43:252–265. https://doi.org/10.1177/0094582X15579900
35. Romestant F (2020) Sustainability agencing: the involvement of stakeholder networks in megaprojects. Ind Mark Manag 89:535–549. https://doi.org/10.1016/j.indmarman.2019.09.005
36. Ninan J, Mahalingam A, Clegg S (2019) External stakeholder management strategies and resources in megaprojects: an organizational power perspective. Proj Manag J 50:625–640. https://doi.org/10.1177/8756972819847045
37. Ninan J, Mahalingam A, Clegg S (2020) Power and strategies in the external stakeholder management of megaprojects: a circuitry framework. Eng. Proj. Organ. J. 9:1–20. https://doi.org/10.25219/epoj.2020.00104
38. Wang D, Fu H, Fang S (2020) The efficacy of trust for the governance of uncertainty and opportunism in megaprojects: the moderating role of contractual control. Eng Constr Archit Manag 27:150–167. https://doi.org/10.1108/ECAM-09-2018-0409
39. Liu C, Cao J, Wu G, Zhao X, Zuo J (2022) How does network position influence project performance in Chinese megaprojects? An inter-organizational conflict-based perspective. Int J Confl Manag 33:448–474. https://doi.org/10.1108/IJCMA-05-2021-0080
40. Wu G, Zhao X, Zuo J, Zillante G (2018) Effects of contractual flexibility on conflict and project success in megaprojects. Int J Confl Manag 29:253–278. https://doi.org/10.1108/IJCMA-06-2017-0051
41. Zecchin F (2020) Megaprojects contracts: between international practices and Italian law. In: Favari E, Cantoni F (eds) Megaproject Management: A Multidisciplinary Approach to Embrace Complexity and Sustainability. Springer International Publishing, Cham, pp 107–122. https://doi.org/10.1007/978-3-030-39354-0_8
42. Chen Z, Agapiou A, Li H, Xu Q (2021) A TRIZ approach to reliable megaproject sustainability. Front. Built Environ. 7:1–18. https://doi.org/10.3389/fbuil.2021.650699
43. Thees H (2020) Towards local sustainability of mega infrastructure: reviewing research on the new silk road. Sustain 12:1–35. https://doi.org/10.3390/su122410612
44. Deelstra Y, Nooteboom SG, Kohlmann HR, Van Den Berg J, Innanen S (2003) Using knowledge for decision-making purposes in the context of large projects in The Netherlands. Environ Impact Assess Rev 23:517–541. https://doi.org/10.1016/S0195-9255(03)00070-2
45. Köberle AC, Garaffa R, Cunha BSL, Rochedo P, Lucena AFP, Szklo A, Schaeffer R (2018) Are conventional energy megaprojects competitive? Suboptimal decisions related to cost overruns in Brazil. Energy Policy 122:689–700. https://doi.org/10.1016/j.enpol.2018.08.021
46. Sandhu M, Khan A (2017) Benchmarking project management dimensions at the lapse of a century. Benchmark An Int J 24:1675–1689. https://doi.org/10.1108/BIJ-03-2016-0043
47. Alsudiri T, AlKaraghouli W, Eldabi T (2013) Alignment of large project management process to business strategy: A review and conceptual framework. J Enterp Inf Manag 26:596–615. https://doi.org/10.1108/JEIM-07-2013-0050
48. AkbariAhmadabadi A, Heravi G (2019) Risk assessment framework of PPP-megaprojects focusing on risk interaction and project success. Transp Res Part A Policy Pract 124:169–188. https://doi.org/10.1016/j.tra.2019.03.011
49. Zhao N (2019) Managing interactive collaborative mega project supply chains under infectious risks. Int J Prod Econ 218:275–286. https://doi.org/10.1016/j.ijpe.2019.06.008

50. Chapman P, Quang C (2021) Major project risk management: reconciling complexity during delivery with the inside view in planning. Eng Proj Organ J 10:1–28. https://doi.org/10.25219/epoj.2021.00104
51. Xiao-zheng C, Xiang-ming H, Study of social risk assessment on major projects based on risk perception. In: Proceedings of the 2013 International Conference on Management Science and Engineering 20th Annual Conference Proceedings, IEEE, 2013, pp. 1900–1905
52. Yi H, Chan APC, Le Y, Xu Y, Shan M (2016) Developing a program organization performance index for delivering construction megaprojects in china: fuzzy synthetic evaluation analysis. J Manage Eng 32(4):05016007. https://doi.org/10.1061/(ASCE)ME.1943-5479.0000432
53. Christensen TB (2022) Closing the material loops for construction and demolition waste: the circular economy on the island Bornholm. Denmark Resour Conserv Recycl Adv 15:200104. https://doi.org/10.1016/j.rcradv.2022.200104
54. Çimen Ö (2021) Construction and built environment in circular economy: a comprehensive literature review. J Clean Prod 305:127180. https://doi.org/10.1016/j.jclepro.2021.127180
55. Erbach M (2021) Managing the economic sustainability of the belt and road initiative by applying pragmatic identity matching PrIM. J. Innov. Entrep. 10:33. https://doi.org/10.1186/s13731-021-00158-4
56. del Cerro Santamaría G, Megaprojects (2021) Ecology and the Environment: Dimensions of Sustainability in Global Cities, pp 135–144
57. Moggi S, Dameri RP (2021) Circular business model evolution: stakeholder matters for a self-sufficient ecosystem. Bus Strateg Environ 30:2830–2842. https://doi.org/10.1002/bse.2716
58. Arfaoui N, Le Bas C, Vernier MF, Vo LC (2022) How do governance arrangements matter in the circular economy? Lessons from five methanation projects based on the social-ecological system framework. Ecol Econ 197:107414. https://doi.org/10.1016/j.ecolecon.2022.107414
59. Ryu C, Kwon Y (2016) How do mega projects alter the city to be more sustainable? Spatial changes following the seoul cheonggyecheon restoration project in South Korea. Sustainability 8:1178. https://doi.org/10.3390/su8111178
60. Kasimati E (2015) Post-olympic use of the olympic venues: the case of greece. Athens J. Sport. 2:167–184. https://doi.org/10.30958/ajspo.2-3-3
61. Lee JW (2019) A winter sport mega-event and its aftermath: a critical review of post-olympic PyeongChang. Local Econ 34:745–752. https://doi.org/10.1177/0269094219889608
62. Morseletto P (2020) Targets for a circular economy. Resour Conserv Recycl 153:104553. https://doi.org/10.1016/j.resconrec.2019.104553
63. Shenhar A, Holzmann V (2017) The three secrets of megaproject success: clear strategic vision, total alignment, and adapting to complexity. Proj Manag J 48:29–46. https://doi.org/10.1177/875697281704800604
64. Kardes I, Ozturk A, Cavusgil ST, Cavusgil E (2013) Managing global megaprojects: complexity and risk management. Int Bus Rev 22:905–917. https://doi.org/10.1016/j.ibusrev.2013.01.003
65. Kemp D, Worden S, Owen JR (2016) Differentiated social risk: rebound dynamics and sustainability performance in mining. Resour Policy 50:19–26. https://doi.org/10.1016/j.resourpol.2016.08.004
66. Ranjbari M et al (2021) Three pillars of sustainability in the wake of COVID-19: a systematic review and future research agenda for sustainable development. J Clean Prod 297:126660. https://doi.org/10.1016/j.jclepro.2021.126660
67. Ranjbari M, Morales-Alonso G, Carrasco-Gallego R (2018) Conceptualizing the sharing economy through presenting a comprehensive framework. Sustainability 10:2336. https://doi.org/10.3390/su10072336

Mega-projects and Social Impact Evaluation: The Difficult (Un)Balanced Inclusion of Social Needs in Current Practices

Andrea Caccialanza, Ernesto De Nito, Paolo Canonico, and Edoardo Favari

Abstract The understanding of the interactions between megaproject and social impact assessment disciplines is crucial and an effective implementation of a multi-dimensional perspective has been positively correlated with project success and failures avoidance. Current literature aims to encompass this role, extending the analysis of the impacts to the broader concept of stakeholders and local communities impacted by the project. The aim of the Authors is that these preliminary findings can inspire further and deeper research on these topics, looking for and integrated approach to include all of them into a cohesive framework for managing the social pillar in megaprojects management. The literature review leads to the identification of three different research areas related to the issue of the evaluation of the megaprojects from a social perspective: a first one related to the issue of power and equality that looks coherent with the critical management agenda both from a methodological and theoretical point of view; a second one related to the concept of social space as a construct to adopt in order to enlarge the alternatives in the evaluation process; the third one that depicts the role of technology and social media to manage stakeholders.

Keywords Megaprojects · Social impact · Sustainability · Literature review

1 Introduction

There is always more attention with reference to Megaprojects, sometimes referred as "Major Projects/Programs" or "Large Engineering Projects" [35, 36], that are

A. Caccialanza (✉)
Università Cattolica del Sacro Cuore, Piacenza, Italy
e-mail: andrea.caccialanza@unicatt.it

E. De Nito
Università degli Studi di Salerno, Fisciano, Italy

P. Canonico
Universita' degli studi di Napoli Federico II, Napoli, Italy

E. Favari
Politecnico di Milano, Milano, Italy

E. Favari and F. Cantoni (eds.), *Complexity and Sustainability in Megaprojects*, Lecture Notes in Civil Engineering 342,
https://doi.org/10.1007/978-3-031-30879-6_2

identified by Flyvbjerg [16] as "large-scale, complex ventures that typically cost more than 1 billion US Dollars, take many years to build, involve multiple public and private stakeholders, are transformational, and impact millions of people", even if some scholars noted how the economic constraint (1 billion US Dollars) could be not an universal standard, including in the mega-project label also smaller projects [26, p. 1351].

The increasing number of large capital investments worldwide makes megaprojects a field of rising interest both for researchers and practitioners. The underlying reasons for this growth lie predominantly both in the overall growth of the world's population [54] and the increased complexity of large investments [7, 8] that encompass the economic viability of a project [52, 61].

In particular, megaprojects are usually large-scale interventions, such as infrastructures, that require billions of dollars investments [4] and long-term perspective to be analysed [17].

The respect of the constraints in terms of budget, time and benefit associated to the "Iron law" [18], in the last decades has been associated with sustainability issues. The sustainability perspective requires a multi-dimensional approach not limited to the triple-bottom line (TBL) approach proposed by Elkington [14] that focuses on the economic, environmental and social pillars.

The economic pillar has been widely analysed not only as determinant to succeed in the on time and on budget implementation, but also considering the economic related externalities [41]. Research on the sustainability of megaprojects issues is still not only limited to few geographical areas (such as Eastern China, Western Europe, and Eastern United States) [62], but also suffers the frequent underestimation of the other two non-economic pillars [59, 60].

Since the emergence of the environmental dimension in the last few decades [46, 47], recently we are witnessing an increased awareness for issues of social responsibility [58]. This is "the reason why" we consider the better understanding of the interactions between megaproject and social impact assessment disciplines crucial [2, 10, 15].

The effective implementation of a multi-dimensional perspective has been positively correlated with project success and failures avoidance [5, 6, 32], in particular referring to the social one [40]. Moreover, great effort has been played in the understanding of possible causes of megaprojects failures. Megaprojects have often an underestimated effect on local communities and residents that might be affected by a (temporary or permanent) change in terms of their livelihood conditions or life quality [1, 44]. Crisis are more frequently related to project managers responsibility and forecasting ability [56], in the cases of the industrialization development focus, with repercussions on the liveability of the local area [57].

Literature has pointed out that there are even cases where an environmental impact analysis has been carried out, the overall project performance has not been effective due to the exclusion of the other social dimension within the perimeter of analysis: [23] studied the disrespect of procedures or increase the dissatisfaction of local participants as possible causes of failure, but this is particularly evident in also the case of "Not In My Back Yard" (NIMBY) infrastructures or facilities with respect to local residents [34, 58] proposed as research agenda the proposal of "tools that we might call social-design, helping to include social aspects in the project". This issue seems to be still urgent and a current open field for future works, also according to the evolution of the concept of project manager.

The emerging role of the project manager as innovation leader within the organization has been discussed by the literature [27, 28]. Current literature aims to encompass this role, extending the analysis of the impacts to the broader concept of stakeholders and local communities impacted by the project. Recently, [3] propose also to extend this role also to large scale megaprojects focusing on "the dyadic relationship between influence and the social power sources of strategic project managers".

The project managers' social power is correlated with the overall project performance [9, 51] and positively influenced by expertise and communication abilities [31].

The interactions between the megaproject and social impact literature will then be analysed and discussed in this chapter. In the next section, the research methodology used to systematize the existing literature will be introduced, then in the discussion is proposed a classification of different focus around themes discussed by the research.

2 Research Methodology

In order to propose a comprehensive overview of the existing literature of the social impact of megaprojects a review of the existing contributed has been performed.

The applied mixed methodology is the result of the combination of a systematic and inductive approach. This strategy allows us to combine "the findings of qualitative and quantitative studies within a single systematic review to address the same overlapping or complementary review questions." [22] and stimulate the debate within a multidisciplinary topic finding new research gaps. Within the models proposed by literature [39] for a reproducible and comparable selecting process [49], it has been chosen and adapted the one proposed by Mok et al. (2015).

At first we used for the systematic approach the Scopus database to select those works that contains in the title and abstract "Social Impact" AND "Megaproject" as keywords (202). The first filtering phase limited the research to the articles, books, books chapters, and reviews published in the English language and at final stage of publication. 164 documents were found.

The second filtering stage limited the research to the last 5 years of publications to collect the most recent contributions and research issues, obtaining the final list of 82 selected papers. For those papers it has been done an abstract and full text screening, including 27 papers in the final list considered for the analysis. As output of this first step, we defined 3 research areas with the aim of analysing and discussing them.

Within each research area we eliminated the papers that were not really coherent with the mega-project literature and starting from the remaining papers we enlarged our literature review applying an inductive process that aimed to include relevant reference and cross-references present in the papers selected. This phase also comprehended the full text screening of the 4 systematic literature review previously published on the megaproject multi-disciplinary approach. This stage was the most conservative and inclusive in terms of preservation of both classical and contemporary project management literature.

3 Discussion

As introduced in the methodology our first result was to identify 3 different research areas related to the issue of the evaluation of the megaprojects from a social perspective: 1) The role of power and social issues; 2) Infrastructure and social space; 3) People engagement and endorsement.

In this section we will discuss briefly these topics identifying possible inspirational new trends and research gaps (for the managerial literature).

3.1 *The Role of Power and Social Issues*

As argued by del Cerro Santamaria [12] megaprojects are often related to a logic of growth, development, competitiveness and prosperity, and are shaped as public–private institutional arrangements participated by elites and pro-growth coalitions at various spatial scales that are able to provide the necessary funding and expert knowledge to implement these complex structures [12, p. 263].

In this sense they are the expression of large economic forces, which carry out all the influence they can in order to implement the mega project. Looking at this phenomenon from a different point of view, it is possible to argue that Mega projects, are often characterized by political lobbying, privatization and institutional fragmentation which can result in the marginalization of vulnerable communities as well as concerns regarding the transparency and accountability of the decision-making process [25, p. 294).

No matter what the perspective is, what looks clear it is the role power in order to make these mega-projects possible. Mega-project sponsors have all the interest in presenting the positive relevance of the project itself, giving a partial picture of the situation, underlying the short-term effects (especially the positive economic effects) and undermining the less visible long term social impacts. This uneven power distribution could imply a strong conflict between different players (in particular the project sponsor organizations, both private and public, and the local community) [29, 33, 37, 50]. A typical case is related to the choice of presenting the economic effects of the mega-project adopting a quantitative (and technical) approach, without taking into consideration the qualitative side (often related to the social impact), that is much more ambiguous and difficult to "measure" and to represent. A clear example of this kind of behaviour is discussed by [25] who identified four variables that influence this relationship (and this conflict): choice/alternatives, transparency or access to information, integration of local knowledge and power sharing. According to their study, in the official documents there are different choices in order to underline the positive effects of the mega-project, hiding the potential negative social consequences. The language adopted is always positive and offers only one-sided perspective: there is no space for an alternative choice. There is not much information regarding the project, especially in relation to funding and social impacts. the “expert” knowledge does not take into consideration the local knowledge and experience, excluding local communities from participating in the debate [11, 13]. The power distribution is clearly inequal and the linguistic expressions show the real power of the proponents [38, 42, 43].

Another stream of research to include in this area is related to the Policy Effects of Resistance against Mega-Projects in Latin America [45]. Even if they look at the issue from a political perspective, We would argue that considering these ideas also in the more strict managerial debate is really useful in order to reflect on stakeholder management: in some contexts (as the extractive business in America Latina) social movements work in opposition to the dominant players who want to carry out mega-projects. It is really interesting to reflect on [45, p. 28] how do social movement actors with various interests and power resources form contending coalitions around policies and institutions relating to mega-development projects?

This debate on power is not very present in the managerial literature (expect for some papers related to the issue of conflict management or social responsibilities, and the articles are published in journals belonging to different areas such as geography, environmental issues or engineering), even if it is coherent with a more critical approach to management studies. In this sense it would be very useful to take into consideration these issues (politics, uneven power, social movements, etc.) in order to reflect on the mega-projects’ evaluation.

3.2 Infrastructure and Social Space

Another interesting research area could be represented by the effects of mega-projects in urban infrastructure development. According to Lee et al. [30, p. 2] the long-term social outcomes of urban infrastructure development are rarely assessed (ex-ante) or evaluated (ex post), especially in terms of how the respective interests are or will be affected by the multi-scale spatial changes generated by the projects.

In other terms what is really interested in terms of mega-project evaluation is two-fold: first of all, in many sectors there is still a gap in terms of how to evaluate a project (ex-ante and ex-post); the other significant issue is the temporal span (what are the long-term consequences and how it is possible to evaluate them). In their paper they [30] propose a framework in order to give a contribution in terms of evaluation. In their view it is important to take into account micro and macro levels and the interests related to transport development and to urban development. They consider a long term perspective indicating some specific proxy measures. In their perspective, and in coherence with other scholars [21, 55], the concept of social space is a key construct in order to evaluate the development of urban infrastructure, both at a macro scale (i.e., change in spatial structure of a city) and a micro scale (i.e., change in physical environment in a local area), shows the need for an integrated and broader approach to urban transport infrastructure development that addresses social consequences of multi-scale spatial changes induced by projects, including those that are unexpected or unintended. In addition, Lee et al. [30, p. 16] show how the expansion of urban transport networks by itself does not guarantee increased accessibility or quality of life for the whole of the society and underline the need of a more complex system evaluation where all the spatial changes that occur across a city should be considered as well as the unexpected or unintended negative impacts of spatial transformation at the local neighbourhood scale [20, 21, 48].

3.3 People Engagement and Endorsement

Mega projects are typically criticised for their significant impact on communities and on society as a whole. There is then little surprise that they stimulate and challenge public opinion mobilizing wide citizens' reactions. Given their ability to increasingly attract public attention, they may exert an influence on public participation to collective choices and on political decision making.

The diffusion of social media has of course reinforced such phenomenon. Issues about engagement are crucial with reference to the so called Nimby attitude characterizing the negative social response to unwanted facilities that in recent times unfold especially through social media. Social media have brought about new challenges of widescale information diffusion.

According to Wang et al. [58], focusing on how information related to megaprojects may circulate and develop, it is possible to put forward a series of socio-economic implications of the diffusion of information model. Government administration should understand what kind of social conflict may be prompted by the project. Government should therefore follow a progressive update which should include not only the casualties and economic impacts of the event, but also subsequent framing of the phenomenon, such as results of project evaluation. Important challenges are originated by the interactions between the megaproject and engaged citizens. They [58] argue that government administration should enhance its online influence through micro-blog operations, communicating with elites, defined as the important groups who are extremely active in the communication and directly promote the emergence of peaks of public concerns, to guide the positive online discourse. More generally, with reference to megaprojects, social media may represent a powerful communication channel to deliver public voice integrating public participation in the real world [53].

4 Conclusion and Research Agenda

Far from being an exhaustive literature review, this paper has the main goal to look for new insights and research areas in the managerial domain.

Starting from a simple keyword research, we found three interesting areas to develop from a managerial perspective. A first one related to the issue of power and equality that looks coherent with the critical management agenda (see for example [19, 24]) both from a methodological and theoretical point of view. A second one related to the concept of social space as a construct to adopt in order to enlarge the alternatives in the evaluation process. Finally, the third one depict the role of technology and social media to manage stakeholders.

The aim of the Authors is that these preliminary findings can inspire further and deeper research on these topics, looking for and integrated approach to include all of them into a cohesive framework for managing the social pillar in megaprojects management.

The Authors' research agenda includes a systematic literature review on the topic a set of questionnaires about the topic to be submitted to a large set of megaprojects' stakeholders in order to validate the findings.

This paper is a first attempt to open up a discussion on different issues (new frameworks that emphasise the social pillar in sustainable project management, the scarce role of local communities in a context of increasing capitals and political concentration, the development of a common and comparable set of indicators that may help the understanding of the value distribution among (local) communities) that are still marginalized in the mega-project literature.

References

1. Abdullah AN, Rahman S (2021) Social impacts of a mega-dam project as perceived by local, resettled and displaced communities: a case study of Merowe Dam. Sudan Econom 9(4):140
2. Atkinson CL (2021) Hydropower, development, and poverty reduction in Laos: promises realized or broken? Asian J Polit Sci 29(1):67–87
3. Badi S, Rocher W, Ochieng E (2020) The impact of social power and influence on the implementation of innovation strategies: a case study of a UK mega infrastructure construction project. Eur Manag J 38(5):736–749
4. Brookes NJ, Locatelli G (2015) Power plants as megaprojects: using empirics to shape policy, planning, and construction management. Utilities Policy 36:57–66
5. Caccialanza A, Cantoni F, Favari E, Mariani C (2022) Proposing a sustainability-centered approach to overcome critical issues in megaprojects. In: Cantoni F, Favari E (eds) Sustainability and Megaproject Development. Routledge, London, pp 25–44. https://doi.org/10.4324/9781003305750-3
6. Cantoni F, Favari E (2022) Sustainability and Megaproject Development. Routledge, London. https://doi.org/10.4324/9781003305750
7. Cantoni F, Pagnone F (2019) Sowing the seeds to overcome the "Iron Law" in Megaprojects, A multidisciplinary approach to embrace complexity and sustainability, Giappichelli, Torino, 1–10, ISBN/EAN 978-88-921-3163-7

8. Cantoni F, Favari E, Pagnone F (2019) Large Engineering Projects. Favouring resilience in increasingly complex environments grafting an adaptive approach to waterfall technique, Impresa e Progetto. Electron J Manage 2, ISSN 1824-3576
9. Carvalho MM, Rabechini R Jr (2017) Can project sustainability management impact project success? An empirical study applying a contingent approach. Int J Project Manage 35(6):1120–1132
10. Choudhury NB, Dey Choudhury SR (2021) Implications for planning of hydroelectric projects in Northeast India: an analysis of the impacts of the Tipaimukh project. GeoJournal 86(4):1999–2019
11. Daye M, Charman K, Wang Y, Suzhikova B (2020) Exploring local stakeholders' views on the prospects of China's belt & road initiative on tourism development in Kazakhstan. Curr Issue Tour 23(15):1948–1962
12. del Cerro SG (2019) Megaprojects, development and competitiveness: building the infrastructure for Globalization and neoliberalism. Athens J Soc Sci 6(4):263–290
13. Dembicz K, Carrillo X (2021) Guanacaste tourist mega-projects local perceptions for the future solutions. Prace i Studia Geograficzne 66(4):119–134
14. Elkington J, Rowlands IH (1999) Cannibals with forks: the triple bottom line of 21st century business. Altern J 25(4):42
15. Esteves AM, Franks D, Vanclay F (2012) Social impact assessment: the state of the art. Impact Assess Project Apprais 30(1):34–42
16. Flyvbjerg B (2014) What you should know about megaprojects and why: an overview. Proj Manag J 45(2):6–19
17. Flyvbjerg B (2014) Megaproject Planning and Management: Essential Readings. Edward Elgar Publishing, Cheltenham and Northampton
18. Flyvbjerg B (2017) Introduction: The Iron Law of Megaproject Management, The Oxford Handbook of Megaproject Management, Chapter 1, 1–18. Oxford University Press, ISBN 100198831102
19. Fraser T, Chapman AJ (2018) Social equity impacts in Japan's mega-solar siting process. Energy Sustain Dev 42:136–151
20. Fu Q (2022) Too big to succeed: mega neighborhoods, depression, and actually existing urban governance. Cities 126:103711
21. Geurs T, Boon W, van Wee B (2009) Social impacts of transport: literature review and the state of the practice of transport appraisal in the Netherlands and the United Kingdom. Trans Rev 29:69–90
22. Harden A (2010) Mixed-methods systematic reviews: Integrating quantitative and qualitative findings. Focus 25:1–8
23. Ho P, Nor-Hisham BMS, Zhao H (2020) Limits of the environmental impact assessment (EIA) in Malaysia: Dam politics, rent-seeking, and conflict. Sustainability 12(24):10467
24. Hodgson D, Cimcil S (eds) (2006). Palgrave MacMillan, New York
25. Hossain SR, Fuller S (2021) Understanding conflict in transport mega-projects: social impacts and power dynamics in the WestConnex project, Sydney. Austr Geogr 52(3):293–313
26. Invernizzi DC, Locatelli G, Brookes NJ (2017) Managing social challenges in the nuclear decommissioning industry: a responsible approach towards better performance. Int J Project Manage 35(7):1350–1364
27. Kapsali M (2011) Systems thinking in innovation project management: a match that works. Int J Project Manage 29(4):396–407
28. Kenny J (2003) Effective project management for strategic innovation and change in an organizational context. Proj Manag J 34(1):43–53
29. Khodeir LM, Nabawy EM (2021) Responsive human resource framework for design and building of mega housing development projects in Egypt. Ain Shams Eng J 12(2):2371–2383
30. Lee J, Arts J, Vanclay F, Ward J (2020) Examining the social outcomes from urban transport infrastructure: Long-term consequences of spatial changes and varied interests at multiple levels. Sustainability 12(15):5907

31. Lindebaum D, Cartwright S (2010) A critical examination of the relationship between emotional intelligence and transformational leadership. J Manage Stud 47(7):1317–1342
32. Locatelli G, Milos M, Kovacevic M, Brookes N, Ivanisevic I (2017) The successful delivery of megaprojects: a novel research method. Proj Manag J 48(5):78–94
33. Ma L, Fu H (2022) A governance framework for the sustainable delivery of megaprojects: the interaction of megaproject citizenship behavior and contracts. J Constr Eng Manag 148(4):04022004
34. MarcelinoSádaba S, González-Jaen LF, PérezEzcurdia A (2015) Using project management as a way to sustainability. From a comprehensive review to a framework definition. Journal of Cleaner Production 99:1–16. https://doi.org/10.1016/j.jclepro.2015.03.020
35. Merrow E.: Industrial Megaprojects: Concepts, Strategies, and Practices for Success. Wiley (2011).
36. Miller R., Lessard D.: The Strategic Management of Large Engineering Projects: Shaping Institutions, Risks, and Governance. MIT Press (2001).
37. Mulholland C, Chan PW, Canning K, Ejohwomu OA (2020) Social value for whom, by whom and when? Managing stakeholder dynamics in a UK megaproject. Proc Inst Civil Eng Manage Procure Law 173(2):75–86
38. NikBakht M, ElDiraby TE (2020) Beyond chatter: profiling community discussion networks in urban infrastructure projects. J Infrastruct Syst 26(3):05020006
39. Pearson A, White H, Bath-Hextall F, Salmond S, Apostolo J, Kirkpatrick P (2015) A: Mixed-methods approach to systematic reviews. JBI Evid Implement 13(3):121–131
40. Pitsis A, Clegg S, Freeder D, Sankaran S, Burdon S (2018) Megaprojects redefined – complexity vs cost and social imperatives. Int J Manag Projects Bus 11(1):7–34. https://doi.org/10.1108/IJMPB-07-2017-0080
41. Platoni S, Timpano F (2020) The economics of mega-projects. In: Favari E, Cantoni F (eds) Megaproject Management: A Multidisciplinary Approach to Embrace Complexity and Sustainability. Springer International Publishing, Cham, pp 43–54. https://doi.org/10.1007/978-3-030-39354-0_4
42. Rajput A, Maqsoom A, Shah SWA, Ullah F, Munawar HS, Rehman MSU, Albattah M (2022) Impact of Political, social safety, and legal risks and host country attitude towards foreigners on project performance of China Pakistan economic corridor (CPEC). Buildings 12(6):760
43. Sankaran S, Clegg S, Müller R, Drouin N (2022) Energy justice issues in renewable energy megaprojects: implications for a socioeconomic evaluation of megaprojects. Int. J. Manag. Projects Bus. 15(4):701–718. https://doi.org/10.1108/IJMPB-06-2021-0147
44. SanetraSzeliga J (2022) Culture and heritage as a means to foster quality of life? The case of Wrocław European capital of culture 2016. Eur Plan Stud 30(3):514–533
45. Silva E, Akchurin M, Bebbington A (2018) Policy effects of resistance against mega-projects in Latin America. Eur Rev Latin Am Caribb Stud Revista Eur Estud Latinoamericanos y del Caribe 106:25–46
46. Silvius AJG, Schipper R, Planko J, van den Brink J, Köhler A (2012) Sustainability in Project Management. Gower Publishing, Farnham
47. Silvius G, Tharp J (eds) (2013). IGI Global, Hersey
48. Sodiq A, Baloch AA, Khan SA, Sezer N, Mahmoud S, Jama M, Abdelaal A (2019) Towards modern sustainable cities: review of sustainability principles and trends. J Clean Prod 227:972–1001
49. Stern C, Lizarondo L, Carrier J, Godfrey C, Rieger K, Salmond S, Apóstolo J, Kirkpatrick P, Loveday H (2020) Methodological guidance for the conduct of mixed methods systematic reviews. JBI Evid Synth 18(10):2108–2118
50. Talbot A, Carter TF (2018) Human rights abuses at the Rio 2016 olympics: activism and the media. Leis Stud 37(1):77–88
51. Thamhain HJ, Gemmill GR (1974) Influence styles of project managers: some project performance correlates. Acad Manag J 17(2):216–224
52. Thounaojam N, Laishram B (2022) Issues in promoting sustainability in mega infrastructure projects: a systematic review. J Environ Planning Manage 65(8):1349–1372

53. Tumasjan A, Sprenger T, Sandner P, Welpe I (2010) Predicting elections with twitter: what 140 characters reveal about political sentiment. Proc. Int. AAAI Conf. Web Soc. Med. 4(1):178–185. https://doi.org/10.1609/icwsm.v4i1.14009
54. United Nations (2022) Department of Economic and Social Affairs, Population Division, World Population Prospects 2022 Ten Key Messages
55. Vecchio G, Porreca R, Rivera D (2020) Socio-spatial concerns in urban mobility planning: insights from competing policies in Quito. Sustainability 12:2923
56. Wang A, Pitsis TS (2020) Identifying the antecedents of megaproject crises in China. Int J Project Manage 38(6):327–339
57. Wang Z, Wu F (2019) In-situ marginalisation: social impact of Chinese mega-projects. Antipode 51(5):1640–1663
58. Wang Y, Li H, Zuo J, Wang Z (2019) Evolution of online public opinions on social impact induced by NIMBY facility. Environ Impact Assess Rev 78:106290
59. Wang G, Wu P, Wu X, Zhang H, Guo Q, Cai Y (2020) Mapping global research on sustainability of megaproject management: a scientometric review. J Clean Prod 259:120831
60. Wu X, Chong HY, Wang G, Li S (2018) The influence of social capitalism on construction safety behaviors: an exploratory megaproject case study. Sustainability 10(9):3098
61. Yang D, He Q, Cui Q, Hsu SC (2020) Non-economic motivations for organizational citizenship behavior in construction megaprojects. Int J Project Manage 38(1):64–74
62. Zeng SX, Ma HY, Lin H, Zeng RC, Tam VW (2015) Social responsibility of major infrastructure projects in China. Int J Project Manage 33(3):537–548

Megaproject Cost Growth: A Stakeholder Perspective

Nuno Gil

Abstract I deploy a stakeholder perspective to move forward the debate on the reasons behind cost growth in 'megaprojects' - the social tools that are designed by humans to produce capital-intensive technology. Using the case of High-Speed 2, a new railway network that is currently under planning and construction in the UK, to illustrate my claims, I argue that cost growth over the time it takes to plan and deliver a megaproject is not isomorphic with bad management neither with dishonesty. Rather, the root cause of cost growth are the limitations of project appraisal methods to formally recognise the wider social and environmental value that society ('us all') expects these capital investments to produce - value which goes above and beyond the threshold necessary for a megaproject to conform to law and existing regulations. This pattern occurs because we approve capital-intensive projects based on a narrow analysis of the user willingness to pay for the production costs, while ignoring wider benefits (and mitigation of disbenefits) that the project needs to produce in order to gain consent to progress from society. As a result of this limitation of project appraisal methods, before a capital investments is sanctioned, we lack a structure of incentives to encourage collaboration from stakeholders that control essential resources for the project to progress. A stakeholder perspective therefore suggests that cost growth over project time reflects a need to renegotiate the value proposition ex-post project approval towards the production of a socially valuable outcome from the perspective of essential stakeholders, a negotiation that, perforce, frustrates our ability to keep a megaproject within the initial budget that was approved.

Keywords megaprojects · value distribution · new stakeholder theory (NST) · purpose

N. Gil (✉)
Alliance Manchester Business School, Manchester, England
e-mail: nuno.gil@manchester.ac.uk

E. Favari and F. Cantoni (eds.), *Complexity and Sustainability in Megaprojects*, Lecture Notes in Civil Engineering 342,
https://doi.org/10.1007/978-3-031-30879-6_3

1 Introduction

Cost growth in large-scale projects set up to produce capital-intensive assets such as railways, airports, defence systems, cross-border pipelines, and power plants, so-called 'megaprojects', has long been an empirical regularity. This was the case in the 1950s when the project management profession got institutionalised, and continues to be the case today. In the last seven decades, however, the governance of megaprojects has evolved in fundamental ways. There are now way more layers of scrutiny over the process that precedes the decision to sanction a capital investment, and equally, there are now a large number of layers of scrutiny over the decision-making process to allocate capital ex-post project approval. As well as this, in the last decades, there has been significant technological progress in digital models to help us anticipate conflicts between different design specialties, and to help the project participants visualise the project outcomes. This is to the extent that today it is not unusual that the indirect costs of a large project (so the costs of project management, IT systems, control systems, and legal fees, but excluding design costs) to represent almost 20% of the cost of project implementation.

A case in point is the High Speed 2 (HS2) project, a public-sector financed project that aims to equip the UK with a new high-speed railway network before the middle of the XXI century, and which since the idea gained traction back in 2009, has seen dramatic cost growth. In this brief essay, using the case of HS2 to illustrate my claims, I argue that cost growth is not isomorphic with bad management or dishonesty. Rather, I claim that cost growth reflects the complex organizational structure that underpins the process of value creation through a megaproject. Critically, a capital-intensive, project-based organization will not be able to make actual progress towards the achievement of its goals unless the organization agrees - ex-post project approval - to renegotiate the value distribution to encourage cooperation from stakeholders that control essential resources to achieve the project goals. In other words, the managers of a megaproject and their organizational sponsors need to agree, after sanctioning an initial capital investment, to widen the purpose of the investment beyond the maximization of economic returns based on user willingness to pay. As the purpose of the project broadens post hoc to address legitimate stakeholder concerns, pressure to let the cost forecast grow goes up, a dynamic that is sine qua non for the project to forge ahead and create value.

In sum, I argue that cost growth does not reflect project underperformance. Rather, cost growth is a symptom of a much more complex problem that is rooted our inability, as of now, to define, measure, and verify the social and environmental gains that society ('us all') demands from a megaproject in order to enable that megaproject to progress - gains that go above and beyond the law and existing regulations that inform the cost–benefit analysis that is used to justify the capital investment in first place. But let's look into the story of HS2 to sharpen our intuition for this novel explanation of megaproject behaviour and performance, rooted in new stakeholder theory and organization economics more generally.

2 HS2: A Story of Cost Growth and Evolution of Purpose

Back in 2009, HS2 Ltd, the arms-length government body that was charged with delivering the new railway network, was launched to great fanfare with the aim of improving connectivity and capacity in the UK's national railway network. The announcement occurred as the UK economy was suffering the effects of the global financial crisis, and policymakers were eager to launch new capital-intensive initiatives that could help reinvigorate the country's economy.

Because the idea for HS2 had been conceptualised within the railway fraternity, notably inside Network Rail, the public agency in charge of managing the UK railway infrastructure, the initial goal for HS2 Ltd was to develop a high-speed railway to tackle a bottleneck in transport capacity as anticipated by the railway industry. Supporting this goal was extensive mainstream cost–benefit analysis (CBA). This analysis, considered by policymakers "the best game in town" for project appraisal, measured the benefits of HS2 largely in terms of user willingness to pay for the new railway (based on travel time savings, crowding benefits) together with some other measurable wider economic benefits (e.g., noise reduction, carbon impact, air quality, agglomeration, labour supply impacts). At the time, the managers of HS2 Ltd and their organizational sponsor (the UK government) recognised that the new railway project could potentially internalise other negative and positive externalities to create more social and environmental value beyond fiduciary duties. To broaden the purpose of HS2 Ltd would be a choice to go above and beyond the law and regulations that constrained capital investment in railways. For example, HS2 Ltd could invest in grandiose stations in the centre of the cities on the route to catalyse urban regeneration; it could build more and longer tunnels to further mitigate the negative impact on the landscape and townscape and on the welfare of local communities; and it could make extra provisions to tackle societal concerns around loss of biodiversity, growing social inequality, and climate change.

Were HS2 Ltd to broaden the purpose of the project beyond its fiduciary duties, the organization would be attending to issues that were being raised by key stakeholders in the surrounding context at informal consultation events, including local authorities of the cities on the route, environmental agencies, interest groups, and local communities – so-called "nonmarket" stakeholders in that these stakeholders control essential resources for a project to go ahead and create economic value, but these actors do not operate in a regime of efficient markets and their property rights maybe ill-defined. But were HS2 Ltd to agree to meet the claims of these stakeholders and commit to widen the purpose of the project, the company would frustrate its ability to demonstrate through mainstream cost–benefit analysis that the capital investment would be "Value for Money" (VfM). This is because VfM was codified as a benefit-to-cost ratio, and while the costs of internalizing more externalities above and beyond the threshold necessary to conform to the law would be substantial (increasing the denominator), there wereno objective data neither robust methods available to quantify the potential benefits (the numerator). Thus, to keep the benefit to cost ratio well above one, the purpose of HS2 Ltd was restricted to build a new railway, while staying within the law and existing regulations.

However, knowing that the purpose of HS2 would need to broaden in the future, HS2 Ltd. did a sensitivity analysis on the CBA, with and without uplifting the cost forecast to build the railway. Uplifting the cost forecast to account for the unpredictable requirements of stakeholders was recommended policy by the UK Treasury, the so-called "optimism bias" policy. To calculate the budget uplift, HS2 Ltd used a highly imperfect method, Monte Carlo simulation, which assumes that, first, all project tasks can be stochastically modelled; and second, it assumes that the realization of the unpredictable requirements of the stakeholders over project time will not change the project task structure – two assumptions that fundamentally contradict empirical evidence suggesting that not only changes in the project scope are par for course in megaprojects, but also that the realization of uncertainty in stakeholder requirements is not a quantifiable risk. Still, by conducting CBA with and without uplifting the cost forecast, HS2 Ltd. was able to produce substantive variation in the "Value for Money" indicator, including producing scenarios without a budget uplift, which suggested that the HS2 project was 'high' value for money.

By 2013, the year when the last cost–benefit analysis was published, based on a more advanced concept design, HS2 managers started formal consultations with the stakeholders to be materially impacted by the future railway - over 70,000. By then, the capital investment had been sanctioned and substantive capital resources had therefore been committed to move forward with the project. Unsurprisingly, ten years later, between 2013 and the time of this writing, the purpose of HS2 Ltd has evolved radically. To progress with the project, HS2 Ltd had to engage in a major renegotiation of the value distribution. This renegotiation has occurred as essential stakeholders such as the local authorities of the cities on the route; owners of site-specific assets; and environmental agencies started to make legitimate claims (from their perspective, at least) on the HS2 design choices and budget. In a nutshell, these stakeholders have been arguing since 2013 that HS2 Ltd needs to allow for much greater creation of societal value. Enabling these stakeholders to make claims on the resources that de jure are controlled by HS2 Ltd is the fact the UK is an open access society where stakeholders can leverage institutions in the context in order to challenge the decisions from public agencies.[1]

Local authorities along the route, for example, have demanded that their voices should be heard. Manchester is a case in point. To keep costs under control to protect the original cost–benefit analysis, HS2 managers and its sponsor proposed a utilitarian station in Salford. This is a vibrant area not far from the Manchester city centre, but nonetheless at least a 15 min commute by light rail. Accessing into Salford would

[1] 'Open access societies' use entry and competition to order the patterns of social organizations that structure human relationships and organizations. They contrast with 'restricted access societies' that use the political system to regulate economic competition and create economics rents, enabling political groups to manipulate economic interests and abuse their power to obtain resources or to coerce others. In North, D. C., Wallis, J. J., & Weingast, B. R. 2009. ***Violence and social orders: A conceptual framework for interpreting recorded human history***. Cambridge: Cambridge University Press.

require not much tunnelling, which would limit the construction costs. Outrightly, however, the local authorities of Greater Manchester pushed back. The local authorities argued that Manchester needs a grandiose city centre station, one that should be fully integrated with the existing station in Piccadilly. As well as this, the Manchester authorities demanded a long tunnel into the city centre station to avoid disruption to the welfare of the local communities and a second station to be added to the route at Manchester airport. As the parties started to negotiate back in 2013, the cost forecast started to grow concomitantly. After 10 years of conversations, the parties have yet not been able to achieve consensus, and recently, they have deferred conflict resolution to a Parliamentary committee, a conflict-resolution structure in the UK that aims to avoid stakeholders resorting to the public courts to resolve planning disputes.

3 Can We Keep Megaprojects Within Budget?

The HS2 case illustrates a novel theoretical perspective on megaprojects that I have been developing in the last decade, jointly with my doctoral students and collaborators, for a forthcoming book Megaprojects and Society: A Theory of Purpose, Value Creation, and Value Distribution. This perspective posits that cost growth in megaprojects is rooted in a renegotiation of the value distribution.[2] So a renegotiation away from the narrow purpose that informed the initial cost–benefit analysis based on which the project was approved towards a broader purpose that commits to produce a socially valuable outcome from the perspective of the stakeholders that control essential resources for the project to go ahead. Seeing this value renegotiation ex-post project approval is essential to understand the behaviour of capital-intensive projects promoted both by public agencies as well as by privately financed actors.

In a nutshell, this theoretical perspective suggests that capital-intensive projects are approved based on an analysis of the benefits that are measurable against the costs that need to be incurred to produce those benefits. But these projects are not the projects that society will then enable to go ahead. Rather, to progress, capital projects need to gain consent from numerous non-market stakeholders through consensus-oriented negotiations. Today, for example, HS2 is only a railway building project in name. De facto, HS2 has become an instrument of economic development, for better and worse. This is to the extent that improving railway capacity is but one of the

[2] Readers interested in this line work can for example look into Gil, N. A. & Baldwin, C. Y. 2013. Sharing design rights: A commons approach for developing infrastructure. Cambridge, MA: Harvard Business School Working Paper; Gil, N. & Pinto, J. K. 2018. Polycentric organizing and performance: A contingency model and evidence from megaproject planning in the UK. ***Research Policy***, 47(4): 717–734; Gil, N. 2022. Megaprojects: a meandering journey towards a theory of purpose, value creation and value distribution. ***Construction Management and Economics***, 40(7–8): 562–584. Gil, N. & Fu, Y. 2022. Megaproject performance, value creation, and value distribution: an organizational governance perspective. ***Academy of Management Discoveries***, 8(2): 224–251; Gil, N. 2023. Cracking the Megaproject Puzzle: A Stakeholder Perspective? International Journal of Project Management, 41 (3) https://doi.org/10.1016/j.ijproman.2023.102455

seven strategic goals of HS2 Ltd. This wider purpose is what essential stakeholders have consented to.

As major projects face demands to widen their purpose, is it inevitable that costs will grow after the capital investment has been approved? The answer is yes. Given it is unlikely the use of cost–benefit analysis as a prerequisite for sanctioning a capital investment will go away, it is also unlikely that we can create a structure of incentives, before a capital investment is approved, to enable managers to engage in deeper forms of stakeholder enfranchisement and consensus-oriented negotiations with essential stakeholders. In other words, today, capital investments in major projects like HS2 are only approved after initially demonstrating that they are "value for money" from a narrow fiduciary perspective. But while we know how to codify, count, and verify economic returns based on user willingness to pay and some other wider economic benefits, there are many other social and environmental benefits that these investments can create to attend to legitimate societal concerns, but which we do not know how to codify and measure. And yet, unless managers enter into compromises with the stakeholders and agree to some of their claims, projects will struggle to progress ex-post the approval of the capital investment.

For example, we have methods and historical data to forecast the number of passengers who will use HS2 and the impact of the new railway network on user productivity; we also have data and methods to quantify some wider economic benefits, such as the impact of the new railway on property prices. But we still lack data and robust methods to measure many potential benefits from internalising externalities with a view to further mitigate negative impacts on local communities, further improve the health, safety, and wellbeing of construction workers; create "good" local jobs, and enable technological spillovers. We also lack consensus on how to measure in money terms the value of mitigating environmental disbenefits such as carbon emissions; threats to biodiversity; landscape impact; and noise pollution. Put simply, we do not know how to monetize the societal value that megaprojects can create by playing the role of moral actor.

And so, the game starts. Project sponsors have no alternative but to exclude upfront the costs of realising societal gains that go above and beyond the threshold necessary to conform to the law in order to make the project "bankable". Eventually, cost forecasts maybe uplifted with large amounts of unallocated funds using imperfect techniques such as stochastic modelling and Reference Class Forecasting to create more conservative budget envelopes; layers of governance are added to control decision-making on capital allocation; and consultants are brought on board to mitigate the "stakeholder risks" and try to keep the project "within budget", a normative marker of high performance. While these efforts have increased dramatically the costs of major projects, the ability of these efforts to keep projects "within budget" is limited, as HS2 and other cases show. Because unless costs grow, the essential stakeholders will not enable the project to progress.

To make matters worse, as costs grow, many actors in society from opponents to a megaproject to the press start to question the 'character' of the managers of the project and of the leaders of the project sponsors. Critics may argue that if costs are growing, it is because managers and the project sponsors were dishonest. As the trust deficit between a project and society widens, extra layers of scrutiny and control are added, with the leaders of the project sponsors depriving the managers of the project of decision-making autonomy, which hurts efficiency and productivity. As costs continue to grow, the leaders of the project sponsors may find it tempting to launch independent reviews to make go/no go decisions and even sack the top managers of the project, decisions which may further undermine efficiency. As well as this, because the benefit to cost ratio doesn't acknowledge the benefits to be appropriated by non-market stakeholders, over time, the "value for money" indicator will appear to worsen. This deterioration of the VfM indicator may cause support for the project to decline in the court of public opinion, putting pressure on managers and policymakers to slash parts of the project scope that may nonetheless be essential to realise the full economic value of the project - just like we are now seeing with the controversial decision to eliminate the Eastern Leg of HS2. So, major projects get trapped in a vicious circle of incomplete budgets, cost growth, mistrust, and value destruction to sustain a narrative that the project is 'within budget'.

4 Where Does this Leave Us?

I argue that what we need is research towards developing novel methodologies and metrics to inform decisions on project-based capital investment, together with research towards understanding how institutional environments can be created to move away from deep-seated norms where budget overruns are isomorphic with bad management and/or dishonesty. As of now, the benefit to cost ratio just isn't a realistic measure of all the value that megaprojects can potentially create. We need better ways to quantify in money terms the benefits that society will demand from megaprojects, including gains in social welfare, inclusion, equality, job creation, technological spill overs, resilience, redundancy, health, safety, wellbeing, and environmental issues. Interestingly, CBA at its genesis was not developed to inform large-scale capital investment, but rather to facilitate a prioritization of investment among a family of similar projects, eg prioritise investment in local roads. Notwithstanding this, because of a lack of trust on governments, since the 1950s, CBA got institutionalised and became standard practice to inform large capital investment decisions. This created a situation where a project appraisal tool that was not designed to inform major capital investment decisions, has nonetheless become an indispensable tool to enable project-based capital investment.

Complicating matters, there is the deep-seated norm that a high-performing project is one that stays within budget, on time and sees no changes to scope over project time-the so-called "iron triangle" coined in the 1950s, when the "stakeholder" construct had not yet been developed (Insert this footnote here: See Gaddis, PO 1959 The Project Manager, Harvard Business Review, May-June) While evolution in the institutions and technology surrounding megaprojects in the last decades has given stakeholders a "voice" on capital allocation decision-making, dislodging this norm is far from trivial. On the one hand, there is a pragmatic issue in that it is hard for policymakers to sanction a capital investment unless there is a pre-defined budget, and more so when capital resources are scarce. Commitments to capital budgets that are enabled by the institutions in the context, even if the commitments lack credibility because the commitments are premature, succeed nonetheless to reduce ambiguity in the value proposition and thus give legitimacy to choices to allocate scarce resources. It would be inconceivable that HS2 could get political support to forge ahead unless there was a budget with institutional credibility based on which reserves could be estimated. As well as this, a large ecosystem of professional bodies, management consultants, watchdogs, and scholars has mushroomed in the last decades around the project's norm that equates project high-performance to "on time, on budget". The emergence of this ecosystem has created vested interests in perpetuating a narrative that traces cost growth to incompetence and dishonesty. In recent years, scholarly dissent from this narrative has increased. Ideas like strategic misrepresentation and optimism bias have been strongly refuted. Ideas have also been floated about the need to develop heuristics that can help us anticipate the costs of collective action contingent on the institutional context. But norms that are institutionalised take time to change because the powerholders that control those institutions may lack incentives to acquiesce to change, and the mental models behind existing norms and institutions are also ingrained in society. So, it will take time for new narratives to dislodge the old 'iron triangle' narrative, or at least offer a compelling alternative for those who want to listen.

5 Conclusion

In this essay I argue that we need better norms, metrics, and methodologies to define and measure how megaprojects create and distribute value. Megaprojects are complex social tools designed by humans that need to be consented by society to achieve their goals. Their large-scale means that megaprojects are complex, non-linear enterprises that unfold under conditions of high uncertainty, and directly impact many self-interested stakeholders. Progress towards recognising that these stakeholders are unlikely to enable megaprojects to forge ahead unless these capital investments commit to produce socially valuable outcomes from the perspective of these stakeholders will support a search for novel frameworks that enable greater efficiency in megaprojects. But until new norms, tools and policies are available, managers and policymakers will continue to play a game upfront to make projects look bankable while knowing those projects will have to be fully redesigned along the way, just as HS2 has been. If instead we create an institutional environment where the costs and benefits of realising societal gains can be codified earlier, managers could then enfranchise the essential stakeholders earlier and negotiate a project purpose to which society consents. By getting the value distribution right earlier, there would be less disruption during project execution, and thus also less need to adapt capital-intensive buyer–supplier contracts ex-post the contracts have been agreed, which can hypothetically encourage suppliers to act opportunistically to increase their profit - another source of managerial complexity. And crucially, trying to agree a value distribution that stakeholders see as fair earlier woud also be a better basis to not only agree who gets what, but also who pays what.

Researchers and policymakers are working on this. Recently, the G20 has produced new guidance to measure the positive effect of new infrastructure in developing countries on tackling gender inequality. Progress on measuring the benefits of decarbonising projects is also happening. And advances have also been made on the value of negotiating community-benefit agreements or committing public sponsorship funds to realize societal gains. What we saw tackling the Covid-19 pandemic suggests that we are at a tipping point of a paradigm shift as to the values that we cherish as a society. As such, old ways of measuring the value to be created by megaprojects – as well as old norms deployed for assessing megaproject performance– have become anachronic to our times. But until new ways dislodge existing ones, it will be hard to stop the game. Science, the story goes, advances funeral by funeral. So should megaproject policy and practice.

Acknowledgements This essay builds on a blog that was published jointly with Visiting Professor Ian Reeves, CBE, on October 28, 2021 for the Original Thinkers series sponsored by the Alliance Manchester Business School. The contribution of Ian Reeves, CBE, is gratefully acknowledged.

Understanding Project Stakeholders Management as a Key Driver for Managing Complexity Within Mega-projects

Primiano Di Nauta, Cristina Simone, Francesca Iandolo, Stefano Armenia, and Marco Arcuri

Abstract While the existing research has mainly investigated a project management's static view, this contribution aims at discussing the impacts of projects' structure and dynamics on their performances, with a specific focus on the influence of stakeholder management within mega projects.

The chapter starts with the discusses a project as a Complex Adaptive System (CAS). In the second part, it analyzes stakeholder management as a key driver for managing complexity within mega projects. The proposed model integrates several previously developed tested project structures, adding a separate, even if complementary, structure for the negotiation process. Simulations describe the behaviors generated by the interaction of customized development processes in single-phase projects. Project performances are measured with common variables such as time, quality and cost. Originality/value – This research aims at showing how human behavior through stakeholder management is a key factor for the management of complexity within mega projects. With the aim of including the dynamic dimension, the extension of the models used to managing stakeholders within projects requires, for researchers and practitioners, a change of perspective and interpretation, especially regarding the mega projects.

P. Di Nauta (✉)
Department of Economics, University of Foggia, Via Romolo Caggese 1, 71121 Foggia, Italy
e-mail: primiano.dinauta@unifg.it

C. Simone · F. Iandolo
Department of Management, Sapienza University of Rome, Via del Castro Laurenziano 9, 00161 Rome, Italy
e-mail: cristina.simone@uniroma1.it

F. Iandolo
e-mail: francesca.iandolo@uniroma1.it

S. Armenia
Università Telematica degli Studi IUL, Via Buonarroti 10, 50122 Firenze, Italy
e-mail: s.armenia@iuline.it

M. Arcuri
AMPM Consulting s.r.l., Viale Le Corbusier 39, 04100 Latina, Italy
e-mail: m.arcuri@ampmconsulting.it

E. Favari and F. Cantoni (eds.), *Complexity and Sustainability in Megaprojects*, Lecture Notes in Civil Engineering 342,
https://doi.org/10.1007/978-3-031-30879-6_4

Methodology - The system dynamics methodology provides some of the tools for developing and implementing such a wider extension in project models.

Keywords Complex Project Management · Mega Projects · Complex Adaptive Systems · Organizational dynamics · Management for stakeholder · Stakeholder engagement

1 Projects as Complex Adaptive Systems (CASs): The Need for New Lens

In recent years, Project Management has been profoundly influenced by the complexity perspective, absorbing its concepts, principles and methodologies. This led to the progressive abandonment of the traditional waterfall approach, typically reductionist and sequential, towards the adoption of Complex Project Management, enriched by iterations, incrementality, adaptability, contextualization and stakeholder management. Actually, the several contexts (private firms, public institutions, etc.) in which Project Management is applied become more and more complex (i.e. uncertain and characterized by unpredictable feedbacks among their own variables and their environments). This highlighted the need for an enrichment (and perhaps a new conceptualization) of the endowment of information variety [10] provided by the methodologies and tools of Project Management, with respect to the specific requisite variety expected at a local level (i.e. the specific organizational contexts), to lead a project with efficiency, effectiveness and sustainability (Simone, [3, 53–55, 6]. For example, the traditional standard for Project Management can no longer be considered a "comfort zone" (i.e. a set of established and "familiar" frameworks, rules and tools aiming to ensure specific and predictable results). On the contrary, the standard for Project Management should shift towards an open standard, able to consistently co-evolve with the increasingly complex contexts that even more require new tools, creative solutions, and original combinations between exploitative and explorative knowledge.

Originally, we propose to conceptualize methodologies, methods, tools, and standards (such as ISO 21502, PRINCE2, IPMA, PMBOK, etc.) of Project Management as providers of a huge, precious and evolutive endowment of information variety [14]. This conceptualization leads to focus attention on the level of fit between the exploitable information variety provided by the discipline of Project Management and the specific requisite variety needed at a local level for the successful management of the project. In so doing, the proposed conceptualization promotes helpful reflections also on the potential gap between the information variety provided by the extant knowledge (and view) on Project Management and the requisite variety implies by the local needs that, in turn, promote new solutions enriching the Project Management information variety endowment [3].

Moreover, adopting the perspective of complexity leads to original implications in the way to conceive a project itself. Each project can be viewed as a CAS (Complex

Adaptive System), i.e. a reticular, open system whose components operate in parallel and with non-linear interactions. This highlights that a project co-evolves as a system that self-organizes and learns from the experience of positive and negative feedback. Considering a project as a CAS focuses the attention on the requisite variety to effectively lead a project: in an original way, the paper highlights how the requisite variety needed to lead a project will be provided by the project manager according to a tailor-made approach, creating original and emerging connections and combinations among modules of knowledge that heterogeneous standards make available.

Particularly, in a more and more complex and unpredictable world, Project Management will be characterized by ever greater context variability and by a growing variability of the scope, the requirements, and the constraints to be respected. This will generate an ever-increasing need to improve the capability in leading complexity [7, 12, 15, 37, 38, 46]. Moreover, in complex projects, stakeholder management supports the generation of that perceived business value, which leads to project success by achieving those project goals, which, in turn, fulfill stakeholder expectations [48].

Trying to provide a useful framework for project managers, the following matrix crosses critical organizational dimensions and the related challenges of Project Management in complex contexts (Table 1).

Accordingly, we highlight how urgent requirements for Project Management will be:

1. continuous enrichment with respect to intellectual, methodological and creative solicitations that may arise from the perspective of complexity;
2. tailoring at the local level: project management must increasingly present itself as a meta-platform knowledge, whose modules should be selected, adapted and

Table 1 Critical organizational dimensions and challenges for project managers in leading complexity

Critical organizational dimension	Challenges for project managers in leading complexity
Functional integration and coordination	Increasing need for bridge capabilities
HRM	Increasing need for negentropic human resources
Relationship among vertical and horizontal dimension	Increasing stress on the horizontal organizational dimension
Quality management	Ensuring quality in a service society
Relationship among managerial and entrepreneurial function	Ask for an entrepreneurial role of project manager
Learning	Increasing need to "learn to learn"

Source: [60]

combined according to the variety and variability of each specific local context (contextualization);

3. designing new tools for project managers to be able to tackle the complexity arising in projects [5] that hence demand for an approach to decision making which is based on current project data and models describing its structural dynamics [9].

2 Project Complexity and the Need for a System Dynamics Approach

The bundle of non-linear relationships, and risks of time delays and overbudgets in a project life cycle are among the main causes of the recursive reduction in performances, hence causing projects to be very difficult to manage (Thomas and Napolitan, 1994; [17, 51]. As a matter of fact, the overall complexity of projects seem to be poorly understood by managers [18, 47, 50, 62]. In addition, the traditional tools often appear inadequate when dealing with the increasing complexity of projects [56, 12, 1].

In this perspective, the modeling and simulation methodology known as System Dynamics (SD) [21, 61, 39] can play a key and strategic role in understanding the root causes of projects low performance and hence managing the risks of going overtime and overbudget (with poor quality of the final outcomes). Indeed, several risk assessment techniques are based on the subjective judgement of domain experts but subjective judgement is often human-intensive and error-prone [58]. It is thus clear how entrusting such a key strategic issue to the lone manager's intuition (even when supported by qualified experience) could be a source of poor technicality, potentially generating a proliferation of harmful events which, in a wider perspective, would be much better to avoid in projects.

Additionally, notwithstanding the fact that projects should be really considered as man-made goal-oriented open systems and, thus, not devoid of unpredictability and instability, on the other hand their complexity has increased the disruptive effect of subjective human factors such as bounded rationality, cognitive bias and information asymmetry [12], [3]. The managerial perspective, even when based on extensive experience, often shows its limits to coping with this problem. In this path, the need for a deep understanding of the strategy related to Project Management (PM) emerges, allowing to effectively learn from past failures. This can be achieved through a more formal systemic analysis, and by means of a methodological approach.

System Dynamics (SD) appears to be very useful in describing causal and dynamic complexity arising in complex projects and organizations [4], thus eventually allowing for the design of a new set of tools, which can support managerial decisions, as well as allow decision makers to experiment in risk-free learning environments, checking out their hypotheses and intuitions without necessarily implementing them first, waiting for the, rather often, catastrophic consequences.

Research practice has demonstrated that the methodological approach of SD is more suitable to strategic decisions than to operational ones [61], since it aims at considering not just the single elements of a system, but rather the sum of its parts intended as a whole, by concentrating on their relationships instead of elements considered separately [57]. The most fascinating conclusion is that margins for improvement emerge by considering the synergy of traditional PM techniques with an SD approach, to assign operational planning activities to the first ones, and strategic decisions based on the dynamic and feedback structure intrinsic to managing projects (with reference to greater complexity typical of mega projects) to the second one [8].

3 Stakeholders Management as a Key Driver for Managing Complexity Within Mega Projects

Megaprojects can be defined as large and complex projects involving substantial investment, social relevance, and long-time span [16].

Megaprojects are characterized by complexity, uncertainty, ambiguity, dynamic interfaces, significant political or external influences, and time periods reaching a decade or more [20].

They also involve different stakeholders with a variety of cultural differences, backgrounds, political systems, and languages [59].

The APM (Association for Project Management, 2008) describes a complex project as one which will typically involve interaction between several stakeholders requiring the coordination of the work of several disciplines and involving a wide range of project management methods, tools and techniques.

According to Turner and Cochrane [63], complexity arises not from what is known and under control, but from what is uncertain and unpredictable, then a definition of a complex project warranting examination would be "a complex project is one which exhibits a high degree of uncertainty and unpredictability, emanating from both the project itself and its context".

For the purposes of this paper, the causes of complexity in mega projects can be grouped into three categories: human behavior, system behavior, and ambiguity PMI® [49], Navigating Complexity).

Human behavior is the source of complexity that may arise from psycho-social interactions between stakeholders within a project. These interactions are numerous and non-linear in mega projects, due to factors such as power relationships, leadership, conflicts and political influence, and individuals' experiences and perspectives.

According to the complex project management approach, effective stakeholders' interactions contribute to project success, while the multiplicity and diversity of stakeholders involved in those interactions generate complexity in a project, even if they could be a valuable source of information variety and interdisciplinarity, if effectively managed.

In mega projects the level of differences, influence and number of stakeholders is particularly high, making a mega project very complex.

How is it possible to lever partially unpredictable human behaviors to effectively managing megaprojects in a with the rise in complexity?

To this aim, some crucial issues need to be put under focus:

- the necessity to manage a set of different unrealistic and unachievable expectations from different stakeholders within a project. This grade of unrealizability and unreality increases in mega projects.
- the presence of misunderstanding and/or disagreement with main aspects of projects, as goals, decision processes, priority, etc.
- the lack of support, clear guidelines, and effective decision-making processes from the higher governance subjects.
- the presence of a low maturity level within the performing organization and the people involved in the managing of projects. The required maturity level for a mega project should be higher to manage the greater complexity level.
- the lack of an effective communication process able to generate a clear, updated and shared knowledge of all the main aspects of the project.

Such conditions of complexity may be especially challenging for the managing of a project dealing with external stakeholders with a wide range of disagreements and unwritten or tacit agreements.

When individuals act, they do so both on their own behalf and on behalf of the groups and organizations that they represent or with which they identify.

The behavior of the individual becomes part of the complexity of the project; however, human behaviors are neither always rational nor deliberate.

This individual behavior fosters complexity by inadvertently introducing elements of irrationality, such as:

- Optimism bias and planning fallacy. Because of these elements, senior management tends to underestimate costs and durations and overestimate organizational benefits from projects under consideration.
- Anchoring. Anchoring is a bias that occurs when great significance is attached to information acquired during the initial project phase when there is a lack of knowledge about the project able to affect project elements as estimates, requirements assumptions, and other types of information. Anchoring may prevent the new information developed during the project to be correctly evaluated and incorporated in the plan.
- Framing effect. The way and the person who presents information affects the way in which this information is perceived and understood by other stakeholders and the consequent decision-making processes.
- Sunk cost effect. When a great deal of emotion, energy, and resources are invested in the management of a project, people involve are reluctant to accept negative performance and to terminate it if the business value is not reachable.
- Resistance. The projects generate change and there is a natural resistance to change on the part of the people involved.

Another element regarding how human behavior can generate complexity within a project are the team, organizational and political behavior.

Among this element, we can underline:

- Tribal mindset that involves rivalries with members of other groups.
- Groupthink, when rational decision making is overwhelmed by the group's desire to achieve conformity and harmony. In this situation groups can ignore essential information that are contrary to their way of thinking.
- Groupshift, when discussions among the group lead individuals to take more extreme positions than normal, generating a worst risk management approach.
- Lack of stakeholder commitment. It is not possible for a project to achieve success without the explicit commitment and support of its key stakeholders.

Another element regarding human behavior are aspect of communication and control.

Geographical dispersal of the project team, client, suppliers, and other key stakeholders creates complexities for communication and control within the project, such as:

- Varying legal perspectives from country to country, generating complexities in planning and control, potential misinterpretation, and ethical dilemmas.
- Cultural aspects regarding different from country to country and within different stakeholders may affect common understanding as well as how stakeholders communicate and interact with one another. There are various cultural dimensions that could increase complexity when a project is conducted in a cross-cultural environment: tolerance for inequality in levels of authority in organizations and institutions; the degree to which people tend to operate within a group or autonomously; the distribution of roles between men and women, or the levels of comfort with unstructured situations.

Another element regarding human behavior is opacity. If there are untransparent and unshared communications regarding decision-making processes and people management, it could lead to a lack of trust by a project's stakeholders, with a consequent mistrust of the organization by project team members and mistrust among team members that may lead to complexities regarding the stakeholder's management.

4 Discussion

This chapter introduced the perspective of mega projects as complex adaptive systems (CAS), and proposed System Dynamics as a possible approach to adopt a comprehensive and systemic view of the issues that may arise in managing projects; finally, building on the very same systemic perspective, it focused on stakeholders management as one of the key drivers to address complexity in megaprojects.

Indeed, as described in previous paragraphs, megaprojects constitute a particularly complex environment whose system structure is characterized by the presence of interdependent variables, feedback, delays, and nonlinear relationships, typical of the application of an SD approach: this allows for the description of a project's dynamic behavior by evidencing the key drivers that determine such behavior.

The concept of project development as a set of interactive demand-driven activities may be used to build a rich description of the causal relationships that emerge from previous research. The strong direct and indirect influences of development processes, in fact, may help to explicitly separate development processes from resources, scope, and targets, and by identifying iterations into and among project phases due to the intrinsic structure of projects. Usually, delays due to development iterations are intrinsic to all projects' structures and, even if the leading IT companies are making a daily use of traditional project management techniques and tools, there has not been any improvement until nowadays that brought tangible advantages in order to mitigate (or at least try to assess at the project kick-off) the time and overhead costs. The most advanced managerial practices today use optimization and modelling methodologies or tools that, however, can only consider the direct effects of measured variables related to a project, while indirect effects have been shown to be usually those who impact most on project performances. It is then natural to assess that if the delays and disruption problem persist even by recurring to advanced techniques, logical thinking would bring to consider the following: either the techniques used to assess the expected values on time and costs are wrong or, rather, the considered variables are not sufficient to describe the dynamics, the more complex structure and, ultimately, the behavior of the whole system.

Additionally, The proposed approach to projects defined as CAS may help in improving the methodologies and models used for management and the complexity of project structures. In fact, widening the knowledge and understanding of project dynamics is critical for improving projects' performances and developing new or improved tools, such as simulation (dynamic, discrete and/or hybrid) for communication and management practice, which is also expected to be essential to translating improved knowledge and understanding into improved performance.

Starting from the above, this chapter proposes the need for a new approach to methodology in megaprojects management practice and for research in order to develop new tools able to assess the feedback structures intrinsic to project management systems (as the examples in Fig. 1 shows).

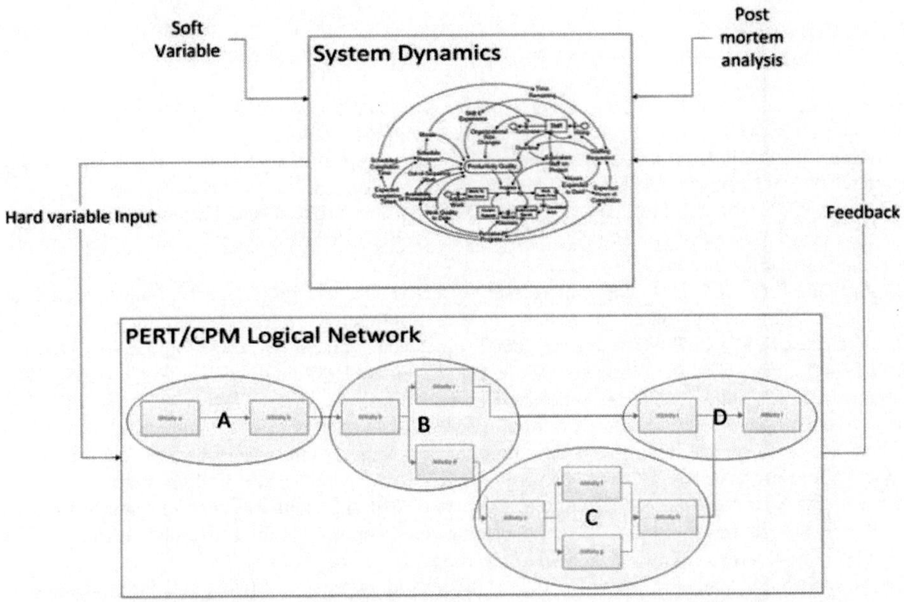

Fig. 1 A possible evolution in Project Management methodology and tools. Source: Armenia et al. [5]

5 Conclusions

This chapter, as a part of an ongoing research path, has led to insights concerning the dynamics of projects according to the Complex Adaptive Systems perspective and to the System Dynamics modeling and simulation approach. It discusses also stakeholders management as a driver for managing complexity. Finally, it proposes a possible evolution in Project Management methodology and tools.

Ultimately, this contribution provides new insights to expand the study of project dynamics in several potentially valuable directions and pushes project management toward a broader image of projects and their role in project performance. It proposes a way to improve performance through an improved understanding of the projects' structures and behaviors. Further research is needed to expand and refine the understanding and the adoption of dynamics approaches to manage projects. Particularly, future lines of research will regard possible applications of the proposed model; the development of standard SD-based project evaluation metrics; the integration at the operational level of our proposed SD model with traditional project modelling approaches.

References

1. Aguiari, R., Di Nauta, P.: Governing business dynamics in complex contexts. Mercati e Competitività, fascicolo. 1, pp 39–59, Franco Angeli, Milano (2012)
2. Arcuri M (2021) Il Project Management. Conoscenze, strumenti e tecniche in accordo con la UNI ISO 21502:2021, Tangram Edizioni Scientifiche Trento
3. Arcuri M, Simone C, La Sala A, Simone C (2018) The PMBOK standard evolution: leading the rising complexity. In: Workshop PMI 2018, Sapienza University of Rome, 14 September, Sapienza University Press
4. Armenia S (2018) Analysing management effectiveness and project performance through a system dynamics approach: the case of software development projects. In: Nonino F, Annarelli A, Gerosa S, Mosca P, Setti S (a cura di), Project Management. Driving Complexity, Atti 3rd PMI® Italian Academic Workshop, Project Management Institute Italy Chapters, pp. 139–144, Roma, 20–21 settembre 2018, Sapienza Università Editrice [ISBN 978-88-9377-086-6]
5. Armenia S, Canini D, Falsini D, Schiraldi MM (2011) Improving management and tools for project performance in software development: a system dynamics approach. Proceedings of the 11th European Academy of Management (EURAM) Conference, Tallinn, Estonia
6. Armenia S, Dangelico RM, Nonino F, Pompei A (2019) Sustainable project management: a conceptualization-oriented review and a framework proposal for future studies. Sustainability 11(9):2664. https://doi.org/10.3390/su11092664
7. Armenia S, Di Nauta P, Simone C, (2019) Understanding project structure and dynamics as key complexity drivers for project management. The case of a software project in a system dynamics perspective. In: Proceedings of the 2019 WOA Congress, Palermo Italy. Associazione Italiana di Organizzazione Aziendale (Assioa)
8. Armenia S, Di Nauta P, Simone C, Pompei A, (2019) La modellizzazione dei progetti come Complex Adaptative Systems (CAS). Il contributo della System Dynamics", in Barile S., Paniccia P. (a cura di), Il fascino della precarietà. Studi sull'evoluzionismo sistemico, Edizioni Nuova Cultura, Roma, pp 243–267. [ISBN 978-88-3365-248-1]
9. Armenia S, Ferreira FE, Mecella M, Onori R (2017) Smart model-based governance: from big-data to future policy making. In: (a cura di): Fabio Nonino, Stefano Armenia, Gandolfo Dominici, Model-based Governance for Smart Organizational Future - BSLab-SYDIC International Workshop - Book of Abstract. Business Systems Laboratory (2017) ISBN: 9788890824258
10. Ashby WR (1985) Requisite variety, and its implications for the control of complex systems. Cybernetica. **1**(2)
11. Bakhshi J, Ireland V, Gorod A (2016) Clarifying the project complexity construct: past, present and future. Int J Proj Manag 34(7):1199–1213
12. Barile S, Di Nauta P, Iandolo F (2016) La decostruzione della complessità. Studi MOA – Collana di Management e Organizzazione Aziendale, Editrice Minerva Bancaria, Roma
13. Bredillet, CN (2006) The future of project management: mapping the dynamics of project management field in action. In: Global Project Management Handbook: Planning, Organizing, and Controlling International Projects. McGraw-Hill, New York
14. Calabrese G, Di Nauta P (2014) Professionalization and dissemination processes for project management practices. A situationist perspective. In: Baglieri D, Metallo C, Rossignoli C, PezzilloIacono M (eds) IS, Management, Organization and Control: Smart Practices and Effects. LNISO, vol. 6, pp 289–300. Springer International Publishing, Switzerland
15. Calabrese M, Simone C, Magliocca P (2016) Going away from the "protocol culture": innovation, complexity and the need for a culture of variety. China-USA Bus. Rev. 15(4):194–204
16. Cerić A, Vukomanović M, Ivić I, Kolarić S (2021) Trust in megaprojects: a comprehensive literature review of research trends. Int J Project Manage 39(4):325–338
17. Cooper KG (1980) Naval ship production: a claim settled and a framework built. Interfaces 10(6):20–36
18. Diehl E, Sterman JD (1995) Effects of Feedback Complexity on Dynamic Decision Making Organizational Behavior and Human Decision Processes. Elsevier, Amsterdam

19. Eppinger SD, Whitney DE, Smith RP, Gebala DA (1994) A model-based method for organizing tasks in product development. Res Eng Des 6(1):1–13. https://doi.org/10.1007/BF01588087
20. Floricel S, Miller R (2001) Strategizing for anticipated risks and turbulence in large-scale engineering projects. Int J Project Manage 19:445–455
21. Forrester JW (1961a) Industrial Dynamics. MIT Press, Cambridge
22. Halpin DW, Woodhead RW (1980) Construct Management. John Wiley & Sons, New York
23. Howick S, Eden C (2001) The impact of disruption and delay when compressing large projects: going for incentives? J. Oper. Res. Soc. 52:26–34
24. Kelly JC (1987) A comparison of four design methods for real-time system. In: Proceedings of the 9th international conference on Software Engineering table of contents, Monterey. IEEE Computer Society Press, Los Alamitos, CA, USA
25. Kerzner H (2006) Project Management: A Systems Approach to Planning, Scheduling, and Controlling. Wiley, New York
26. van Oorschot KE, Sengupta K, van Wassenhove LN (2009) Dynamics of agile software development. In: Proceedings of the International Conference of the System Dynamics Society 2009, Albuquerque NM, USA
27. Kwak YH, Anbari FT (2009) Analyzing project management research: perspectives from top management journals. Int J Project Manage 27(5):435–446
28. IPMA (2018), Individual Competence Baseline for Project Management.
29. ISO (2021) ISO 21500:2021 Project, programme and portfolio management -- Context and concepts
30. ISO (2020) ISO 21502:2020 Project, programme and portfolio management - Guidance on Project Management
31. ISO (2017) ISO 21505:2017 Project, programme and portfolio management - Guidance on Governance
32. ISO (2018) ISO 21508:2018 Earned Value Management in project and programme management
33. ISO (2018) ISO 21511:2018 Work Breakdown Structures for project and programme management
34. ISO (2018) ISO/TR 21506:2018 Project, programme and portfolio management – Vocabulary
35. Leach LP (1999) Critical chain project management improves project performance. Proj Manag J 30(2):39–51
36. Lindstrom L, Jeffries R (2003) Extreme programming and agile software development methodologies. In: Brown C, Topi H (eds) IS Management Handbook, 8th Edition. Auerbach Publications. https://doi.org/10.1201/9781420031393.ch41
37. Martinez M, Di Nauta P, Merola B (2016) Dynamic Capabilities and T-Shaped Knowledge in Project Management. A Viable Systems perspective. In: eBook of Abstracts of the 4rd Business Systems Laboratory International Symposium
38. Martinez M, Di Nauta P, Sarno D (2017) Real and Apparent Changes of Organizational Processes in the Era of Big Data Analytics. Studi Organizzativi, n. 2/2017, Franco Angeli
39. Meadows DH (2009) Thinking in systems : a primer. Earthscan, London, Sterling, VA
40. Meso P, Jain R (2006) Agile software development: adaptive systems principles and best practices. Inf Syst Manage 23(3):19–30. https://doi.org/10.1201/1078.10580530/46108.23.3.20060601/93704.3
41. Miscia S (1997) in "Introduzione" a Archibald R.D., Project Management. La gestione di progetti e programmi complessi (ed. it.) Franco Angeli, Milano
42. Moder JJ, Phillips CR, Davis EW (1983) Project Management with CPM, PERT and Precedence Diagramming. Van Nostrand Reinhold, New York
43. Ostvold M, Magne Jorgensen K (2005) A comparison of software project overruns-flexible versus sequential development models. IEEE Trans Softw Eng
44. Morelli MD, Eppinger SD, Gulati RK (1995) Predicting technical communication in product development organizations. IEEE Trans Eng Manage. **42**(3)
45. Nevins JL, Whitney DE (1989) Concurrent Design of Product and Process. McGraw Hill, New York

46. Ng ICL, Badinelli R, Polese F, Di Nauta P, Löbler H, Halliday S (2012) S-D logic research directions and opportunities: the perspective of systems, complexity and engineering. Mark. Theory. **12**
47. Paich M, Sterman JD (1993) Boom, bust, and failures to learn in experimental markets. Manage Sci 39(12):1439–1458. https://doi.org/10.1287/mnsc.39.12.1439
48. Pirozzi M (2018) The Stakeholder management perspective to increase the success rate of complex projects, featured paper. PM World J. **7**(I)
49. PMI® (2014) Navigating Complexity: A Practice Guide. Project Management Institute. Newton Square, PA
50. Rechtin E (1991) Systems Architecting: Creating and Building Complex Systems. Prentice Hall, Hoboken
51. Reichelt KS (1990) Halter Marine: A Case Study in the Dangers of Litigation, Technical Report D-4179, System Dynamics Group. MIT Sloan School of Management, Cambridge, MA
52. Rodrigues AG, Williams TM (1996) System dynamics in software project management: towards the development of a formal integrated network. Eur J Inf Syst 6:51–56
53. Saviano M, Bassano C, Piciocchi P, Di Nauta P, Lettieri M (2018) Monitoring viability and sustainability in healthcare organizations. Sustainability 10(10):3548. https://doi.org/10.3390/su10103548
54. Saviano M, Di Nauta P, Montella MM, Sciarelli F (2018) Managing protected areas as cultural landscapes: the case of the Alta Murgia National Park in Italy. Land Use Policy 76:290–299. https://doi.org/10.1016/j.landusepol.2018.03.052
55. Saviano M, Di Nauta P, Montella M, Sciarelli F (2018) The cultural value of protected areas as models of sustainable development. Sustainability 10(5):1567. https://doi.org/10.3390/su10051567
56. Saviano M, Di Nauta P (2011) Project Management as a compass in complex decision making contexts. A viable systems approach. In: Caivano D, et al PROFES 2011, 12th International Conference on Product Focused Software Development and Process Improvement. ICPS, pp 112–119. ACM
57. Senge P (1992) The Fifth Discipline: The Art and Practice of the Learning Organization. Sperling & Kupfer
58. Yacoub SM, Ammar HH (2001) A Methodology for Architectural-Level Reliability Risk Analysis. Publishing Systems and Solutions Laboratory, HP Laboratories Palo Alto
59. Shore B, Cross BJ (2005) Exploring the role of national culture in the management of large-scale international science projects. Int J Project Manage 23:55–64
60. Simone C, Arcuri M, La Sala A (2017) Be vicarius: the challenge for project management in a service economy. IConference Proceedings of the 20th Excellence in Services International Conference, 7–8 September 2017, Verona, pp 853–867
61. Sterman J (2000) Business Dynamics: Systems Thinking for a Complex World. Irwin/McGraw-Hill
62. Sterman JD (1994) Learning in and about complex systems. Syst Dyn Rev 10(2–3):291–330
63. Turner JR, Cochrane RA (1993) Goals-and-methods matrix: coping with projects with ill defined goals and/or methods of achieving them. Int J Project Manage 11(2):93–102. https://doi.org/10.1016/0263-7863(93)90017-H
64. Wheelwright SC, Clark KB (1992) Revolutionizing Product Development, Quantum Leaps in Speed, Efficiency, and Quality. The Free Press, New York
65. White D, Fortune J (2002) Current practice in project management an empirical study. Int J Project Manage 20(1):1–11
66. Womack JP, Jones DT, Roos D (1990) The Machine that Changed the World. Rawson Associates, New York.

The Concept of Sustainable Infrastructure: A Content Analysis of Construction Companies Reports

Silvia Gordano, Daniel Torchia, and Laura Corazza

Abstract In 2015, the 17 Sustainable Development Goal (SDGs) were introduced by the United Nations. Among them, SDG 9 (*Industry, innovation and infrastructure*) refers to the notion of 'resilient' and 'sustainable' infrastructure to promote the transition towards sustainable industrialisation. The notion of 'sustainable infrastructure' has been subject of debate over time. The variety of definitions and assessment frameworks used to categorise sustainability components by academics and professional bodies, call for the need to further investigate and reflect on the concept. Hence, this exploratory study aims to provide first evidence on how the term is conceived and interpreted by companies. To this goal, a manual content analysis on non-financial disclosures published by the top-10 European construction companies, is conducted. Findings suggest that firms largely refer to SI in relation to the early phases of the project life-cycle, while neglecting the ultimate stage (i.e., dismantling). Furthermore, there is not a consensus towards a unifying representation of the components enclosed in the notion of SI, although all companies appear to agree with its 'green' attributes (e.g., energy intensity, emissions, materials). Consistently, a prevailing use of environmental assessment criteria is testified. The emphasis placed on other sustainability-related issues (i.e., social and economic) and the broadness of stakeholders' interests addressed vary considerably across corporate reports, thus supporting that the concept of SI is still fragmented and in evolution.

Keywords Sustainable Infrastructure · Resilient Infrastructure · Sustainable Construction · Sustainable Building · Sustainability Reporting · Content Analysis

1 Introduction

The 2030 Agenda by the United Nations has introduced 17 Sustainable Development Goals (SDGs) striving for societies to reach peace and prosperity (United Nations, 2015). In particular, SDG 9 (*Industry, Innovation and Infrastructure*) focuses on

S. Gordano (✉) · D. Torchia · L. Corazza
Università degli Studi di Torino, Turin, Italy
e-mail: silvia.gordano@unito.it

E. Favari and F. Cantoni (eds.), *Complexity and Sustainability in Megaprojects*, Lecture Notes in Civil Engineering 342,
https://doi.org/10.1007/978-3-031-30879-6_5

infrastructure and promotes the concept of 'resilient' and 'sustainable' infrastructure to drive the transition towards sustainable industrialisation.

The construction sector is largely recognised as one of the most environmentally impactful industrial sectors. It is responsible for 30% of global energy consumption, 40% of raw material extraction, and 30% of carbon emissions [7]. However, public expenditure in infrastructure is crucial to promote the economic wellbeing, growth, and social development of countries, by increasing employment, connectivity and ensuring adequate provision of critical utilities.

Given the relevant implications of infrastructural projects, and considering the profound, wide-ranging and lasting impacts that they generate on society and economy (Zeng et al., 2014), sustainability issues in infrastructure have drawn considerable academic attention in recent times.

Overall, there is no unequivocal consensus on how to cope with the sustainability assessment of infrastructure (Chang et al., 2022). As testified by the definition provided by the Inter-American Development Bank [16], the concept of sustainable infrastructure (SI) is connected with the evaluation of a wide range of sustainable impacts over the entire life-cycle of the project. Therefore, sustainable project management of infrastructure requires reconciliating the conflicting, even though interconnected, dimensions of sustainability, in a multistakeholder and long-term context. Such complexity is made apparent by the large number of different assessment frameworks developed by professional bodies and authorities (including, CEEQUAL, NYSDOT, Greenroads, ISI, ISCA), as well as by scholars [9, 28], Ugwu and Haupt, 2007, [30, 6], to measure the sustainability impacts and performance of infrastructure.

To deal with infrastructure through the lens of sustainability and sustainable development, further reflection on the still blurred concept of 'sustainable infrastructure' is required (Chang et al., 2022).

Therefore, the purpose of this study is to provide evidence on how the notion of SI is conceived and interpreted by companies in the construction industry, in order to contribute to the lively and ever-growing debate on sustainable infrastructure. To this aim, we carry out a content analysis on the latest-available non-financial disclosures published by the top 10 European construction companies.

Findings suggest that companies largely refer to SI in relation to the early phases of the project life-cycle (i.e., planning and execution), whilst the end of life stages are neglected (i.e., dismantling). Furthermore, there is not a widely shared representation of the components enclosed in the notion of SI, although all companies seem to agree with its environmental attributes (e.g., energy intensity, emissions, materials). To exacerbate such differences, the emphasis placed on social and economic issues and the breadth of stakeholders' interests addressed vary across corporate reports. Nonetheless, a prevailing use of environmental assessment criteria is observable.

2 Literature Review and Theoretical Framework

When engaging with infrastructure and construction projects, it must be underlined the high degree of complexity that surrounds them (Zeng et al., 2014). Infrastructural projects impacts are wide-reaching, encompassing environmental, social and governance matters, which require urgent sustainability management (Flyvberg, 2014; [20], Zeng et al., 2014).

The environmental impacts enclose climate change, pollution, biodiversity, natural resources, and ecosystems, calling for the sustainable ways of managing waste and greenhouse gas emissions (GHG) across supply chains, designing infrastructure with increased resource-use efficiency and greater adoption of green technologies and processes (SDG 9). Such impacts are widely acknowledged by civil society, so that traditionally the notion of sustainability in infrastructure projects was (almost) exclusively associated with 'green' attributes of buildings and infrastructure [3]. Indeed, the review work carried out by Thomé et al. [33] on the literature in sustainable infrastructure from 1995 to 2015, highlights that most studies conducted in the early stages of the field focused on materials for the greening of buildings, energy efficiency, and water management.

The social facet of infrastructure not only concerns the social impact of the construction activities on the community, but also the construction organisations' engagement with all stakeholders through the whole project life-cycle, such as suppliers and workforce (Valdes-Vasquez & Klotz Leidy, 2013). Health and safety and the high reliance on labour are hallmarks of the construction industry. Even though infrastructure can extensively improve the conditions of local community, especially in developing economies, negative externalities can stem, among others, from prolonged construction projects and the hazardous nature of activities on construction sites. Today, various social issues remain prominent, such as unequal treatment of workers and low working environment standards [26].

Furthermore, governance has long been a recurrent topic in the construction sector, due to the size and inherent complexity of projects, competitive bidding processes, and the necessity to engage with both public and private stakeholders, along with the need of preventing bribery and corruption in supply chains.

Based on the above, it appears now widely acknowledged that the sustainable management of infrastructure requires consideration of a balanced development of sustainability along the 'triple bottom line' (TBL) (i.e., environmental, social, and economic sustainability) [6]. This requires a marked shift from traditional conceptualisations embracing solely an environmental perspective, to account for the broad range of environmental, social and economic impacts generated, the multitude of stakeholders directly and indirectly involved, and the whole project life-cycle [3, 24].

To try evaluating and improving the environmental, social, and economic outcomes of infrastructure projects, a number of assessment frameworks have been therefore developed by professional bodies in the last two decades. Among them, it is worth mentioning the Civil Engineering Environmental Quality Assessment and Award Scheme (CEEQUAL), Envision, GreenLITES, Greenroads, and the IS rating scheme. However, while sharing several common components, noticeable differences and/or complementarities are observable across frameworks. Furthermore, several studies have criticised them for the overemphasis placed on environmental aspects, while neglecting projects' social return on investment [29, 31].

By reviewing the major existing frameworks developed by professional bodies and assessment frameworks proposed by scholars across the academic literature, Chan et al. [6] recently come up with a comprehensive assessment framework for sustainable infrastructure (see Table 1). Once again, the authors conclude that social and economic sustainability of infrastructure projects are not perceived as importantly as environmental matters.

3 Methodology

In order to investigate how companies interpret the concept of SI, we conduct a content analysis of corporate disclosures on sustainability-related topics published by the top 10 companies in the 2021 list of European Contractors by Construction Europe[1] (see Table 3, in Appendix). We choose this method of analysis as corporate narratives have been suggested to represent meaningful means for detecting how the discourse around social issues is constructed by organisations [5, 19, 34],O'Connor and Shumate 2010; [1].

In particular, the study applies both a qualitative and quantitative content analysis, to both facilitate the interpretation of data and objectively present facts from the text in the form of frequencies. Furthermore, the nature of our content analysis is both deductive and inductive. Indeed, we draw upon Chan et al.'s framework (2022), to analyse the components and subcomponents associated with the notion of sustainable infrastructure, to find out if those or other components of sustainable infrastructure emerged from corporate disclosures.

To carry out our analysis, we first screen sustainability-related information in corporate reports by entering a search query based on the following keywords: "*sustainable infrastructure*", "*sustainable construction*", "*sustainable building*", "*resilient building*", and "*SDG 9*". Subsequently, we extract information located in the same paragraphs where the identified keywords are mentioned and carry out a manual coding. Specifically, based on previous works in the SI field, the coding procedure aims to identify the following key aspects:

1) the project life-cycle phase (s) addressed [24],
2) the components (i.e., sustainability issues/topics) and sub-components (i.e., sustainability sub-topics) related to SI [6],
3) the metrics employed to assess SI.

---

[1] Available at: https://www.construction-europe.com/Files/Download/20210927-112301-CE-09-2021-CE100.pdf.

Table 1 Assessment framework for sustainable infrastructure [6]

Components of sustainable infrastructure	Subcomponents
Materials	Material quality control
	Materials reuse
	Local materials
	Innovative materials
	Material intensity
Waste	Divert from landfill
	Construction and operational waste
	Deconstruction
Energy and ecology	Energy use
	Air quality control
	Greenhouse gas emission control
	Protection of biodiversity
	Ecological connectivity
	Renewable energy
Water	Water use
	Protection of water quality
	Stormwater management
Community engagement	Tangible heritage management
	Public opinion
	Future visions
	Level of engagement
	Intangible heritage management
	Skill development opportunities
Health and safety	Occupational mental health
	Occupational physical health
	Public physical health
	Public mental health
	Safety
Corporate social responsibility	Anti-discrimination
	Anti-corruption
	Fair wage
Project management	Procurement and supply chain governance
	Type of contract

(continued)

Table 1 (continued)

Components of sustainable infrastructure	Subcomponents
	Life-cycle cost management
	Environmental management
Environmental impact on local community	Vibration control
	Noise control
	Light pollution control

4 Findings

The keyword-based search resulted in nine companies out of ten showing some kind of information associated with sustainable infrastructure. For such companies, therefore, key aspects related to SI are discussed in the following.

4.1 Project Life-Cycle Phases

From the analysis, it emerges that the concept of SI is largely applied by companies in the *design* and *execution* phases of projects life-cycle (see Fig. 1). Considering sustainable impacts in the genesis of infrastructure project, is crucial, since decisions related to the early phases are highly risky and have profound and irreversible consequences (Ma et al., 2017). Even though project processes involve progressively decreasing levels of risk and uncertainty through time (Winch, 2001), the dismantling phase need to be addressed nonetheless, to implement circular economy practices.. However, recent works point out a substantial lack of tools and standards related to the ultimate life-cycle stages (Corazza et al., forthcoming).

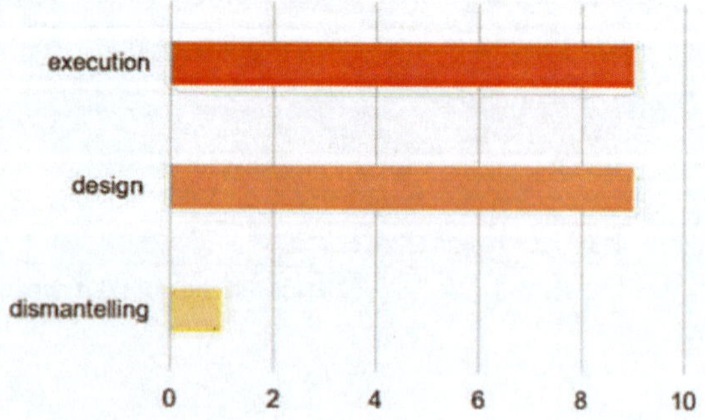

Fig. 1 Project life-cycle phases addressed

4.2 Components and Sub-components of SI

Frequencies of components detected from our analysis are shown in Fig. 2. The component 'Energy' records the highest relative importance, appearing in all corporate reports. Companies widely attach sustainability attributes to low-carbon emission projects, capable of producing fewer GHG emissions compared to traditional infrastructure and positively contributing to climate neutrality. This is not surprising, given that the construction sector has been traditionally put under the spotlight by civil society for its high levels of energy consumption and GHG emissions. Energy-based criteria represent the most commonly employed measures to assess sustainability in buildings [2], to the extent that the terms 'sustainable' and 'energy efficient' have been largely used as synonyms [3]. Accordingly, the analysis also shows that the majority of subcomponents concerns energy-efficiency instead of renewable energy investment.

The component 'Materials' is also recurrently anchored to the concept of sustainable infrastructure. Previous studies show that materials-related issues are noticeable in almost all existing assessment frameworks [6]. Specifically, the analysis highlights that companies focus on reusing and recycling, turning to certified and low-energy-embodied materials. By contrast, the use of local materials appears not to be a prevailing option, as this practice may lead to higher costs and delays in building projects [14].

Following, topics related to 'Water', 'Biodiversity', 'Community', and 'Innovation' appear to be equally associated with the notion of SI. Water management is usually connected to water use reduction and stormwater management, in line with the focus on storm water runoff shown by studies conducted in the first decade of the 2000s [33]. 'Biodiversity' is linked to SI mostly in terms of prevention of damages

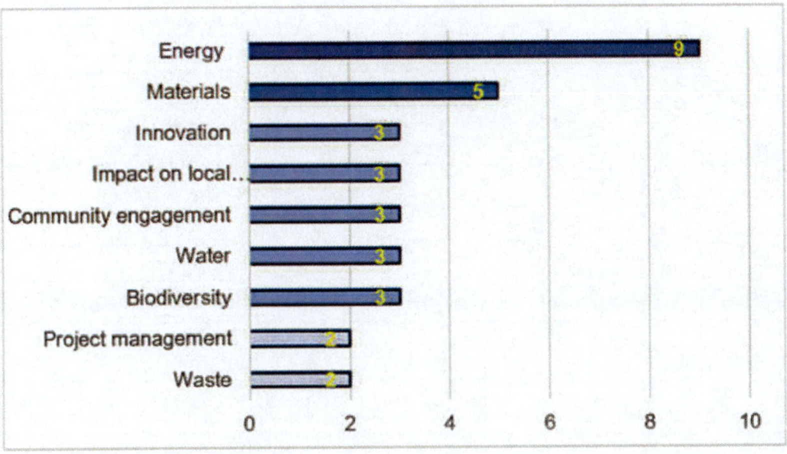

Fig. 2 Components and sub-components of SI

caused by natural forces, and ecosystem protection and restoration thus recalling the notion of 'green' infrastructure promoted by the European Union (EU) (EU, 2013).

The component related to community can be categorised into two categories, i.e., 'Impact on local community', which refers to reducing negative impacts falling over local communities (such as, noise, dust, and traffic), and 'Community engagement', which instead relates to company's engagement in initiatives for community wealth development (such as, information-sharing endeavours, seminars, workshops, educational events dedicated to youngsters, apprenticeship and work placement opportunities). In this light, intangible impacts of infrastructure, which have been indicated as a white spot in many existing assessment frameworks [6], appears to be addressed to some extent by companies. Furthermore, SI is also linked to innovation, in terms of ground-breaking project design and innovation sharing to improve people's life.

By contrast, consistently with the scarce consideration devoted to the disposal phase of infrastructure projects, waste management is rarely associated with SI, as well as economic issues related to project management.

In Table 2, a re-elaboration of Chan et al.'s [6] framework is presented, based on the components and subcomponents that emerged from the content analysis. Components and subcomponents marked in bold indicate labels attributed by the authors that differ from Chan et al.'s [6] framework.

Table 2 Components and subcomponents of sustainable infrastructure recognised by the authors

COMPONENTS	SUBCOMPONENTS
Materials	
	Material quality control
	Materials reuse
	Materials recycling
	Local materials
	Innovative materials
	Material intensity
Waste	
	Divert from landfill
	Construction and operational waste
Energy	
	Energy use
	Renewable energy
	Greenhouse gas emission control

(continued)

Table 2 (continued)

COMPONENTS	SUBCOMPONENTS
Biodiversity	
	Protection/**restoration** of biodiversity
	Ecological connectivity
	Water use
Water	
	Stormwater management
Community engagement	
	Future visions
	Level of engagement
	Skill development opportunities
	Quality relationship
Project management	
	Procurement and supply chain governance
	Type of contract
	Environmental management
Impact on local community	
	Noise control
	Traffic issues
Innovation	
	Design for change
	Innovation sharing

4.3 Assessment Metrics

The analysis supports that companies manly rely on environmental assessment criteria to assess the sustainability of infrastructure, including revenues from projects with environmental certifications (such as, LEED, BREEAM, etc.), indicators drawn from the Global Reporting Initiative (GRI) 302 (*Energy*) and 305 (*Emissions*), the percentage of sustainable materials (i.e., with an environmental certification) on sales, the R&D expenditures related to eco-projects.

These findings support scholars' criticism on the prevailing environmental focus of extant SI assessment frameworks and further advocate that companies adhere to such (mis)conception. Indeed, we found that the social sphere is seldomly recalled, with just few companies making references to the social value of projects in terms of its impact on local community. Yet, the economic pillar of sustainability is quite neglected, with sporadic use of assessment metrics drawn from the GRI 201 (*Economic performance*) and 203 (*Indirect economic impacts*).

5 Conclusions

Sustainable infrastructure has the potential to play a primary role in delivering long-term socio-economic benefits, as well as environmental ones [24]. Over time, the sustainable infrastructure field has gradually broadened its scope from green infrastructure to the TBL of economic, social, and environmental sustainability [3]. However, the variety of components and the still prevailing environmental façade of SI in extant assessment frameworks [6], call for further investigation on the concept. Therefore, this exploratory study has aimed to add a company-centric perspective to the SI discourse, by analysing SI-related contents as disclosed by companies of the construction industry in their corporate reports.

Our findings support that firms largely refer to SI in relation to the early phases of the project life-cycle (i.e., planning and execution), whereas very little linkages to the final project stages are detectable. This can be explained by the paucity of assessment tools related to the deconstruction/demolition phase (Corazza et al., forthcoming). Furthermore, the analysis supports that the components and subcomponents associated with the concept of sustainable infrastructure/construction/building are not unanimously shared by companies. In particular, the components mostly associated with the notion of SI are those pertaining to the environmental dimension and involving actions with a restricted focus (predominantly internal to the organisation). Such components reflect corporate commitments that are aligned with business aspects largely acknowledged by the public as the most critical in the industry (for instance, the high levels of CO2 emissions, energy consumption and material use) [26]. More rarely, the SI concept embraces the social dimension and involves a broader spectrum of stakeholders, including natural ecosystems (Haigh and Griffiths, 2007), local community and society at large. Ultimately, the SI concept is very seldomly represented through an economic perspective, by referring to sustainable project management, mainly in terms of sustainable procurement and non-financial performance-based types of contracts.

The study is exploratory and, therefore, presents several limitations, which can also be intended as inputs for future developments. First of all, the empirical analysis has been conducted on a small sample of companies to provide first insights into the topic. Expanding the sample would certainly contribute to the generalisability and enrichment of our findings. Moreover, the manual text classification could be accompanied and complemented by deploying automatic text mining techniques, to analyse terms associations based on frequency support and statistical confidence [37]. Further research is suggested to better define the concept of SI, which is still fragmented and deserves more attention from academics and practitioners.

Appendix

Table 3 List of companies considered for the analysis

Ranking by revenues	Company
01	Vinci
02	ACS
03	Bouygues
04	Hochtief
05	Eiffage
06	Skanska
07	Strabag
08	TechnipFMC
09	Balfour Beatty
10	Saipem

References

1. Beauchamp LL, O'Connor A (2012) America's most admired companies: a descriptive analysis of CEO corporate social responsibility statements. Publ Relat Rev 38(3):494–497
2. Berardi U (2012) Sustainability assessment in the construction sector: rating systems and rated buildings. Sustain Dev 20(6):411–424
3. Berardi U (2013) Clarifying the new interpretations of the concept of sustainable building. Sustain Cities Soc 8:72–78
4. BoschRekveldt M, Jongkind Y, Mooi H, Bakker H, Verbraeck A (2011) Grasping project complexity in large engineering projects: The TOE (technical, organizational and environmental) framework. Int J Project Manage 29(6):728–739
5. Buhr N, Reiter S (2006) Ideology, the environment and one world view: A discourse analysis of Noranda's environmental and sustainable development reports. Environmental Accounting. Emerald (MCB UP), Bingley, pp 1–48
6. Chan M, Jin H, van Kan D, Vrcelj Z (2022) Developing an innovative assessment framework for sustainable infrastructure development. J Clean Prod 368:133185
7. Choi J-H (2019) Strategy for reducing carbon dioxide emissions from maintenance and rehabilitation of highway pavement. J Clean Prod 209:88–100
8. Corazza L, Cottafava D, Torchia D (forthcoming) Toward Sustainable Infrastructural Megaprojects. In: Leal Fihlo W, et al (eds), SDGs in the European Region, Implementing the UN Sustainable Development Goals – Regional Perspectives. Springer Nature, Switzerland AG
9. Dasgupta S, Tam EK (2005) Indicators and framework for assessing sustainable infrastructure. Can J Civ Eng 32(1):30–44
10. Graneheim UH, Lundman B (2004) Qualitative content analysis in nursing research: concepts, procedures and measures to achieve trustworthiness. Nurse Educ Today 24(2):105–112
11. Gupta H, Kumar A, Wasan P (2021) Industry 4.0, cleaner production and circular economy: an integrative framework for evaluating ethical and sustainable business performance of manufacturing organizations. J Clean Prod. **295**, 126253

12. Flyvbjerg B (2014) What you should know about megaprojects and why: an overview. Proj Manag J 45(2):6–19
13. Haigh N, Griffiths A (2009) The natural environment as a primary stakeholder: the case of climate change. Bus Strateg Environ 18(6):347–359
14. Hayles CS, Kooloos T (2005) The challenges and opportunities for sustainable building practices. Benefits 2:1–9
15. Hsieh HF, Shannon SE (2005) Three approaches to qualitative content analysis. Qual Health Res 15(9):1277–1288
16. Inter-American Development Bank (2018) What is sustainable infrastructure? A framework to guide sustainability across the project cycle. IADB, Washington, DC
17. KPMG (2020). The sustainable infrastructure opportunity. https://home.kpmg/xx/en/home/insights/2020/06/covid-19-recovery.html
18. KPMG (2022). Emerging trends in infrastructure 2022. https://assets.kpmg/content/dam/kpmg/xx/pdf/2022/01/emerging-trends-in-infrastructure-2022.pdf
19. Laine M (2009) Ensuring legitimacy through rhetorical changes? A longitudinal interpretation of the environmental disclosures of a leading Finnish chemical company. Acc Audit Accountabil J. 22:1029–1054
20. Levitt RE (2007) CEM research for the next 50 years: Maximizing economic, environmental, and societal value of the built environment. J Constr Eng Manag 133(9):619–628
21. MarcelinoSádaba S, GonzálezJaen LF, PérezEzcurdia A (2015) Using project management as a way to sustainability. from a comprehensive review to a framework definition. J Clean Prod 99:1–16
22. Moldavska A, Welo T (2017) The concept of sustainable manufacturing and its definitions: a content-analysis based literature review. J Clean Prod 166:744–755
23. Nilsson M, Griggs D, Visbeck M, Riegler C, McCollum D (2017) Introduction: a framework for understanding SDG Interactions. In: A guide to SDG interactions: From science to implementation. International Council for Science (ICSU)
24. OECD (2019) Sustainable Infrastructure for Low-Carbon Development in Central Asia and the Caucasus. Paris
25. O'Connor A, Shumate M (2010) An economic industry and institutional level of analysis of corporate social responsibility communication. Manag Commun Q 24(4):529–551
26. Opoku A, Deng J, Elmualim A, Ekung S, Hussien AA, Abdalla SB (2022) Sustainable procurement in construction and the realisation of the sustainable development goal (SDG) 12. J Clean Prod 376:134294
27. Reed B (2007) Shifting from 'sustainability'to regeneration. Build Res Inf 35(6):674–680
28. Sahely HR, Kennedy CA, Adams BJ (2005) Developing sustainability criteria for urban infrastructure systems. Can J Civ Eng 32(1):72–85
29. Shaw G, Kenny J, Kumar A, Hood D (2012) Sustainable infrastructure operations: a review of assessment schemes and decision support. In: Proceedings of the 25th Australian Road Research Board Conference, pp 1–18. Australian Road Research Board (ARRB)
30. Shen L, Wu Y, Zhang X (2011) Key assessment indicators for the sustainability of infrastructure projects. J Constr Eng Manag 137(6):441–451
31. Surbeck, C. Q., & Hilger, H. (2014). Social sustainability and important indicators in infrastructure. In *World Environmental and Water Resources Congress 2014* (pp. 2078–2093).
32. Thacker S et al (2019) Infrastructure for sustainable development. Nature Sustainability 2(4):324–331
33. Thomé AMT, Ceryno PS, Scavarda A, Remmen A (2016) Sustainable infrastructure: A review and a research agenda. J Environ Manage 184:143–156
34. Tregidga H, Milne MJ (2006) From sustainable management to sustainable development: a longitudinal analysis of a leading New Zealand environmental reporter. Bus Strateg Environ 15(4):219–241
35. Ugwu OO, Kumaraswamy MM, Wong A, Ng ST (2006) Sustainability appraisal in infrastructure projects (SUSAIP). Part 1: development of indicators and computational methods. Autom. Constr. 15(2):239–251

36. Valdes-Vasquez R, Klotz LE (2013) Social sustainability considerations during planning and design: framework of processes for construction projects. J Constr Eng Manag 139(1):80–89
37. Wang X, Yuen KF, Wong YD, Li KX (2020) How can the maritime industry meet sustainable development goals? An analysis of sustainability reports from the social entrepreneurship perspective. Transp Res Part d: Transp Environ 78:102173
38. Zeng SX, Ma HY, Lin H, Zeng RC, Tam VW (2015) Social responsibility of major infrastructure projects in China. Int J Project Manage 33(3):537–548

Integrating Risk and Stakeholder Management in Complex Mega-projects: A Multilayer Network Analysis Approach

Gabriel Castelblanco, Enrico Maria Fenoaltea, Alberto De Marco, Paolo Demagistris, Sandro Petruzzi, and Davide Zeppegno

Abstract The complexity of megaprojects requires methodological methods able to analyze the interaction between multiple systems in an integrated way rather than analyzing isolated systems (e.g., either stakeholders or risks). However, although project management has devoted significant efforts to employing classical unidimensional networks to analyze either stakeholders or risks in an isolated way, there is lacking multidimensional analyses that integrate both perspectives to unravel complexity. By integrating the risks and the economic transaction between stakeholders in a multilayer network analysis, this study aims to constitute the first effort to develop multilayer analysis to unravel behaviors that are hidden in unidimensional analysis. Using a 220 million € Italian education megaproject, this study exposes the risk allocation patterns and potential suboptimal risk allocation patterns resulting in stakeholders' underperformance and project issues. The analysis exposed that public stakeholders with low interdependence demonstrated a good performance due to their active role in the monetary transfers but low-risk exposition, which highlights risk transfer to the private sector. Complementary, the analysis of the betweenness and eigenvector centrality in the aggregated multilayer network demonstrates that when is significantly lower than these metrics in the unidimensional money flow network, there is a potential suboptimal risk allocation that may result in underperformance of these stakeholders.

Keywords Network Analysis · Risk Propagation · Management for stakeholder · Project success factors · New metrics for performance management · Change of paradigm

G. Castelblanco (✉) · A. De Marco · P. Demagistris
Politecnico di Torino, Turin, Italy
e-mail: gabriel.castelblanco@polito.it

S. Petruzzi · D. Zeppegno
Università Degli Studi di Torino, Turin, Italy

E. M. Fenoaltea
University of Fribourg, Fribourg, Switzerland

E. Favari and F. Cantoni (eds.), *Complexity and Sustainability in Megaprojects*, Lecture Notes in Civil Engineering 342,
https://doi.org/10.1007/978-3-031-30879-6_6

1 Introduction

Megaprojects are socio-technical systems involving long execution periods (including in some cases long-term operation and maintenance) and huge investments (often involving private and public funds) [1]. The complexity of these projects involves a myriad of interconnected impacts on the economy, society, and environment [2].

Project Management literature has devoted significant efforts to employing classical unidimensional networks to analyze either stakeholders or risks in an isolated way [2–7]. However, these unidimensional classical networks are restricted to one-dimension analysis, neglecting the multidimensional complex of megaprojects. Moreover, complex systems can not be fully understood just by knowing their elements (dimensions) in an isolated way but it is required to analyze them from a multidimensional perspective [8].

Most network analyses in extant literature rely on Social Network Analysis for identifying, analyzing, and prioritizing stakeholders and their interests and impact on the project's outcomes without considering both economic transfers between stakeholders and risks [3, 5, 9]. Conversely, a few papers are focused on applying Social Network Analysis to risks [3]. However, these networks rely on semi-qualitative data resulting from workshops involving a few participants (i.e., less than 10 persons) instead of building the networks based on non-subjective quantitative data [3].

To address the multidimensional complexity of megaprojects, this paper proposes a multilayer network that integrates the risks and the economic transaction between stakeholders in a multilayer network analysis. This analysis is applied in a 220 million € Italian education megaproject.

2 Background

2.1 Limitations of Traditional Network Analysis

Over the past two decades, network science has emerged as the main quantitative approach to investigating complex systems in a variety of scientific fields such as physics [10], biology [11], sociology [12], and economics [13].

In its basic form, a network describes a system as a set of simplified entities, called nodes, that interact with each other through so-called links (which can be indirect, i.e., without a direction, or direct, i.e., with a specific direction). This representation provides a complete and holistic view of the system under consideration and, in addition, reveals novel features that would be invisible with different approaches. Moreover, the structure of the network, i.e., the pattern of interaction between nodes, is crucial to the properties and evolution of a system [14].

Often, however, such a representation turns out to be a too reductive description of the system, and one risks finding the same problems typical of reductionist approaches that network science aims to avoid [15]. For example, consider stakeholders' systems: although the description of stakeholders as a set of nodes (i.e., individuals/organizations) and their relationships (i.e., links) can be useful in understanding the mechanisms underlying important phenomena such as polarization [16] or information spreading [17], it is too simplistic and stylized. Indeed, all possible types of relationships between two individuals/organizations (e.g., collaboration) cannot be represented by the same object (i.e., a link between two nodes) and, moreover, they can be the result of different dynamic processes (such as negotiations or information sharing) that increase their complexity [18–20]. To complicate the picture, the evolution of a social network depends not only on the social network itself, but there are other social dimensions (cultural, economic, or historical factors, for example) that can influence its dynamics.

Similarly, in the field of project management, an analysis based on traditional networks has the shortcoming of being reductive: the relationships between stakeholders in a project are multiple and, for example, the interaction (e.g., economic transaction, collaboration, trust) between two stakeholders may depend on exogenous factors to the network of stakeholders or even to the project itself. It is therefore necessary to add more information to traditional network models while keeping them tractable and powerful.

2.2 Multilayer Networks

To overcome the limitations of traditional networks described above, so-called multilayer networks have recently been introduced [15]. They can be seen as multi-dimensional networks in which each dimension describes a different layer of complexity of the system. Specifically, a multilayer network is a collection of M traditional networks, called layers. Different layers can also communicate: two layers are connected if at least one node in one layer has a link to at least one node in the other layer. Links between two nodes belonging to the same layer are called "intra-layer" links, while those between two nodes belonging to different layers are called "inter-layer" links (see the left of Fig. 1 for an example of a multilayer network). In general, there can be an overlap between nodes of different layers, that is, there can be multiple copies of the same node in different layers. A classic example is the multilayer description of online social networks: each layer represents the network of a different online platform (e.g., Facebook, Instagram, Twitter); each individual may have an account on different platforms, and therefore they are a node present in different layers of the multilayer. However, as shown in the next section, in the multilayer network of the system considered in this paper, the nodes are all different, that is, no node has a copy of itself.

Multilayer networks are a direct generalization of traditional networks. Indeed, traditional networks are multilayer networks with only one layer and thus are also called unidimensional or monoplex networks. however, in addition to the explicit advantage of being able to incorporate more information and thus describe the system

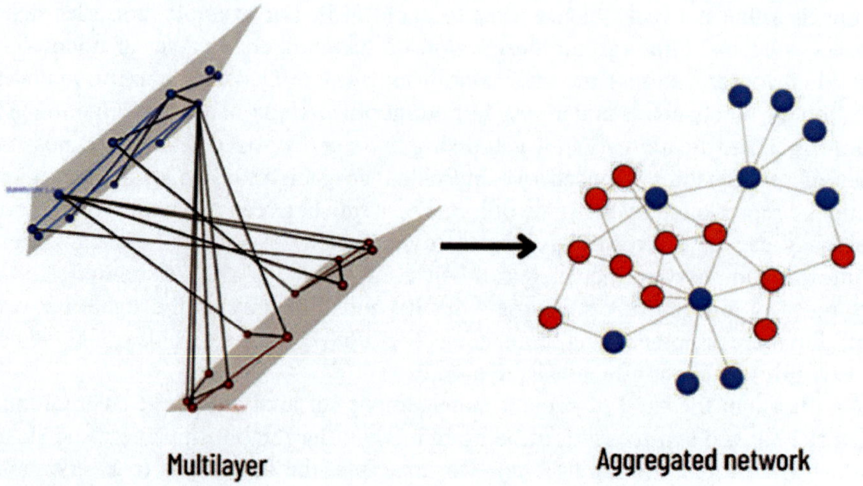

Fig. 1 Representation of Multilayer Network and Aggregated Network

more accurately, the analysis of multilayer networks can reveal novel properties of the system that would have been undetectable with an analysis based on unidimensional networks. For example, navigability and diffusion properties in a multilayer network can be very different from those obtained by studying its one-dimensional version [21, 22]; also, a topic relevant to this paper concerns node centrality, that is, the importance of a node's geographic location within the network: multilayer structures have significant effects on centrality measures, and ignoring them can lead to different ranking results than what one observes for multilayer networks [23, 24].

So, especially in the field of project management, where misleading analyses can lead to large losses of money and time, it is increasingly necessary to adopt the multilayer networks formalism to obtain reliable descriptions and predictions.

3 Methodology

3.1 Data Collection

To apply the proposed approach, a megaproject in the education sector in Italy was selected as the real case for implementing a multilayer network approach. This case was selected for this study because it gathers the following features required for a complex megaproject:

- Involves huge investments: more than 220 million €.
- Comprehends a long execution period: one year for design, three years for the construction period, and twenty years of operation.

- There are multiple risks involving a huge monetary valuation.

To gather the relevant information from the case study, multiple data sources were triangulated. Firstly, the legislation was reviewed in order to get all the baseline information established in the adjudication decree concerning stakeholders' transactions, responsibilities, and financial sources. Next, the contractual information was collected including the risk valuation considered and the specific financial transactions projected between the owner and the contractor. Lastly, the internal project management reports were gathered to contrast the intended risk and money transfer with the real ones during the execution of the project.

3.2 *Development of the Multilayer Network*

Based on the data collected (i.e., the economic transactions between stakeholders and risk factors of the project), a 2-layer network was designed. In the first layer, the nodes represent the stakeholders and there exists a direct link between node i and node j if there is an economic transaction from stakeholder i to stakeholder j. In the second layer, on the other hand, the nodes are the risk factors and there exists a direct link between node a and node b if there is a potential cause-and-effect relationship between the occurrence of risk a and that of risk b. The two layers are interrelated through "inter-layer" links, which connect stakeholders to risk factors. Specifically, a link exists between a node in the first layer i and a node in the second layer a if the stakeholder i is responsible for the risk factor a. It is worth mentioning that inter-layer links are not treated as direct links. Indeed, inter-layer links are considered bidirectional, since this study assumes that the cause-and-effect relationship is always possible in both directions: the stakeholder i, responsible for risk factor a, may cause risk occurrence because of its negligence; and simultaneously, the occurrence of risk a affects stakeholder i economically.

The multilayer network defined above belongs to a particular class of multilayer networks: the "networks of networks" [25]. This specific typology consists of networks that are interlaced to each other but formed by different types of nodes (i.e., the set of nodes of a layer has no overlap with the set of nodes of all the other layers). These have the advantage that they can be reduced to an aggregate network without loss of information, as demonstrated in [26]. The aggregate network is simply the reduction of the multilayer network into a single layer in which all nodes in the original multilayer network are present. As a result, there is no distinction between inter-layer links and intra-layer links (i.e., links that connected either two stakeholders or two risk factors). A graphical representation of how to reduce the multilayer network of the project to the corresponding aggregate network is shown in Fig. 1.

Figure 1 (left) represents a multilayer network of the project. In this network blue nodes are stakeholders and blue links are intra-layer links of the stakeholder layer; while red nodes represent risk factors, red links are intra-layer links of the risk layer; and black links are the inter-layer links. Figure 1 (right) shows the corresponding aggregated network. In this network, the nodes and links are drawn on the same plane, without distinction between intra-layer and inter-layer links.

The main advantage of working with the aggregated network rather than with the multilayer is that it is not necessary to rely on the tensor formalism that is normally used for the analysis of multilayer networks (and which is often tedious and ambiguous) [25] but the same metrics that are used to describe traditional networks can be implemented for the analysis of the aggregated network.

3.3 Metrics Meaning and Calculation

The centrality of stakeholders in the aggregate network is studied and compared with that obtained in the single-layer network composed of only stakeholders. The most intuitive measure of centrality is the so-called *node's degree centrality.* For node *i*, it is defined as the number of *i*'s neighbors divided by the total number of nodes in the network. Figure 1 shows that the node's degree of stakeholders in the aggregated network is different from that in the single-layer network. Indeed, it is evident the difference in the number of neighbors of each node when considering only blue nodes (single layer) versus considering all nodes (aggregated network). However, while the node's degree centrality is a simple measure of centrality, it is also uninformative. Indeed, it only gives information about the number of neighbors of each node but not about the node's strategic influence within the network. Thus, in this paper, the following metrics as measures of centrality are considered:

- The *eigenvector centrality*: Computes the transitive influence of a node within the network [27]. It is based on the recursive concept that nodes connected with highly influential nodes are also highly influential. Formally, eigenvector centrality is derived by solving the eigenvalue equation $\lambda x = xA$. Here, x *is* a vector whose element x_i is the eigenvector centrality of node *i; A* is the network adjacency matrix whose element A_{ij} is one if there is a directed link from node *i* to node *j,* and is zero otherwise; finally, λ is the largest eigenvalue of the matrix *A*. Hence, *x* corresponds to the eigenvector relative to the largest eigenvalue of the network. From the above eigenvalue equation, it is possible to write explicitly the eigenvector centrality of node *i* as $x_i \sum_k a_{k,i} x_k / \lambda$.
- The *betweenness centrality:* Measures how many shortest paths between any pair of nodes pass through a given node. Nodes with large betweenness may have considerable control over information passing [28]. Indeed, their removal from the network would highly affect communications between other nodes because they lie on the largest number of communication channels.

  Formally, the betweenness centrality of node *i* is expressed as $b_i = \sum_{s,t} \frac{n^i_{s,t}}{n_{s,t}}$, where $n^i_{s,t}$ is the number of shortest paths between node *s* and node *t* that pass by node *i,* while $n_{s,t}$ is the total number of shortest paths between node *s* and node *t.*

Unlike nodes' degree centrality, it is not obvious how these two metrics vary in the presence of multiple layers of networks (in our case, in the presence of the risk network layer). However, looking at the aggregated network of Fig. 1, for example, it is clear that the number of paths between two stakeholders in the aggregated network (i.e., considering both blue and red nodes) is larger than that in the single layer network (i.e., considering only blue nodes). As shown later, this means that the calculation of the eigenvector and betweenness centralities of stakeholders in the aggregated network will lead to different insights than those following from the same calculations in the single layer network.

Beyond the centrality of a stakeholder in the project, it is also interesting to know how important the addition of the risk layer is for extracting information regarding the reachability (in terms of information or economic costs, for example) of a stakeholder. In other words, a metric is needed to measure the dependence of a node on the multilayer structure in terms of reachability. To have a quantitative measure of this, interdependence metrics for a general multilayer network with undirected links were proposed in [29]. In the following, to adapt this metric to our network with directed links, a generalized definition of interdependence is proposed:

- I*nterdependence:* Measure how much a stakeholder's reachability varies from the single-layer network to the multilayer network. A stakeholder with a high value of interdependence can influence and be influenced by other stakeholders mainly through paths that pass into the risk layer network (thus through risk propagation); on the other hand, a stakeholder with a low value of interdependence can influence and be influenced by other stakeholders mainly through paths that pass into the stakeholders' layer network itself (thus through money transfer). Formally, the interdependence of a stakeholder i is defined as $\lambda_i \sum_{i \neq j} \frac{\psi_{ij}+\psi_{ji}}{\sigma_{ij}+\sigma_{ji}}$, where σ_{ij} is the total number of shortest paths between stakeholders i and j and ψ_{ij} is the number of shortest paths between stakeholders i and j that makes use of nodes in the risk layer.

3.4 Risk Network Simulations and Metrics Calculations

As mentioned above, data on potential risk factors in the project and economic transactions between stakeholders were collected. However, cause-effect relationships among the risk factors have not been identified. There are various approaches in the literature for determining such relationships, particularly based on probabilistic inference and, mostly, on expert judgment [30–32]. In this paper, a different approach is implemented: the network of risk factors was assumed to be a random network, i.e. a là Erdos-Reny [33], with connection probability *p*. Specifically, in the risk layer network, a link exists between node *a* and any node *b* with probability *p*, while such a link does not exist with probability *1-p*. The resulting network has a binomial degree distribution and average degree $(k) = p(N-1)$, where *N* is the number of nodes. In the extreme cases where $p = 1$ and $p = 0$, the result is a complete network or an empty network, respectively.

The above assumption is reasonable since, given the high complexity of the underlying phenomena, precise identification of cause-and-effect relationships among risk factors is often not possible, and even expert judgment can prove to be a misleading approximation. Hence, the assumption of a random network is the best guess. Moreover, within this framework, one can perform a Monte Carlo simulation to allow the description of a more general picture beyond the specific case of this paper, as is explained below.

To compute the metrics described above (i.e., x_i, b_i and λ_i) and compare the results for stakeholders between the single-layer network and the aggregated network, the following steps are performed:

1. Considering the single-layer network of stakeholders, the following metrics are calculated for each stakeholder *i*: x_i and b_i (interdependence only makes sense in the multilayer structure). These are then normalized such that $\sum_i x_i = \sum_i b_i = 1$.
2. Given the risk factors and a connection probability *p*, the risk network layer was constructed: as explained above, a link exists between risk *a* and risk *b* with probability *p*, while such a link does not exist with probability *1-p*. The resulting network is aggregated to the layer of the stakeholders' network, as in Fig. 1.
3. Considering the aggregated network constructed in Step 2, the metrics x_i, b_i and λ_i are computed for each stakeholder *i*. Again, they are normalized such that $\sum_i x_i = \sum_i b_i = \sum \lambda_i = 1$. Note that the sum is only over stakeholders and not risks.
4. *Montecarlo simulations:* steps 2 and 3 are repeated for 1000 realizations, and the values of the metrics are averaged over all the realizations. Hence, the value of any metric *m* is defined as $m_i = \sum_{j=1}^{1000} m^j_{i/1000}$, where $m^j{}_i$ is a general metric obtained in the *j*-th realization for the stakeholder *i*.

In the tables and figures of the next section, the results of the metrics obtained in step 1 are compared with those obtained in step 4. In the following, $p = 0.1$ is set as a reasonable value for the specific project presented in this paper. Computations with different values of *p* can be easily implemented to better describe projects with different properties.

4 Findings and Discussion

To standardize the representation of the networks, the nodes represent the stakeholders in all the cases. An analytical categorization based on a top-down approach was conducted for obtaining ten relevant stakeholders across the public and private sectors: Four stakeholders within the public sector (i) the Ministry of Education (MoE), which approved the project; (ii) the public university (owner) that act as the contractor of the project; (iii) the municipality (Mun) where the project is located, which plays a relevant role for the approval of the permits; and (iv) the Ministry of Finance, which provided funding for the feasibility phase of the project. The remaining stakeholders belong to the private sector composed. In this regard, the contractor is made by (v) the Responsible for Financing (RfF) that provides short-term private finance to the remaining contractor members and will be reimbursed in the long-term by the owner; (vi) the Responsible for Maintenance (RfM) that has to conduct the maintenance of the facilities during the 20 years maintenance period; (vii) the Responsible for the Execution (RfE) that must develop the construction of the project; (viii) the Responsible for the Design (RfD) that must develop the detailed design previous to the construction phase; and (ix) the Furniture provider (Forn) that supply the equipment and furniture for the labs and facilities. Lastly, (x) the Project Management Office (PMO) is responsible for managing the project on behalf of the owner.

4.1 Interdependence Analysis

To obtain fine-grained insights regarding the interaction between risks and stakeholders, this study conducted an interdependence analysis of stakeholders. In this regard, stakeholders with large interdependence can affect and be affected mostly through risk propagation. Conversely, stakeholders with low interdependence are less affected by risk but mostly because of the monetary transfers between stakeholders.

The results of the interdependence analysis are shown in Fig. 2 and Table 1. According to the network and its metrics, two stakeholders within the contractor present the highest risk exposure, namely, the Responsible for Execution and the Responsible for Maintenance. Conversely, the public sector's stakeholders demonstrated the lowest risk exposure, namely, the Owner, the Municipality, the Ministry of Education, and the Ministry of Finance. These stakeholders within the public sector demonstrated playing a more active role in the monetary transfers but transferring most of the risks to the private sector.

4.2 Comparison between Unidimensional Social Network Analysis and Aggregated Network

Social Network Analysis was incorporated into the analysis to provide more insights at the stakeholder level based on their roles according to quantitative metrics. Therefore, based on the contractual documents, the financial exchange between stakeholders was transformed into networks. The direction of the links establishes the source and the recipient of the financial transfers between stakeholders.

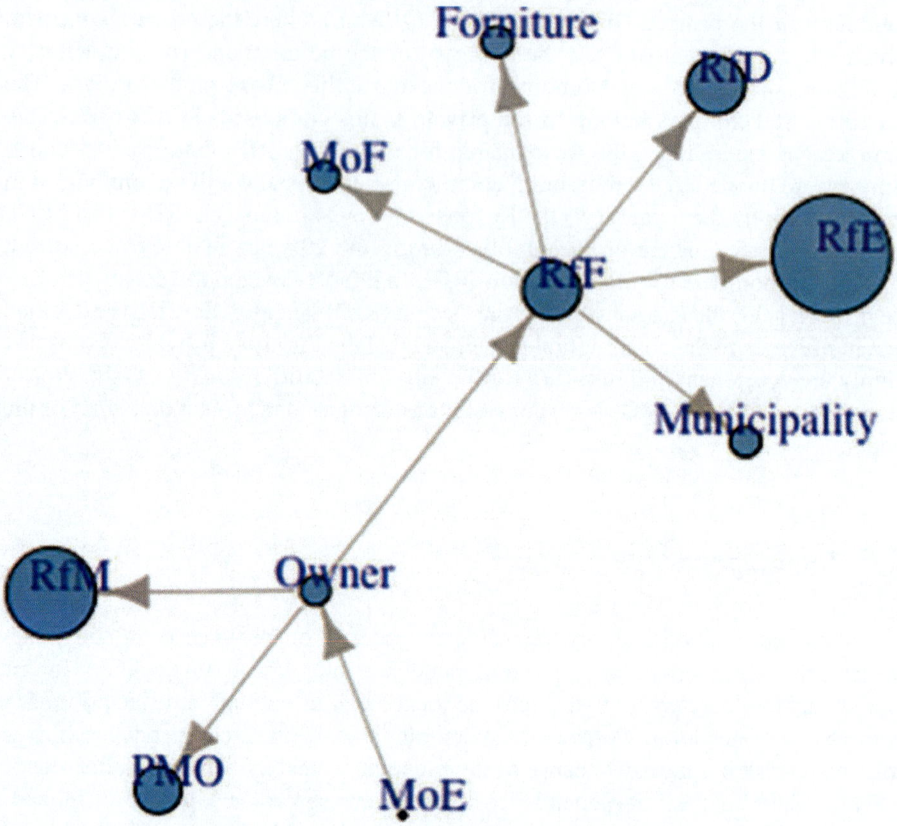

Fig. 2 Aggregated Network

Table 1 Interdependence Metrics

Stakeholder	MoE	Owner	Mun	RfF	RfM	RfE	RfD	PMO	Mof	Forn
Metric	*0.00*	*0.15*	*0.16*	*0.30*	*0.47*	*0.61*	*0.30*	*0.24*	*0.16*	*0.16*

Note: The size of the node is proportional to the stakeholder's interdependence value.

The comparison between traditional unidimensional social networks and aggregated multilayer networks through the betweenness is presented in Table 2 and Fig. 3. Interestingly, there is a misalignment regarding the preponderance of the public owner and the Responsible for Finance between the unidimensional network (before aggregation) and the multidimensional network (after aggregation). In the former network, the Responsible for Finance has the leading role because of its role to provide finance to the contractor companies and pay taxes to the municipality. However, in the latter network, when the risk is incorporated into the analysis, the Responsible for Finance loose preponderance due to its low exposure to risk. In effect, despite the significant flow of money that is mediated through the Responsible for Finance, there is no significant risk exposure considering that the disbursements made have to be approved by the Owner, which means that the owner bears the risks in case of any future dispute due to cost overruns. Additionally, the contractual clauses establish that the public owner bears macroeconomic risks. Consequently, the Responsible for Finance is only a financial mediator with no risk exposure.

This misalignment between both networks in regards to both stakeholders exposes a relevant issue in the scheme of the contractor. While project delivery methods that incorporate the finance and operation within the contract (e.g., Design-Finance-Build-Operate-Transfer) use to rely on special purpose vehicles that do not include financial corporations but are composed mainly of private firms with expertise in building, managing, and operating specific megaprojects; the case study analyzed incorporated within the contractor group a financial corporation that did not add any value for the development of the project but only provides financing. As a result, this contractor member demonstrated significant disinterest in the performance of the project due to the risk allocation pattern that neglected to transfer any risk to this company.

This finding also demonstrates that traditional unidimensional networks are not able to capture misalignments between money flow (and even profitability) and risk exposure properly, against the recommended practice in risk allocation and management [34–45]

Simultaneously, the comparison between traditional unidimensional social networks and aggregated multilayer networks through the eigenvector centrality is presented in Table 3 and Fig. 4. Besides the misalignment previously explained,

Table 2 Betweenness before and after aggregation

	MoE	Owner	Mun	RfF	RfM	RfE	RfD	PMO	Mof	Forn
Before aggregation	*0.00*	*0.44*	*0.00*	*0.56*	*0.00*	*0.00*	*0.00*	*0.00*	*0.00*	*0.00*
After aggregation	*0.00*	*0.51*	*0.00*	*0.23*	*0.02*	*0.22*	*0.02*	*0.00*	*0.00*	*0.00*

Note: The size of the node is proportional to the stakeholder's betweenness centrality computed before (left) and after (right) the aggregation.

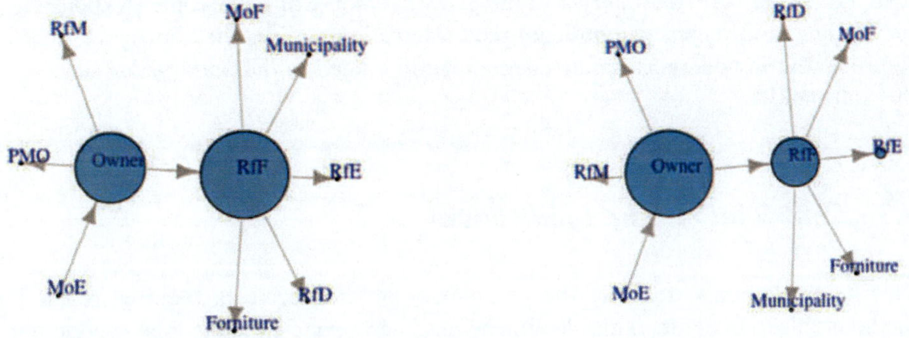

Fig. 3 Betweenness before and after Aggregation

Table 3 Eigenvector before and after aggregation

	MoE	Owner	Mun	RfF	RfM	RfE	RfD	PMO	Mof	Forn
Before aggregation	*0.06*	*0.16*	*0.09*	*0.23*	*0.06*	*0.09*	*0.09*	*0.06*	*0.09*	*0.09*
After aggregation	*0.06*	*0.38*	*0.02*	*0.11*	*0.10*	*0.18*	*0.04*	0.06	*0.02*	*0.02*

Note: The size of the node is proportional to the stakeholder's eigenvector centrality computed before (left) and after (right) the aggregation.

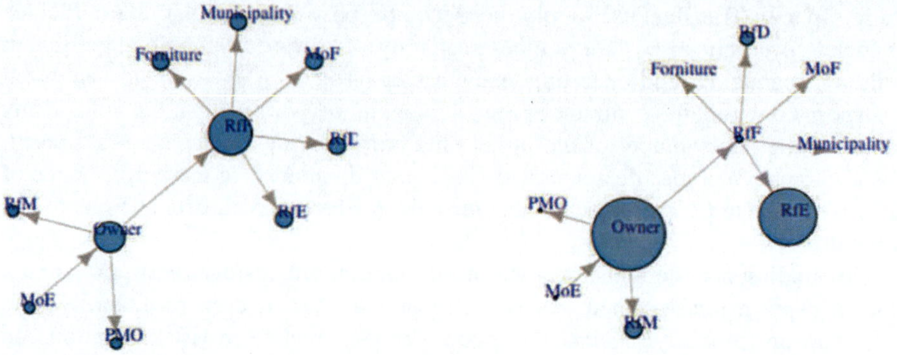

Fig. 4 Eigenvector before and after Aggregation

this metric allows for identifying the significant change in the role of the Responsible for Execution and the Responsible for Design between the unidimensional network of money flow (before aggregation) and the multidimensional network (after aggregation). When the risk is not considered (unidimensional network) both stakeholders had similar representativeness among the stakeholders within the private sector. However, when the risk is incorporated into the analysis (multidimensional network), the Responsible for Execution doubles its representativeness while the Responsible for Design halve its eigenvector value.

This disagreement demonstrates, on one hand, that the Responsible for Execution bear a significant proportion of the risks, following the risk allocation pattern of transferring the construction risk to the private sector. Conversely, the Responsible for Design is not bearing significant risk. Interestingly, the project analyzed had a significant time underperformance significantly related to the poor performance of this stakeholder.

4.3 Contributions and Implications

This study demonstrates that the complexity of megaprojects requires research methods able to integrate multiple dimensions interconnected rather than conducting unidimensional analysis. Particularly, multidimensional network analysis demonstrated the ability to unravel complex behaviors integrating traditional dimensions of Social Network Analysis and risk propagation.

Moreover, this study revealed that whenever the multilayer betweenness or eigenvector decreases for a company in the private sector in the multilayer network in comparison with the unidimensional money flow network, there is a potential suboptimal risk allocation that may result in underperformance of these stakeholders.

5 Conclusions

The complexity of megaprojects requires methodological methods able to analyze the interaction between multiple systems in an integrated way rather than analyzing isolated systems (e.g., either stakeholders or risks). However, although project management has devoted significant efforts to employing classical unidimensional networks to analyze either stakeholders or risks in an isolated way, there is lacking multidimensional analyses that integrate both perspectives to unravel complexity.

By integrating the risks and the economic transaction between stakeholders in a multilayer network analysis, this study aims to constitute the first effort to develop multilayer analysis to unravel behaviors that are hidden in unidimensional analysis. Using a 220 million € Italian education megaproject, this study exposes the risk allocation patterns and potential suboptimal risk allocation patterns resulting in stakeholders' underperformance and project issues.

The analysis exposed that public stakeholders with low interdependence demonstrated a good performance due to their active role in the monetary transfers but low-risk exposition, which highlights risk transfer to the private sector. Complementary, the analysis of the betweenness and eigenvector centrality in the aggregated multilayer network demonstrates that is significantly lower than these metrics in the unidimensional money flow network, there is a potential suboptimal risk allocation that may result in underperformance of these stakeholders.

References

1. Priemus H, Flyvbjerg B, van Wee B (2008) Decision-making on mega-projects: cost-benefit analysis, planning and innovation. Edward Elgar Publishing Limited, Cheltenham, UK
2. Zheng X, Chen J, Han Y, Ren L, Shi Q (2021) Unveiling complex relational behavior in megaprojects: a qualitative-quantitative network approach. Int J Proj Manag 39(7):738–749. https://doi.org/10.1016/j.ijproman.2021.07.001
3. Yang RJ, Zou PXW, Wang J (2016) Modelling stakeholder-associated risk networks in green building projects. Int J Proj Manag 34(1):66–81. https://doi.org/10.1016/j.ijproman.2015.09.010
4. Mok KY, Shen GQ, Yang RJ, Li CZ (2017) Investigating key challenges in major public engineering projects by a network-theory based analysis of stakeholder concerns: a case study. Int J Proj Manag 35(1):78–94. https://doi.org/10.1016/j.ijproman.2016.10.017

5. Mok KY, Shen GQ, Yang RJ (2017) Addressing stakeholder complexity and major pitfalls in large cultural building projects. Int J Proj Manag 35(3):463–478. https://doi.org/10.1016/j.ijproman.2016.12.009
6. Shen W, Wang Y, Lee SH (2022) Formation of inter-project ties from the sender–recipient perspective: Roles of task interdependence and functional interdependence. Int J Proj Manag 40(5):577–586. https://doi.org/10.1016/j.ijproman.2022.05.005
7. Wen Q, Qiang M, Gloor P (2018) Speeding up decision-making in project environment: The effects of decision makers' collaboration network dynamics. Int J Proj Manag 36(5):819–831. https://doi.org/10.1016/j.ijproman.2018.02.006
8. Meadows DH (2009) Thinking in systems. Earthscan, London. https://doi.org/10.1016/j.aeue.2014.08.008
9. Rojas R, Bennison G, Gálvez V, Claro E, Castelblanco G (2020) Advancing collaborative water governance: unravelling stakeholders ' relationships and influences in contentious river basins. Water (Switzerland) 12(3316):1–25. https://doi.org/10.3390/w12123316
10. Strogatz SH (2001) Exploring complex networks. Nature 410(6825):268–276
11. Gosak M et al (2018) Network science of biological systems at different scales: a review. Phys Life Rev 24:118–135. https://doi.org/10.1016/j.plrev.2017.11.003
12. Watts DJ (2004) The 'new' science of networks. Annu Rev Sociol 30:243–270. https://doi.org/10.1146/annurev.soc.30.020404.104342
13. Jackson MO (2010) Social and Economic Networks. Princeton University Press, Princeton
14. Albert R, Barabási AL (2002) Statistical mechanics of complex networks. Rev Mod Phys 74(1):47–97. https://doi.org/10.1103/RevModPhys.74.47
15. Aleta A, Moreno Y (2019) Multilayer networks in a nutshell. Annu Rev Condens Matter Phys 10(1):45–62. https://doi.org/10.1146/annurev-conmatphys-031218-013259
16. Li L, Scaglione A, Swami A, Zhao Q (2013) Consensus, polarization and clustering of opinions in social networks. IEEE J Sel Areas Commun 31(6):1072–1083. https://doi.org/10.1109/JSAC.2013.130609
17. SîrbuA, Loreto V, Servedio VDP, Tria F (2017) Opinion dynamics: models, extensions and external effects. In: Loreto V, et al. Participatory sensing, opinions and collective awareness. understanding complex systems. Springer, Cham. https://doi.org/10.1007/978-3-319-25658-0_17
18. Baybusinov IB, Fenoaltea EM, Zhang YC (2022) Negotiation problem. Phys A Stat Mech Appl 592:126806. https://doi.org/10.1016/j.physa.2021.126806
19. FenoalteaEM, Baybusinov IB, Na X, Zhang YC (2022) A local interaction dynamic for the matching problem. Phys A Stat Mech Appl 604. https://doi.org/10.1016/j.physa.2022.127690
20. Fenoaltea EM, Baybusinov IB, Zhao J, Zhou L, Zhang YC (2021) The stable marriage problem: an interdisciplinary review from the physicist's perspective. Phys Rep 917:1–79. https://doi.org/10.1016/j.physrep.2021.03.001
21. De DomenicoM, Granell C, Porter MA, Arenas A. (2016) The physics of spreading processes in multilayer networks 12, August 2016. https://doi.org/10.1038/NPHYS3865
22. SalehiM, Sharma R, Marzolla M, Magnani M, Siyari P, Montesi D (2015) Spreading processes in multilayer networks 2(2):65–83
23. De Domenico M, Solé-Ribalta A, Omodei E, Gómez S, Arenas A (2015) Ranking in interconnected multilayer networks reveals versatile nodes. Nat Commun 6:1–6. https://doi.org/10.1038/ncomms7868
24. De DomenicoM, Solé-Ribalta A, Omodei E, Gómez S, Arenas A (2013) Centrality in interconnected multilayer networks, pp 1–12. arXiv:1311.2906. https://doi.org/10.1038/ncomms7868
25. Boccaletti S, Bianconi G, Criado R, Genio CI (2014) The structure and dynamics of multilayer networks. Phys Rep 544(1):1–122. https://doi.org/10.1016/j.physrep.2014.07.001
26. De Domenico M, Nicosia V, Arenas A, Latora V (2015) Structural reducibility of multilayer networks. Nat Commun 6:1–9. https://doi.org/10.1038/ncomms7864
27. Ruhnau B (2000) Eigenvector-centrality - a node-centrality. Soc Netw 22(4):357–365. https://doi.org/10.1016/S0378-8733(00)00031-9

28. Barthélemy M (2004) Betweenness centrality in large complex networks. Eur Phys J B 38(2):163–168. https://doi.org/10.1140/epjb/e2004-00111-4
29. BattistonF, Nicosia V, Latora V (2014) Structural measures for multiplex networks. Phys Rev E - Stat Nonlinear, Soft Matter Phys 89(3):1–14. https://doi.org/10.1103/PhysRevE.89.032804
30. Ojha R, Ghadge A, Tiwari MK, Bititci US (2018) Bayesian network modelling for supply chain risk propagation. Int J Prod Res 56(17):5795–5819. https://doi.org/10.1080/00207543.2018.1467059
31. Fang C, Marle F (2013) Dealing with project complexity by matrix-based propagation modelling for project risk analysis. J Eng Des 24(4):239–256. https://doi.org/10.1080/09544828.2012.720014
32. Gu J, Xia X, He Y, Xu Z (2017) An approach to evaluating the spontaneous and contagious credit risk for supply chain enterprises based on fuzzy preference relations. Comput Ind Eng 106:361–372. https://doi.org/10.1016/j.cie.2017.02.012
33. Gómez-GardeñesJ, Moreno Y (2006) From scale-free to Erdos-Rényi networks. Phys Rev E - Stat. Nonlinear, Soft Matter Phys 73(5):1–7. https://doi.org/10.1103/PhysRevE.73.056124
34. CastelblancoG, Guevara J, Mesa H, Flores D (2020) Risk allocation in unsolicited and solicited road public-private partnerships: Sustainability and management implications. Sustain 12(11). https://doi.org/10.3390/su12114478
35. Castelblanco G, Guevara J (2022) Crisis driven literature in PPPs: a network analysis. In: IOP Conference Series: Earth and Environmental Science. IOP Publishing, Melbourne. https://doi.org/10.1088/1755-1315/1101/5/052002. https://oxford-abstracts.s3.amazonaws.com/bbb900d1-74cf-4dbb-b573-81550cf6ef90.pdf
36. Castelblanco G, Guevara J, Salazar J (2022) Remedies to the PPP crisis in the COVID-19 pandemic: lessons from the 2008 global financial crisis. J Manag Eng 38(3):1–18. https://doi.org/10.1061/(ASCE)ME.1943-5479.0001036
37. Castelblanco G, Guevara J, Mesa H, Hartmann A (2022) Social legitimacy challenges in toll road PPP programs: analysis of the colombian and chilean cases. J Manag Eng 38(3):1–15. https://doi.org/10.1061/(ASCE)ME.1943-5479.0001010
38. Marcellino M, Castelblanco G, De Marco A (2022) Multiple linear regression model for project's risk profile and DSCR. In: IOP Conference Series: Materials Science and Engineering
39. Marcellino M, Castelblanco G, De Marco A (2022) Contract renegotiation in PPPs: evidence from Italy. In: IOP Conference Series: Materials Science and Engineering
40. Castelblanco G, Guevara J (2020) Risk Allocation in PPP unsolicited and solicited proposals in Latin America: pilot study in Colombia. Constr Res Congr 2020:1321–1329. https://doi.org/10.1061/9780784482889.141
41. Castelblanco G, Guevara J, Rojas D, Correa J, Verhoest K (2023) Environmental impact assessment effectiveness in public-private partnerships: study on the Colombian road program. J Manag Eng. https://doi.org/10.1061/JMENEA/MEENG-5015
42. Castelblanco G, Guevara J (2022) Building bridges: unraveling the missing links between public-private partnerships and sustainable development. Proj Leadersh Soc. https://doi.org/10.1016/j.plas.2022.100059
43. CastelblancoG, Guevara J, Mendez-Gonzalez P, PPP renegotiation flight simulator: a system dynamics model for renegotiating PPPs after pandemic crisis. https://doi.org/10.1061/9780784483978.011
44. CastelblancoG, Guevara J, Mendez-Gonzalez P (2022) In the name of the pandemic: a case study of contractual modifications in PPP solicited and unsolicited proposals in COVID-19 Times. https://doi.org/10.1061/9780784483978.006
45. Castelblanco G, Guevara J, Mesa H, Sanchez A (2021) Semantic network analysis of literature on public-private partnerships. J Constr Eng Manag 147(5):1–16. https://doi.org/10.1061/(ASCE)CO.1943-7862.0002041

Stakeholder Involvement in the Development of Megaprojects: A Dedicated Framework for Transport Infrastructure

Francesco Cellerino and Mauro Mancini

Abstract Megaprojects have emerged as the preferred way to deliver results in a wide range of industries characterized by high complexity. Lately, due to the failure of several megaprojects because of poor stakeholder management, there was an increasing recognition of the need to involve the "non-market stakeholders" since the early stages. However, this awareness was not followed by the development of an applicable model for stakeholder inclusion in transport infrastructure, the subject of this work. After a review of specific literature, existing guidelines on citizen consultation were integrated with specific individual cases to balance the two contributions. At the end of the process, a framework composed of questions was proposed to guide the stakeholder involvement process. The questions were paired with cases related to transport infrastructure, to overcome the problem of the inherent generality of the guidelines. Finally, two megaprojects were analyzed through the proposed model, namely the enhancement of the A66 in the United Kingdom, and the preliminary stages of a similar Italian project. The Italian case study will then also be followed in later stages, allowing real-time comparison between the model and a megaproject, which will lead to the refinement of research questions and the integration of data.

Keywords Megaprojects · Stakeholder management · Stakeholder engagement · Transport infrastructure

1 Introduction

Megaprojects are increasingly being used to deliver results, not only in the field of infrastructures and construction projects, but also in a wide range of industries where the stakeholder and the local community involvement, the political interest,

F. Cellerino (✉) · M. Mancini
Department of Management, Economics and Industrial Engineering, Politecnico di Milano, Via Lambruschini 4B, 20156 Bovisa, Milan, Italy
e-mail: francesco.cellerino@polimi.it

M. Mancini
e-mail: mauro.mancini@polimi.it

E. Favari and F. Cantoni (eds.), *Complexity and Sustainability in Megaprojects*, Lecture Notes in Civil Engineering 342,
https://doi.org/10.1007/978-3-031-30879-6_7

the completion timeline, the complexity and the definition of success play a big role. [1] Civil engineering, public infrastructures and construction megaprojects often raise attention within communities because of the peculiar nature of their outcomes. For instance, they may involve the radical transformation of the landscape, both in terms of new venues created or interventions on nature. [2, 3] Therefore, each megaproject has a history of its own, and the so-called "lessons learned" rarely stem from experience.

Specifically, this work is mainly focused on transport infrastructures, which are a peculiar example in the field of megaprojects because of their spatial extent, that affects many different local communities and territorial realities. The development of such infrastructure was not treated from a technical point of view, as the objective was to frame the issue from the point of view of governance and involvement of the so-called "non-market stakeholders". Their involvement is particularly crucial and can boost the generation of benefits precisely because of their knowledge of the territory, extremely valuable to optimize project plans. Over the years, partly due to the failure of several megaprojects because of poor stakeholder management, there has been an increasing recognition of the need to involve stakeholders at various levels in the development of project plans. However, the literature review revealed the absence of a comprehensive framework dedicated to non-market stakeholder consultation in the context of transport infrastructure.

This finding thus gave rise to several research questions, which were addressed by comparing and integrating generic guidelines and more specific documents and papers. Their combination led to the generation of a framework composed of focused research questions, meant to address all the steps of stakeholders' involvement specifically in the case of transport infrastructure megaprojects, and thus to provide a support for managers in the consultation process. Finally, two case studies are analyzed more deeply to provide useful comments generated by an assessment against the provided model. To further assess the validity of the model, one of the two case studies will then be followed in the forthcoming months, as it is still in its early phases.

2 Megaprojects and Stakeholders

Megaprojects were initially defined as projects which had a total cost of more than 1B$, a threshold which has been widely accepted as a universal reference. [4, 5] But the economic relevance itself is not enough to define megaprojects, that by nature present a wider variety of factors scattered throughout their life cycle. The involvement of public and private stakeholders and their relationships create a complicated institutional field that often generates delays and huge cost overruns, where local communities play an important role, too. Thus, because of the impact that they have on different levels, many studies suggest to involve local communities in the early phases of a project to decrease the risk of conflicts emergence, improving project performance. Di Maddaloni and Davis [6] reported that often local communities

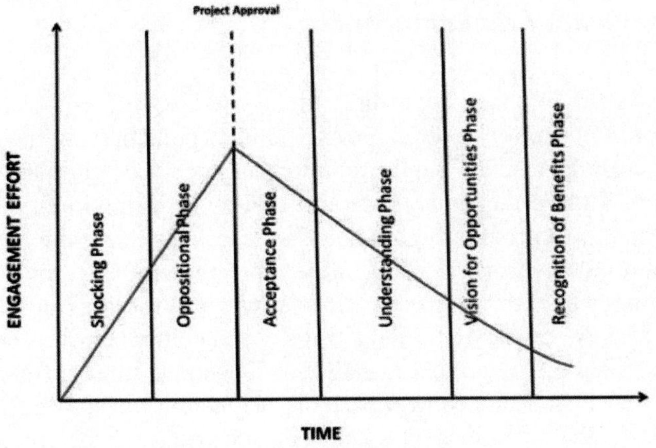

Fig. 1 Engagement effort at the local level through the major public infrastructure and construction projects (MPIC) life cycle (from Di Maddaloni F., Davis K. [6])

experience different behavioral phases throughout the project life cycle, that are a shocking phase, an oppositional phase, an acceptance phase, an understanding phase, the vision for opportunities phase and the recognition of benefits phase. From the perception of project managers, the engagement effort of the local communities is different in every phase, and Fig. 1 provides a summary of the engagement trend throughout the project.

1. The **shocking phase** begins when the project becomes real, and a scheme is proposed. The uncertainty generated by the announcement in the local community triggers the engagement effort and an initial opposition.
2. In the **oppositional phase**, the local communities try to influence the decision-makers before the project approval, often causing delays. The outcome of the negotiations and the conflicts will define the boundary for the acceptance phase, requiring effort and investments to achieve the approval.
3. The **acceptance phase** begins as soon as the project is approved. The influence of secondary stakeholders drops together with strong opposition.
4. In the **understanding phase**, project managers perceive that the emotional attachment of local communities weakens, as they start to understand the purpose of the project.
5. As the first results become tangible, the local communities usually enter the **vision for opportunities phase**, recognizing the future benefits that the project can deliver.
6. In the **recognition of benefits phase**, the tangible asset has been delivered. The local communities can finally analyze the project as a whole and recognize the benefits delivered by it.

2.1 *Involvement of Stakeholders*

Megaprojects remarkably impact the daily life of citizens, so they must be managed with the utmost caution on every level. From a political point of view, their outcome is often seen as a material result of public policy and is perceived as important to consolidate the legacy. Thus, policymakers are attracted by investing in infrastructures to create employment, to provide high quality services, to reduce the environmental impact and many other reasons. [3, 5, 7] Furthermore, the temporary nature of projects requires a higher level of effort to ensure trust between authorities and stakeholders, generating a "sense of ownership", and the social and cultural environment has to be carefully considered by the project managers. The general trend instead is to focus on technical aspects and procedures, partially neglecting the stakeholder management. To improve the performance, a better balance between the "management of stakeholders" and "management for stakeholders" approaches should be sought. [8, 9] Another aspect that cannot be ignored is that often times communities are involuntary stakeholders, and in their position, they lack the possibility of effectively negotiating their relationship with their counterpart. From the literature review emerged that there is a limited understanding of the involvement of secondary actors in public infrastructures and construction projects, which jeopardizes the achievement of the final expected benefits [4].

2.2 *Non-market Stakeholders and Project Governance*

Megaprojects remarkably impact the daily life of citizens, so they must be managed with the utmost caution on every level. Furthermore, the temporary nature of projects requires a higher level of effort to ensure trust between authorities and stakeholders, generating a "sense of ownership", and the social and cultural environment has to be considered carefully by the project managers. [10] The general trend instead is to focus on technical aspects and procedures, partially neglecting the stakeholder management. Again, a more balanced approach between the "management of stakeholders" and "management for stakeholders" approaches would sensibly improve the effectiveness of the whole organization. [8, 11] Another aspect that cannot be ignored is that often times communities are involuntary stakeholders (Table 1), and in their position they lack the possibility of effectively negotiating their relationship with the counterpart. Though, from the literature review emerged that there is a limited understanding of the involvement of secondary players in public infrastructures and construction projects, which jeopardizes the achievement of the final expected benefits [6].

In the case of infrastructure development and improvement, the promoter's commitment to the redistribution of the value generated by the project is necessary to create an inclusive and fair project environment. The more involved the stakeholders are, the more willingly they interact and cooperate towards a shared better solution,

Table 1 Market and Non-Market stakeholders

	Examples	Involvement
Market stakeholders	Owners, partners, employees, shareholders, suppliers, creditors, lenders, investors, competitors	Directly and voluntarily involved on a financial level
Non-market stakeholders	Regulators, local communities, interest groups, landowners, general public, media, NGOs, other businesses, governments, activists, political groups	Not involved in the organization, but directly affected by its output

unified by a higher-order objective. As it is hard to obtain the essential resources that they control, the promoter can share a part of the authority over the resources with them, in order to involve them. In such a context, the stakeholders are enticed to collaborate rather than competing for the resources, thus seeking compromises. The stakeholders might be tempted to act opportunistically and seize the highest portion of the generated benefits, so the promoter must supervise the negotiations among them, showing that through a compromise, in the long-term the outcome will benefit all of them. In terms of governance framework, the more the promoter adds local structures of shared decision-making and joint generation of value, the more project governance becomes "polycentric". [12, 13] The megaproject promoter can choose whether to decentralize the governance or not, and which degree of freedom to leave the involved stakeholders. Inviting them to create value together is a way to finally access the essential resources held by non-market stakeholders.

2.3 *Behavior and Value Distribution of Megaprojects*

The social surplus is the difference between the expected economic value and the total generated value. It is then redistributed in many different forms to the stakeholders, who concede the resources that they own and that are needed. Thus, the infrastructure can be seen as an intermediate good which enables many downstream activities to arise and flourish, through a spill-over effect. [14, 15] It is hard to quantify the entity of the benefits provided, in particular for what concerns the comparison between costs and benefits. [16] In fact, the social gains may be hard to include because they are often vague and hardly quantifiable. It is common to have a negative outcome from the traditional Cost–Benefit Analysis, so it must be extended to consider all of the benefits which are hard to describe in monetary terms. An arbitrary weight has to be assigned to each one of them, and thus the perception of the decision-maker can prove decisive for the future of the megaproject [17].

2.4 Lessons Learned and Case Studies

One of the main problems when talking about megaprojects is the typical lack of lessons learned. Every project is peculiar, often site-specific and brought on in particular contexts, that make it hard to derive general and suitable best practices. However, some cautions can turn out to be decisive for the project's success. Thus, some relevant case studies were analyzed in order to highlight some critical factors for the success of the megaproject, especially when it comes to stakeholder involvement or to the recognition of less immediate benefits. Table 2 summarizes the valuable lessons learned that can be derived from the study of some specific cases identified.

3 The Model

3.1 Premises

The objective is to fill the gaps in the existing guidelines, generating a specific model for transport infrastructure through the analysis of specific literature and proven practices. The first step was indeed an assessment of the as-is situation in literature, to identify the main shortcomings in terms of stakeholder involvement. For the purpose of this work, many different Italian and European guidelines have been analyzed and integrated with more specific documents and papers [26], [39–45]. In fact, it was observed that many of the suggestions provided in the proposed model through this work were already present in literature only in a rather scattered way, being too general or too specific. Many guidelines were only aimed at public, non-economic entities, and did not explore compensation elements. Moreover, stakeholders were often mentioned, but in rather general terms, without providing more specific suggestions for their identification and ignoring their differences.

The proposed model is structured as a list of research questions, which follow an increasingly defined thread of logic. They indicate step by step the issues to be addressed and the means to be considered in a context of sharing project plans with stakeholders and are accompanied by specific examples.

3.2 The Framework

The questions chosen to guide the process are:

1. What is needed from non-market stakeholders?
2. What can the stakeholders be offered in exchange for those assets?
3. How can information be collected from stakeholders?
4. What is needed to obtain an informed and fruitful debate?

Table 2 Lessons learned from case studies

Case study	Lesson learned
High-Speed 2 [18] [19 [20] [21] [22] [23] [24] [25] [26] [27] [28]	The escalation of costs could make it harder to recognize and keep in mind the benefits initially identified, as they lose strength in terms of Cost–Benefit Analysis. Promoters could be tempted to reduce the scope of the megaprojects to reduce costs, but this could significantly limit the capacity of the infrastructure to seize the expected benefits
Algeria East–West Highway [29] [30]	If the owners of lands (or assets in general) needed for the project are not involved in the early phases, they could halter the development later on, increasing costs and time delays. Furthermore, non-market stakeholders perceived as hardly influential must not be underestimated, as they can suddenly gain influence and impact the project development
Public debate over a third airport in Paris [17] [31] [32] [33]	The willingness to involve non-market stakeholders is not enough to guarantee a successful cooperative process, as if this process is not well managed, the result would be the opposite. Moreover, the promoters must consider and accept the possibility of receiving negative feedback over the project itself
Barcelona's Sagrera high-speed station [34] [35]	Sometimes megaprojects are taken hostage by politics, and so their life cycle is more determined by the political situation than by technical reasons
Hallandsås tunnel [36] [37] [38]	The involvement of non-market stakeholders, and specifically of citizens, is a valuable means to obtain site-specific suggestions and feedback. The choice not to involve them since the early phases can generate mistrust, that can lead to violent opposition, accompanied by huge time delays and cost overruns, whenever an accident occurs

5. Who are the non-market stakeholders and what might they be interested in?
6. When should stakeholders be consulted?
7. Disclosure of responses.

3.3 *What is Needed from Non-market Stakeholders?*

Site-specific knowledge is incredibly valuable, especially for what concerns the early identification of possible conflicts and criticalities, improving the project outcome.

The collected feedback should also make emerge which are the performance indicators and dimensions that interest the most the stakeholder. Finally, the creation of a positive, cooperative environment would also bring benefits on the governance side, that would be strengthened by coherence and transparency. In the final phases, the collected feedback can be used to evaluate the effectiveness of the implemented decisions. They can be assessed in terms of:

- Timely and relevant input for decision-making and priority-setting
- Organizational learning
- Transparency and accountability towards stakeholders
- Efficient resources allocation.

3.4 *What can the Stakeholders be Offered in Exchange for those Assets: Possible Compensations*

Depending on the stakeholder or group, the bargaining chip to maximize the usefulness of the discussion can vary widely. The first obvious compensation is monetary, as in the case of expropriated land. In addition, stakes of the project may be offered to the stakeholders, or the government may commit to reinvesting a portion of the royalties into works useful for local communities. Another option is sharing part of the decision-making authority, a choice that can bring many benefits, but whose possible downsides must be carefully controlled. Another key theme of discussion is that of job creation, both internal to the project and external. The latter are less immediate to quantify and present to stakeholders, thus falling among those benefits that must be assigned a partially arbitrary weight at the time of the final project decision. Finally, another major element of interest is the possibility to perform works fostered by an intervention in the area [12, 27, 28, 46].

3.5 *How can Information be Collected from Stakeholders?*

Proper consultation windows need to be defined in compliance with existing regulations and the organization's needs. They should not last less than four weeks, but eight to twelve weeks could represent a better time frame for collecting feedback. After setting them, a strategy for the consultations must be chosen, in terms of structure of the consultations, means of interaction and other details.

Questionnaires

Questionnaires are one of the most immediate ways to collect feedback, because of their versatility and the acquaintance that people have with them. According to the phase of the project and its needs, questionnaires can be more or less structured, balancing close-ended and open-ended questions. Open-ended questions are particularly useful when researchers do not know how participants might respond

Table 3 Advantages and disadvantages of open and close-ended questions

Questions	Advantages and disadvantages
Open-ended	Harder processing and analysis of responses, they allow the stakeholders to provide their suggestions, concerns and feedback extensively, less biased on the side of the organization, do not need a priori response options, generally more reliable
Close-ended	Easier processing of responses, easier storage, provide a tool to obtain categories of stakeholders, much quicker, immediate feedback over general perception and issues of the project, lower quality of data, easier to involve a higher number of respondents

or want to avoid influencing their responses. Furthermore, this type of question is often more valid and more reliable, allowing the respondents to use their own words and to emphasize their main priorities. They are thus powerful in the early stages of a project, as there are no response options to worry about. Closed-ended questions may restrict the depth of participant response and thus the quality of data collected may be diminished or incomplete. On the other hand, they are suitable when an organization is interested in a well-defined aspect, and their analysis is way quicker and easier. Questionnaires allow to deepen the discussion topics at different levels and to gather information in different forms, exploiting the stakeholders' knowledge. Table 3 summarizes the advantages and disadvantages of the two possible types of questions [47–50].

Face-to-Face and Telephone Interviews

Another way to collect more extensive feedback is to interview a sample of stakeholders, asking questions both on the general perception of the project and on specific topics, according to the competencies of the specific stakeholder. Face-to-face interviews offer many advantages, especially for what concerns the level of detail of the feedback collected. Though, the choice of a representative sample is critical, to avoid the risk of having biased information due to poor variety of the respondents. Another option which can be followed is to conduct telephone interviews, as they allow to save costs and time. On the other hand, it is debated the fact that they have to be kept shorter, and so the depth of the information collected could be reduced [51, 52].

Other Collective Methods: Focus Groups, Workshops, Forums, Platforms, Public Debates

Focus groups put the emphasis on social interactions, not only as a way to produce data, but to provide data itself. Each focus group has to be considered as a single entity within a sample of groups, so it is not just an interview with a higher number of individuals to speed up the data collection from several sources at the same time. They can be less inhibiting for respondents, but they must be properly moderated to be effective. Similarly, workshops can be conducted together with the stakeholders, and if properly managed, they should generate trust and willingness to cooperate through the development of shared thoughts. Other collective methods can then

Table 4 Advantages and disadvantages when offering decision-making rights

Advantages	Disadvantages
Deeper access to know-how, willingness to volunteer the resources harder to acquire otherwise, involvement leads to higher level of cooperation	Risk of collective action problems (give-some, take-some dilemmas [58]) on one-off design choices, longer time needed to take decisions

emerge or be suitable from time to time, according to the characteristics of stakeholders and the needs of the project. A possible issue called "under-reporting bias" could emerge, that is a concept usually related to online reviews on platforms. It has been studied indeed that usually online reviews mainly capture extreme rather than moderate opinions. Customers with a moderate view on the product are less likely to provide reviews, if compared to highly satisfied or dissatisfied customers. For the same reason, stakeholders highly interested in the project, and so with a strong positive or negative perception of the work, will be much more likely to contribute. The moderate stakeholders will not perceive instead the participation in the debate as a priority, and so they could decide not to provide their feedback [53–57].

Offering Decision-Making Rights
As already discussed in Section "Non-market stakeholders and project governance", the more involved the stakeholders are, the more willingly they will work towards a shared solution, unified by a higher-order objective. The main objective is mainly related to the incorporation of specific expertise valuable for the continuation and development of the project, or to obtain essential resources controlled by them. The basic principle here is indeed to allow people or groups who possess a particular expertise to actively decide on the topics related to their knowledge. This is especially useful in the case of categories of everyday users or people working in that field, because they can provide the absolute most up-to-date and timely knowledge. The promoters must assume a neutral role supervising the discussion, while leaving the right spaces for the stakeholders. They will positively perceive the neutrality of the promoter and will thus be incentivized to behave more coherently and correctly during the discussion, collaborating rather than competing for the resources [12], (Table 4).

3.6 *What is Needed to Ensure an Informed Debate?*

It is important to guarantee the opportunity for all non-market stakeholders to participate through an adequate publicization of the process. If this publicization is not well performed, the main risk is the "capture" of the consultation process by a limited group of actors, which would represent only a small portion of the needs of the ecosystem of stakeholders. In the worst-case scenario, this could lead to lobbying behavior. Then, a general consultation document is needed, to contain the most important information related to the consultation process. This document should include:

- consultation dates and deadline
- an explanation about the object of the consultation
- an in-depth explanation about who the recipients should be
- the presentation of the forthcoming ways of consultation, including for instance the summary of the research questions, the structure of an eventual questionnaire, and so on
- a clear explanation of the mode of sending of the responses
- the required privacy policy
- useful contacts to obtain further information
- the terms of use
- information about the next steps and phases, possibly accompanied by general time estimates at least in terms of months and years

Other project documents must be provided, possibly both in technical and non-technical versions. The non-technical version of the documents should be designed to give all stakeholders the broadest possible understanding of the aspects explored by the promoters, but in an understandable way. As the project plans develop, up-to-date versions with a higher level of detail must then be provided, as for example for what concerns the technical drawings and the environmental report.

3.7 *Who are the Non-market Stakeholders and What Might they be Interested in?*

The main non-market stakeholders and their main needs were identified and assessed, as reported in Table 5. It is of paramount importance to remember that every megaproject is unique and site-specific, so the "other stakeholders" have to be properly assessed. Time, resources and attention must be devoted to the identification of the characteristic stakeholders, through stakeholder mapping, brainstorming and other techniques. The list provided below indeed represents just a general overview of the main non-market stakeholders who are easily found in transport infrastructure megaprojects, but it is impossible to provide a complete list of them.

3.8 *When should Stakeholders be Consulted?*

In the early phases, the main objective is to present the "grand idea" to the stakeholders, so as to get an idea of their perception of the work. The results of this debate will then guide the development of future, more in-depth plans. In the second round of consultations, the goal is to provide a narrower but more defined set of options.

Table 5 Summary of main non-market stakeholders' needs in transport infrastructure

Stakeholder	Main interests and needs
Citizens	Services, jobs, infrastructures which benefit the community, regeneration of the area, increase of real estate value
Landowners	Money, compensations or useful infrastructures, regeneration of the area, increase of land value
NGOs	Investments to reduce some specific impact of the project (e.g. environmental)
Local authorities	Infrastructures, renewal of existing public properties, job creation
SMEs	Improved connection for goods transportation, savings on time and fuel due to improved transport network
Local shops	Improved connection for tourists and visitors
Trade associations	Improved connectivity
Other stakeholders	Their identity and needs should result from site and project-specific brainstorming

It is crucial to make it clear from the outset that there will be less room for radical changes in this new phase, in order not to create false expectations. Finally, before the project is officially submitted to the authorities in its final form, there may be room for an additional phase of non-market stakeholder consultation, to further optimize the identified solutions.

3.9 Disclosure of Responses

The return of consultation results is extremely important and critical for several reasons that go beyond the simple act of informing. In fact, in the respondents' perspective, the outcome will really impact their daily lives and therefore they need to be reassured about the level of attention dedicated as the project develops.

4 Case Studies

Two case studies were analyzed through this framework: the A66 enhancement in the UK, and an Italian megaproject currently in its earliest phases.

4.1 *A66 Northern Trans-Pennine*

The megaproject aims at upgrading the whole road between the two locations of Penrith and Scotch Corner to dual carriageway standard, improving at the same time the junctions along the route on this 80 km long section, shown in Fig. 2. The estimated cost of the upgrading project stands around 1B£ (around 1.22B$, as of August 2022). In terms of quantity of stakeholders involved, the 80 km of A66 that divide Penrith and Scotch Corner present a wide variety of local communities, heritage sites, areas of natural interest such as the North Pennines and Lake District.

According to the United Kingdom legislation (Planning Act, 2008), such a project is classified as Nationally Significant Infrastructure Project (NSIP). This means that an application for a Development Consent Order (DCO) is needed to obtain the permission to proceed with the work. Figure 3 reports the complete path followed for approval, and the steps directly involving non-market stakeholders are highlighted in green. It provides a meaningful example of an inclusive process, which was used to better define some sections of the aforementioned model.

The analysis of this project depicted it as one of the most suitable proven practices in terms of stakeholder involvement, providing rather useful insights. First of all, throughout the development of the project, the promoters have engaged and consulted with stakeholders and the public, applying a variety of solutions to encourage broad participation and properly mapping the stakeholders. It was done in an iterative way, as the stakeholders were mapped more in depth also through the analysis of their responses to the early questionnaires. The material provided was usable and understandable, allowing the interviewees to approach the consultations with adequate preparation, crucial to obtain valuable feedback. Furthermore, the open-ended questions provided an outline to guide the responses through different topics, making easier their analysis later on. Finally, the disclosure of responses was very well performed, returning an understandable overview of the results to the stakeholders through schematic but clear reports [59–66].

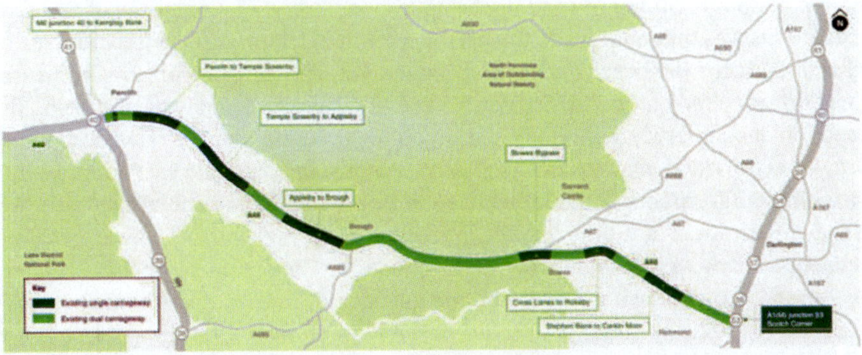

Fig. 2 The single carriageway sections interested by the A66 enhancement project (from National Highways [140])

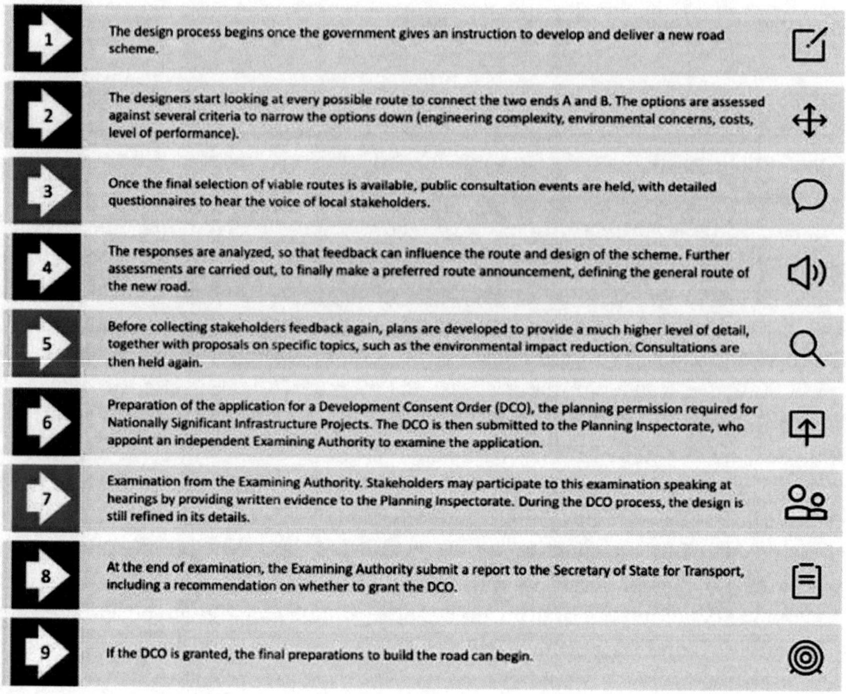

Fig. 3 Approval process in the UK. Non-market stakeholders are directly involved in the green steps

4.2 *Italian Case Study*

During the development of the model, there was the opportunity to follow the initial stages of a megaproject related to a transport infrastructure in Italy, in which a prestigious motorway authority has begun consultations for the widening to four lanes of a section of about 150 km of road under its responsibility. The outcome of the early consultations was thus analyzed through the presented framework, to generate useful feedback both for the organization and the author. The early comments addressed the performance indicators' identification, provided a critical analysis of the positive perception encountered, and finally discussed the broadness of the consultation and the material provided. Due to confidentiality reasons, the outcome was not presented in its complete form, but some indications were shared. A stakeholder register was built, and it showed that only highly influential local stakeholders were involved, leaving non-market stakeholders out of the equation. The organization was criticized by some stakeholders, who considered the quantity of material provided absolutely inadequate. In their opinion, it was not possible to comment on the project due to the lack of information, and so many of them initially decided not to comment and provide feedback and suggestions. Furthermore, some stakeholders did not want to contribute because a similar project recently performed in a nearby area generated controversies and environmental issues. Thus, an effort was made to better present and explain the process of involvement. As indeed reported by the corresponding section of the model, the quantity of material should vary from one phase to another. The low quantity of material provided, and the low level of detail demonstrated a willingness to share the next steps with the involved entities, allowing them to really contribute. It was not instead a sign of low commitment on the side of the organization, and following the suggestions provided they were able to bring on board the stakeholders who initially challenged them on the topic. Finally, the main needs and wants, and the main concerns of the interviewed stakeholders were collected and analyzed, in order to optimize the communication strategy for the next phases. Some examples are reported in Table 6 without entering into detail because of confidentiality reasons, as they are common topics that could emerge when talking about transport infrastructure.

Table 6 Main needs and wants and concerns emerged from the early consultations

Theme	Needs and wants	Concerns
Traffic	Reduce the saturation of heavy goods vehicles (HGV), reduce the saturation of local beltways, incentivize the use of electronic toll collection systems to reduce queues at toll stations, improve viability around a local airport and a hospital, upgrade of tool stations	Increase in motorway tolls would pour traffic onto urban roads, benefits not seized because of the low reliability of the surrounding road system or to inadequate toll stations
Environment and landscape	Consider the possibility to insert architectural artworks in the project to reduce the impact on landscape, reduce the environmental impact of the rest areas along the motorway	Excessive increase in terms of pollution and noise, potential protests from local communities against this increase of noise and pollution, low attention to some nearby infrastructure of hydraulic nature, valuable vineyards put in jeopardy
Other themes	Enhancement of surrounding road system to fully exploit the benefits of the fourth lane, upgrade of the nearby beltways, construction of a depot on the site of a former roadman's house, extension of the project for a fourth lane beyond the section of the motorway considered	Expropriations management must be well planned, nearby railway could be upgraded, some nearby infrastructures have to be demolished, some toll stations cannot be expanded because of space constraints

5 Conclusions

The generality of the existing guidelines does not allow a proper application of their indications in the specific case of transport infrastructure, as they only provide a broad idea of the steps to follow. The literature analysis and comparison with some "proven practices" led to the generation of new research questions, this time to be submitted to the manager interested in stakeholder engagement. Indeed, this framework is intended to provide an overview of the steps to be followed in the process, pairing each of the questions to be answered by the manager with concrete examples, to facilitate understanding. In this case, an attempt was made to achieve a balanced compromise between generality and specific real-life examples. The analysis of the European guidelines, which can be considered a state-of-the-art model, served to further emphasize the need for a more specific model. Currently, the model is being used to support and evaluate the stakeholders' engagement activities in the context of a transport infrastructure enhancement megaproject in Italy. This opportunity allows a constant assessment of the framework against a megaproject under development, thus integrating the observation of past case studies with a state-of-the-art endeavor.

5.1 *Future Developments*

The Italian case study is in its early phases, thus there will be the possibility to follow its development step by step, assessing in real time both the decisions of the promoters against the model, and the model itself against a complete real project. In the meanwhile, more and more transport infrastructure projects will be analyzed through the framework, generating a more complete list of examples linked to the research questions that compose the proposed model, and possibly providing new questions and directions to follow.

Once the gap that was identified in literature is filled, the next massive step could then be to further adapt the guidelines to the different types of transport infrastructure, or at least to the most relevant ones, namely road and rail infrastructure. In particular, railways are also becoming increasingly important from an environmental point of view, given the growing interest governments are showing in shifting passenger and freight traffic from road to rail. Moreover, especially for long-distance routes, high-speed rail tends to reduce road saturation, making it more economically viable for individual passengers to travel by train. That said, regardless of all these reasons, the generation of this model was driven by the desire to provide specific guidelines that could give practical help to managers, including providing examples from which to draw. Further studies devoted to different specific transport infrastructure could therefore only increase the applicability of the framework itself and the relevance of the concrete examples brought.

References

1. van Marrewijk A, Clegg SR, Pitsis TS, Veenswijk M (2008) Managing public–private megaprojects: Paradoxes, complexity, and project design. Int J Project Manage 26(6):591–600
2. Kardes I, Ozturk A, Cavusgil ST, Cavusgil E (2013) Managing global megaprojects: complexity and risk management. Int Bus Rev 22:905–917
3. Van Marrewijk AH, Veenswijk MB (2006) The culture of project management: understanding daily life in complex megaprojects
4. Flyvbjerg B (2014) What you should know about megaprojects and why: an overview. Proj Manag J 45:6–19
5. Mancini M, Derakhshanalavijeh R (2017) Uncertainty in megaprojects: opportunities for the future, pp 368–379
6. Di Maddaloni F, Davis K (2017) The influence of local community stakeholders in megaprojects: Rethinking their inclusiveness to improve project performance. Int J Project Manage 35(8):1537–1556
7. Choudhury B (2014) Aligning corporate and community interests: from abominable to symbiotic. BYU L Rev 257
8. Eskerod P, Huemann M (2013) Sustainable development and project stakeholder management: what standards say. Int J Manag Proj Bus 6:36–50
9. Derakhshan R, Turner R, Mancini M (2019) Project governance and stakeholders: a literature review. Int J Project Manage 37:98–116
10. Grabher G (2002) Cool projects, boring institutions: Temporary collaboration in social context. Reg Stud 36(3):205–214
11. Eskerod P, Huemann M, Ringhofer C (2016) Stakeholder inclusiveness: Enriching project management with general stakeholder theory. Proj Manag J 46:42–53
12. Gil N (2021) Megaprojects: a meandering journey towards a theory of purpose, value creation and value distribution. Constr Manage Econ
13. Ostrom E (2009) Beyond markets and states: Polycentric governance of complex economic systems
14. Ge X, Huang Y, Zhou Y (2019) Transportation infrastructure investment and economic growth: Based on the evidence of quasi-natural experiment. Syst Eng Theory Prac 39(4):922–934
15. Lakshmanan TR (2011) The broader economic consequences of transport infrastructure investments. J Transp Geogr 19:1–12
16. Korytárová J, Papežíková P (2015) Assessment of large-scale projects based on CBA
17. Sébastien D, Roy B (2009) The uses of cost–benefit analysis in public transportation decision-making in France. Transp. Policy 16:200–212
18. Douglas Oakervee. Oakervee review of HS2 (2022). https://assets.publishing.service.gov.uk/government/uploads/system/uploads/attachment_data/%20file/870092/oakervee-review.pdf, August 2019. Accessed 15 Aug 2022
19. HS2 (2022). https://www.hs2.org.uk/, July 2022. Accessed 15 Aug 2022
20. HS2. London to West Midlands (2022). https://www.hs2.org.uk/the-route/london-to-west-%20midlands/, July 2022. Accessed 15 Aug 2022
21. BBC. Route of HS2 (2022). https://www.bbc.com/news/uk-16473296, November 2021. Accessed 15 Aug 2022
22. New Civil Engineer. Mace and dragados sign HS2 Euston station contract (2022). https://www.newcivilengineer.com/latest/mace-and-dragados-sign-hs2-euston-%20station-contract-11%E2%80%9303%E2%80%932019/, March 2019. Accessed 16 Aug 2022
23. Fu Y (2019) Organizational governance and value creation: an empirical Study in megaproject Contexts. University of Manchester
24. HS2. Old Oak Common (2022). https://www.hs2.org.uk/building-hs2/stations/old-oak-%20common/, July 2022. Accessed 16 Aug 2022
25. Department of Transport. High Speed Two phase two, strategic case. Moving Britain ahead (2022). https://assets.publishing.service.gov.uk/government/uploads/system/uploads/attachment_data/file/629393/high-speed-two-phase-two-strategic-case.pdf, 2017. Accessed 05 Sep 2022

26. HM Treasury. Central government guidance on appraisal and evaluation (2022). https://assets.publishing.service.gov.uk/government/uploads/system/uploads/attachment_data/file/1063330/Green_Book_2022.pdf, 2022. Accessed 21 June 2022
27. National Audit Office. Investigation into land and property acquisition for phase one (London - West Midlands) of the High Speed 2 programme (2022). https://www.nao.org.uk/wp-content/uploads/2018/09/Investigation-into-land-and-property-acquisition-for-the-Phase-One-Full-report.pdf, September 2018. Accessed 10 Sep 2022
28. Department for Transport. Government overview of case for phase one of HS2 and its environmental impacts - update for the House of Lords (2022). https://assets.publishing.service.gov.uk/government/uploads/system/uploads/attachment_data/file/918871/HS2-lords-statement-of-reasons-command-paper-web-accessible-version.%20pdf, January 2017. Accessed 10 Sep 2022
29. Iwabuchi A, Shimizu T. The Algeria east-west highway: an interim report
30. Zidane YJ-T, Johansen A, Ekambaram A, Hald LC (2015) When stakeholders shape successes or bring failures - a case study of an Algerian megaproject 64:844–851
31. Barraqué B (2002) Quels indicateurs de gêne et quelle démarche participative pour quelle politique territoriale d'aéroports? Ann Min 26:4
32. BBC. Paris plans third airport (2022). http://news.bbc.co.uk/2/hi/europe/1658259.stm, November 2001. Accessed 05 Sep 2022
33. Paul Michaud. French government cancels plan for third Paris airport (2022). https://www.arabnews.com/node/223245, August 2002. Accessed 05 Sep 2022
34. Baron N (2019) Who killed Barcelona's high-speed station promises? Infrastructure political capture, scaling narratives and the urban fabric. Eur J Geogr 10:6–22
35. Beria P, Grimaldi R, Albalate D, Bel G (2018) Delusions of success: costs and demand of high-speed rail in Italy and Spain. Transp Policy 68:04
36. Global Railway Review. Constructing the complicated Hallandsås tunnel (2022). https://www.globalrailwayreview.com/article/21524/constructing-the-complicated-hallandsas-tunnel/, September 2014. Accessed 11 Sep 2022
37. Sjölander-Lindqvist A (March 2004) The effects of environmental uncertainty on farmers' sense of locality and futurity: a Swedish case study. J Risk Res 7(2):185–197
38. Boholm Å, Löfstedt R (1999) Issues of risk, trust and knowledge: the Hallandsås tunnel case. Ambio 28:556–561
39. Ufficio per la valutazione della performance Presidenza del Consiglio dei Ministri, Dipartimento della Funzione Pubblica. Linee guida sulla valutazione partecipativa nelle amministrazioni pubbliche (2022). https://performance.gov.it/system/files/LineeGuidaeRifNorm/LG_Valutazione_partecipativa_28-11_.pdf, November 2019. Accessed 19 Aug 2022
40. RegioneEmilia-Romagna. Partecipare e decidere. Insieme è meglio. Una guida per amministratorie tecnici (2022). http://partecipazione.formez.it/sites/all/files/Guida%20Partecipazione%20RER.pdf, November 2009. Accessed 21 Aug 2022
41. ParteciPa. Linee guida sulla consultazione (2022). https://partecipa.gov.it/%20processes/linee-guida-consultazione/f/8/proposals?component_id=8&locale=it&participatory_process_slug=linee-guida-consultazione, 2020. Accessed 19 Aug 2022
42. Nanz P, Fritsche M (2012) La partecipazione dei cittadini: un manuale. Metodi partecipativi: protagonisti, opportunità e limiti
43. De Toffol F. Dizionario di democrazia partecipativa (2022). https://centrostudi.alumbria.it/sites/default/files/diz_democrazia_partecipativa.pdf, 2012. Accessed 20 Aug 2022
44. RegioneEmilia-Romagna. Partecipazioni: sostantivo, plurale. Guida metodologica per la gestione di processi di partecipazione integrati (2022). https://partecipazione.regione.emilia-romagna.it/tutte-le-pubblicazioni/pubblicazioni/partecipazione-1/partecipazioni-guida-metodologica-processi-partecipativi-integrati, February 2016. Accessed 20 Aug 2022
45. European Commission. Better regulation guidelines (2022). https://ec.europa.eu/info/sites/default/files/swd2021_305_en.pdf, November 2021. Accessed 27 Aug 2022
46. Shamsa K, Pitafi AH, Rasheed MI, Pitafi A, Iqbal J (2020) Assessment of residents' perceptions and support toward development projects: a study of the China-Pakistan economic corridor. Soc Sci J 59(1):102–118

47. Price P (2015) Research methods in psychology, 2nd Canadian Edition. Online access: center for open education open textbook library. BCcampus
48. Bowling A (2009) Research methods in health: Investigating health and health services. Open University Press, Buckingham. ISBN 0335206433
49. Rattray J, Jones M (2007) Essential elements of questionnaire design and development. J Clin Nurs 16:234–43
50. DiCicco-Bloom B, Crabtree BF (2006) The qualitative research interview. Med Educ 40:314–321
51. Novick G (2008) Is there a bias against telephone interviews in qualitative research? Res Nurs Health 31:391–398
52. Sturges JE, Hanrahan KJ (April 2004) Comparing telephone and face-to-face qualitative interviewing: a research note. Qual Res 4(1):107–118
53. Barbour R, Kitzinger J (1999) Developing focus group research: Politics, theory and practice. Sage Publications Limited, Thousand Oaks
54. Johnson JM (2001) In-depth interviewing. in handbook of interview research, pp 103–119. SAGE Publications, Inc.
55. Silverman D (2006) Interpreting qualitative data: Methods for analyzing talk, text and interaction
56. Caillaud S, Nikos K, Doumergue M (2022) Designing focus groups, pp 684–699
57. Duggleby W (2055) What about focus group interaction data? Qual Health Res 15:832–840
58. Barney JB (2018) Why resource-based theory's model of profit appropriation must incorporate a stakeholder perspective. Strateg Manag J 39:3305–3325
59. NationalHighways. A66 Northern Trans-Pennine project, draft project design report (2021). https://highwaysengland.citizenspace.com/he/%20a66-northern-trans-pennine-design-consulta-tion/supporting_documents/A66NTP%20Preliminary%20Design%20Consultation%20Sept%202021%%2020Project%20Design%20Report.pdf. September 2021. Accessed 28 June 2022
60. National Highways. Environmental statement (2022). https://a66-environmental-statement.virtual-engage.com/Statement.html?entry=1, 2022. Accessed 28 June 2022
61. National Highways. Environmental statement non-technical summary (2022). https://a66-environmental-statement.virtual-engage.com/Statement.html?entry=16, 2022. Accessed 28 June 2022
62. National Highways. Virtual exhibition (2022). https://a66ntp.virtual-engage.com/, September 2021. Accessed 28 June 2022
63. National Highways. Summary of the autumn 2021 statutory consultation (2022). https://highwaysengland.citizenspace.com/he/a66-northern-trans-pennine-design-consultation/results/a66-tc-summary-of-the-autumn-2021-statutory-consultation-online.pdf, March 2022. Accessed 28 June 2022
64. National Highways. Digital preliminary environmental information (PEI) report (2022). https://virtualengage.arup.com/A66-NTP/pei-report/, 2022. Accessed 28 June 2022
65. NationalHighways. A66 Northern Trans-Pennine project, local traffic report (2022). https://highwaysengland.citizenspace.com/he/%20a66-northern-trans-pennine-design-consulta-tion/supporting_documents/A66NTP%20Preliminary%20Design%20Consultation%20Sept%202021%20Local%%2020Traffic%20Report.pdf. September 2021. Accessed 28 June 2022
66. National Highways. A66 northern trans-pennine project. the story so far (2022).https://s3.eu-west-2.amazonaws.com/assets.nationalhighways.co.uk/%20roads/road-projects/A66+Northern+Trans-Pennine/A66+NTP+Project+%20Story+So+Far.pdf, 2022. Accessed 09 Sep 2022

Overtaking the Traditional Leadership Style in Megaprojects

Roberta Virtuani, Barbara Barabaschi, and Franca Cantoni

Abstract The inability to effectively manage structural, social and emergent complexity is one of the main causes of megaprojects failure. The leader's technical skills are beyond necessary but not sufficient to deal with situations of increasing complexity. High levels of complexity can only be addressed with radical new approaches. Specifically, this paper aims to shift the focus from a traditional concept of leadership acknowledged to a specifical individual within a project to an organization-wide, aware and sustainable (WAS) leadership exerted by key people. All key people involved in the implementation of the megaproject (designers, analysts, project managers, ...) are invited to exercise a new leadership style that, in the model we propose, has three main characteristics. To give a practical imprinting to the research, a case study has been selected to describe the traits and skills of the WAS leader in megaprojects.

Keywords complexity · shared leadership · boundary spanners · project manager' reconciliating role · Management for stakeholder · project success factors · new iron triangle · new competencies · value creation · leader-megaproject fit

Addressing the Issue

Megaprojects represent an unstable, non-linear, irregular, uncertain, unpredictable and highly dynamic environment. In other words, they come hand in hand with complexity (Li, Han et al., 2019; [31] which – if not properly managed - may lead to unsuccess and failure [32] (Larson, Gray, 2014) [25]. Therefore, it takes rigorous effort, more than in ordinary projects, to manage a megaproject (Li, Han

R. Virtuani (✉) · B. Barabaschi · F. Cantoni
Università Cattolica del Sacro Cuore, Piacenza, Italy
e-mail: roberta.virtuani@unicatt.it

B. Barabaschi
e-mail: barbara.barabaschi@unicatt.it

F. Cantoni
e-mail: franca.cantoni@unicatt.it

E. Favari and F. Cantoni (eds.), *Complexity and Sustainability in Megaprojects*, Lecture Notes in Civil Engineering 342,
https://doi.org/10.1007/978-3-031-30879-6_8

et al., 2019,Locatelli, Mikic, Kovacevic, Brookes & Ivanisevic 2017). To ensure a megaproject's success this study recognizes an indispensable value to the leadership style adopted by all the key HRs involved in planning and implementation (He, Luo, Hu & Chan, 2015; [19, 35].

1 The "WAS" Style of Leadership

Leadership on megaprojects and other temporary organizations has been extensively researched (Drouin et al., 2021; [39, 41]. However, like project success, leadership research on projects has not kept the pace in addressing emerging issues of project complexity (Al-Subaie et al., 2021; [8], Wan et al., 2020; [12, 43]. Indeed, temporary complex organizations - such as megaprojects - require approaches that differ from those used in permanent organizations [33]. Traditional leaders tended to oversimplify the central role of HRs in terms of engagement and motivation, direction and empowerment and frequently this oversimplification has been one of the main causes of project failure [44]. In a faster-paced and unpredictable world it would be beyond unthinkable not to reformulate the definition and consequently the exercise of leadership (He et al., 2015) [23].

For this reason, we propose an innovative theoretical approach that seeks to define the characteristics of the leader in megaprojects to overcome the limiting views of traditional and classical approaches (Andersen & Polkinghorn, 2008). This new theoretical approach includes three aspects toward a new management model based on a Wide, Aware and Sustainable leadership (WAS) [41]. These three aspects derive from what is possible to observe in megaprojects management and from research that analyses the long-term impact and influence of megaprojects on the environment, society and economy. In the following analysis we will deepen the three aspects to identify the characteristics of a management style that faces the complexity and dynamic evolution of a megaproject [29].

1.1 Organization-Wide Leadership

Organization-wide leadership is a radical paradigm shift that contrasts with the idea of leadership to which we have traditionally been accustomed.

In this new perspective, the figure of the traditional leader no longer exists, but the idea of collaborative, empathic and shared leadership emerges. The leader will therefore not only have to lead but above all to motivate and enhance the merits of individual contributors. This will enable people to feel free to make their own contributions and exercise their initiative depending on the situation and the task required. In this way, through organization-wide leadership, one can be both a leader at one time and a follower at another. Talking about organization-wide leadership we refer to the literature on shared leadership that is based on the sharing of power, influence and decision-making among a set of individuals rather than centralized in one single individual (Pearce. et al., 2007) [20]. The collaborative effort can take place inside a group with one person that can be still in charge. It differs from a more traditional management style where leadership is linked to the vertical hierarchy (Sweeney A., 2022) [40]. On the other hand, shared leadership isn't strictly limited to a team context. It can be broadly applied to the entire company building a business culture based on a cohesive environment with well-understood goals and a strong atmosphere of mutual support. Transparency, autonomy support, a safe environment and the openness to others' ideas are essential aspects of a shared leadership strategy to boost innovation and employee engagement. The smartest company is no longer the one in which some smartest managers have the authority to manage the work of others (Fernandes, 2022) [16]. An entirely different management model emerges based on the possibility to access to the collective knowledge of the company. This is a rapid change in business management from old, authoritarian models to more open methods. A management model based on organization-wide leadership encourages and values personal initiative. Employees are empowered to contribute with their knowledge and ideas instead than just waiting instructions about what to do. Company's performance benefits through the increase of productivity and job satisfaction thanks to the employees' higher sense of ownership (Qiu et al., 2019) [38]. When power and responsibility increase the impact on organization is higher because decision-making gets spread across multiple individuals [17]. In the case we analyze all these aspects are emerging. This organization-wide leadership strategy is already applied and followed as an organizational principle that boost performance, innovation, employee's engagement through an open-door management practice. Megaprojects are complex and include several stakeholders. Recent studies have introduced a novel framework of leadership activities in megaprojects based on boundary spanners [8] as a structural pattern of shared leadership in megaprojects. Share leadership is the expected form of leadership in megaprojects and boundary spanners are shared leadership roles for planning and delivery megaprojects (Pearce et al, 2007) (Sweeney, 2022) [34, 40].

The organizational complexity of megaprojects derives from the collaboration of several teams that form a complex network of stakeholders evolving and changing over the megaproject life cycle. The consequence is that leadership is transferred and shared with great frequency among actors that are always different (McCauley, Palus, 2021) [30]. In this context boundary spanners are interorganizational interfaces that align teams taking various leadership roles because, especially in megaprojects, leadership cannot be attributed to a single person nor single team [47]. Team members become horizontal leaders with complementary leadership tasks for social interaction among project teams. This view overcomes the traditional leadership perspective that cannot answer to the requirements for the integration of different experts and knowledge to lead the megaproject activities [26].

1.2 Aware Leadership

Increasing awareness of self, others and situations is at the heart of most leadership development—be it in leadership training, executive education courses or coaching, feedback, and reflection activities. These leadership development initiatives use many processes where individuals discover more about themselves, the perceptions of their behaviors and performance, and then attempt to create a working agenda of priorities to address to maximize their leadership performance for the future. These processes are designed to provide leadership and personal insights. Where these are deep, effective and fully integrated, we use the term "insightful awareness" to encompass a profound understanding by the leader—about their strengths, weaknesses, drivers and blockers [45]. [18] Relate self-awareness to self-reflection. According to them, "by reflecting through introspection, authentic leaders gain clarity and concordance with respect to their core values, identity, emotions, motives and goals" (p.347). Increasing self-awareness entails attempting to understand oneself and one's surrounding predicated on self-reflection.

1.3 Sustainable Leadership

The goal of improving the quality of megaproject management is not only to deliver more projects on time and on budget, but to realize the change organizations and society need to develop in a sustainable way. [5, 6]. Sustainable leadership is a process of influence that delivers direction, alignment, commitment, and aims to address social, environmental, and economic issues to create a better world [37, 41]. Sustainable leadership is about adopting a responsible approach to "*the way that leaders lead*", stopping to think about the wider impact of actions on society and the environment (Al-Subaie, Talib et al., 2021) (Cantoni, Pagnone, 2020) [2, 7]. This might mean considering the wider stakeholder group: sustainable leaders do not manage stakeholders, they need to include them in order to understand a wide range of points of view and to drive decision-making with all those stakeholders. In particular, the contemporary global purpose to act a sustainable development calls for paying attention to specific values typical of management-for-stakeholders (instead of management of stakeholders), such as transparency, fairness, trust (Al-Subaie, Jabeen et al., 2021) [1]. This approach offers an inclusive perspective which aims to involve a broader group of stakeholders, by meeting or exceeding their needs and expectations in coherence with a project socially responsible governance which finally attempts to balance its economic, ecologic, and social interests [13] (Cantoni & Pagnone, 2020) [7].

Hard-to-satisfy stakeholders or in conflict with the project guidelines can be in any case sources of value creation, if approached with the "no trade-offs" mindset of managing-for-stakeholders. The leader must be able to learn from these difficult situations also by means of a strong personal motivation and purpose, looking up and out beyond the role, organization, and sector (Anderson and Polkinghorn, 2008) [3].

In this context an aspect of the sustainable leadership relates to the ability to exercise one's role and function in fluid and hybrid modes, in-presence and remotely [4]. During and after the pandemic, leadership structures, work systems and technology adoption have undergone great changes (Han et al., 2022). Shared leadership, as an emergent approach to face complex challenges, has become a powerful tool to manage virtual teams [24]. In fact, virtual teams are different from face-to-face teams [10] with research showing that virtual teams are more difficult to lead and manage [9] due to the risk of information lack or dispersion. One of the missions of virtual human resource development is to build virtual teams as integral part of the digital workforce [21] according to a sustainable leadership perspective, trying to reduce the risks and, on the contrary, using digital tools to reinforce teams'cohesion.

2 The WAS Leader: Traits and Competencies

Larson and Gray (2008) state that leadership in project management is characterized by the ability to cope with change. In megaprojects the project leader must recognize the need to change, initiate change, provide direction and motivation and, finally, innovate and adapt as necessary. In mega projects, the project leader faces several dilemmas, including:

- seeing the big picture, while keeping an eye on the important details,
- innovating and maintaining stability at the same time,
- stressing the team while motivating individuals,
- adapting a hands-on as well as a hands-off approach,
- being firm and flexible at the same time.

This research - recognizing the crucial role of technical skills to overcome structural complexities - analyzes the personal and interpersonal skills and traits that all key HRs involved in a megaproject must apply and studies the reconciliation role a PM must demonstrate. The ability to practice healing any conflicts generated by leadership with the characteristics listed above.

Indeed, personal, and interpersonal skills are extremely important in coping with social and emerging complexities.

By social complexities we mean the difficulties arising because of multiple stakeholders, while emerging complexities refer to all the complications that can arise, develop and amplify during the different phases of a megaproject life cycle. These complications are not easy to predict and depend largely on unforeseen developments in the context and scenarios (e.g., Covid-19, commodity price speculation related to the war in Ukraine, ...).

In traditional megaproject literature, technical and hard skills prevailed in the megaproject leader profile (Fernandes et al., 2022) [16]. Empirical studies and professional practices reveal the crucial role of social and relational abilities to face the complex external and internal environments, both influenced by stakeholder diversity and frequent unexpected events. Our work aims to contribute to the broader literature by focusing on the set of social and interpersonal skills enabling megaproject leaders to improve their possibilities to succeed.

Table 1 shows our attempt to systematize the leader's traits and skills, considering also social and relational dimensions along with the technical one.

Among the most relevant social skills we detect communication, negotiation, self-awareness, emotional intelligence [15]. Within this set, communication has revealed to be the most important skill in the knowledge society we live in. Megaproject

Table 1 IPMA ICB4 framework model adapted to megaprojects

Complexity	Knowledge Areas (according to PMBOK)	Traits and Skills
Structural	Technical (Practice): - integration management - scope management - time management - cost management - quality management - risk management - knowledge management - procurement	https://ipma.it/ipma_/images/IPMA_Italy_ICB4_PfM.pdf
Social	Personal and interpersonal (People): - HR management - Communication management - Stakeholder management	- proper communication - negotiation - collaboration among multi-party teams - problem solving - self-awareness and discipline - adaptability - complexity management - to be focused in a complex environment - trust and team-work - control and follow-through - empathy and emotional intelligence - risk-taking - team building
Emergent (dynamic aspects)	Technical + Personal and interpersonal (Perspective)	- strategic thinking - crisis management - flexibility - change management - procedural planning and control - alignment and integration - knowledge and experience to exploit current and available resources

managers who are good communicators can foster stakeholders' engagement, to help them understand the scope of their duties and solve problems as they arise. They can also benefit from listening to their team members to promote dialogue overall. Negotiation requires some give and take, to be willing to compromise to earn the reputation of being fair and respectful: this skill requires frequent consultation, aligning the responsibilities and goals making sure all parties are satisfied, so reinforcing the project "social capital" [11, 36].

Awareness means being detail-oriented and able to see the project in its entire lifecycle, being able to spot potential problems along the way and take steps to avoid them interacting with others through empathy. This last skill has a closed connection with emotional intelligence since emotions can easily get in the way of project success, not only considering employees or other internal stakeholders, but also taking into consideration secondary stakeholders such as local communities or social movements engaged in the environment or human rights protection.

3 Purpose and Methodology

The purpose of the chapter is to propose a new perspective on leadership styles in megaprojects management. Our view comes out from the recent theorical contributions and from our empirical work based on interviews of managers involved in managing teams in a megaprojects' setting. We are particularly interested in their life stories that we collect through their storytelling.

Stories are based on personal experience and the storyteller communicates his/her personal world to other people. In fact, storytelling is considered an important soft skill especially for those practitioners in senior leadership positions. Stories in organizational life help in building an understanding of self and organizational identities. They are a powerful tool that through its informality allows to project images to others. They have a social and emotional content that can create empathy and connection when constructing identities.

The way in which project leaders express themselves and communicate a point of view influence and shape reactions of the team members to problems. For example, "*whether a leader categorizes an event as an opportunity or a threat influence how others respond*" [22], p. 974) The organizational reality becomes constituted by the sense-making expressed by project leaders through the messages and words that they use in their communication. Leaders can use both narratives, more formal, or stories, more informal, to communicate their intents according to different purposes and audiences.

In the context of organizational life, stories can concern specific events or people. They can also take the form of biographies. We are interested in gaining insights on how leadership is expressed by managers in a megaproject context through managers' life journey. Our approach is based on a narrative inquiry that helps in grabbing lessons learnt from managers' experience in a megaproject context. Organization studies have used the narrative analysis and storytelling that a few authors started also to apply to project management research. That's why this paper contributes to the growing literature on megaprojects using novel approaches of research and analysis [39, 14]. In fact in megaprojects research stories and storytelling are getting more and more attention especially for what concern their nature and role as far as leadership is concerned. The human and social aspects in managing projects are becoming increasingly important. Stories and narratives have a key role to construct new working practices and to face organizational change processes.

In our future research we plan to expand the number of managers to interview to collect their life stories learning new lessons on leadership in a megaproject context. Complex projects have wide implications for the three pillars of sustainability: society, economy and environment. The analysis of leadership traits and attributes required to managers allows to show the human side of the megaproject adopting a multistakeholders' perspective with people related and societal issues implications (Li et al., 2019) [27].

Through the narrative approach with our analysis, we would like to contribute to the megaproject management literature. Extracting lessons on leadership from life stories of managers can be helpful for development and for managing change and innovation in megaproject contexts [46].

The research question that guides us is to deep which leadership traits and styles help managers to face growing levels of complexity and uncertainty that affect individuals, organizations, governments, and society alike.

4 The Case

The manager interviewed is Matteo Zilocchi, internal project department manager in Emerson Automation Solutions, Isolation Valves business unit. Emerson was founded in 1890 in St. Louis, Missouri, as a manufacturer of electric motors and fans. Over the past 100-plus years, they have grown from a regional manufacturer into a global technology solutions powerhouse. This business unit is operating in a niche sector and manufactures valves made of different materials according to customer specific needs. The customer-centric approach adopted by Emerson involves the design, construction, and after-sales services according to the specific needs of the buyers. Therefore, the company is very focused on end-customer satisfaction, which only happens because of the great professionalism and technical skills of the work teams.

Matteo leads 16 colleagues: 2 of Matteo's hierarchical reports are Project Manager (PM) team leaders, managing 5 PMs each. The other 4 Matteo's direct reports are accountable of Order Entry (2 head counts) and Customer Care (remaining 2 head counts). On all the managed projects, PMs are coordinating engineering, quality, production, witness test (by customer and/or third-party agency), and the documentary part. Matteo's team is also driving the information on factory order execution to the external PMs who, whose are handling the relationships between Emerson and clients all over the world.

In his position, Matteo believes that for a leader it is essential to have a great knowledge of the company, its structure, its functioning, and key people. Coordinating and managing his work team he behaves as a leader: he is solid from a technical point of view (boasting various experiences in international companies) but at the same time he pays great attention to the human side by recognizing the importance of the person at the center of any organizational discourse.

During Covid-19 pandemic the company suffered on the organizational part (for logistic and supplying processes delays) but also from the perspective of human resource management. It was during this period that Matteo had to master his leadership skills to the fullest by acting in presence and remotely while trying to keep the working group cohesive.

According to his perspective, a leader needs to have a wide view and exercise leadership in a somewhat silent way, trying to make his/her value perceived with no imposition.

According to his view, it is fundamental to guarantee a continuous communication activity. This is what Matteo does for the company: he takes advantage of informal chats, always leaving open doors in encouraging people to have a debate on all the hot topics they are facing.

Matteo adopted a customized approach talking to every member of his team adapting to his/her characteristics: he exchanges ideas and opinions leaving ample room for action and autonomy for group members to organize their work according to assigned and shared goals.

An aspect that distinguishes Matteo's leadership is the way he manages conflicts: when possible, he ran into problems as soon as they have manifested when they are still in their origins. It is through open discussion and constant exchange that he prevents conflicts from escalation. His style of leadership is based on a great value of the emotional part: he tries to manage emotions and reactions reassuring the person allowing him/her to dispose of discomfort. When the discomfort has passed, Matteo intervenes trying to understand the real causes and problems.

Once again, the way to manage conflicts is personalized. Recognizing individuality to the person he intervenes in a kind and gentle way, allowing time for the critical situation to decant. Matteo's preventive action is strong and daily.

Matteo's leadership style is recognized and appreciated by colleagues, employees, and customers: it can be defined as wide, aware, and sustainable. In fact, Matteo knows how to motivate people to give their best by feeling fulfilled in their work. He shares influence and decision-making among his team members, and this gives him a high power and great efficacy.

Moreover, he is extremely inclusive, and this grants him the opportunity to explore multiple viewpoints and perspectives and to drive decision-making with all the internal and external stakeholders.

5 Practical Implications and Conclusions

For practitioners, this study proposes effective leadership competency combinations for managing various complexities of the megaproject.

The leader's technical skills are beyond necessary but not sufficient to deal with situations of increasing complexity. To ensure a megaproject's short and long-term success this study recognizes an indispensable value to the leadership style adopted by all the key HRs involved in planning and implementation of a megaproject.

To keep the pace in addressing emerging issues of project complexity leadership research on projects needs to evolve finding traits and styles that can face that complexity. We propose an innovative theoretical approach that seeks to define the characteristics of the leader in megaprojects to overcome the limiting views of traditional and classical approaches. The WAS style includes the characteristics of the new leader according to a collaborative, empathic and shared leadership. The diffuse leader will therefore not only have to lead but above all to motivate and enhance the merits of individual contributors through a personalized approach. This will enable people to feel free to make their own contributions and exercise their initiative depending on the situation and the task required. In this way, through organization-wide leadership, one can be both a leader at one time and a follower at another. Increasing self-awareness entails attempting to understand oneself and one's surrounding predicated on self-reflection. The goal of improving the quality of megaproject management is not only to deliver more projects on time and on budget, but to realize the change organizations and society need to develop in a sustainable way. Sustainable leadership is a process of influence that delivers direction, alignment, commitment, and aims to address social, environmental, and economic issues to create a better world. According to these three drivers of wide, aware, and sustainable leadership, leaders can afford the changes that are taking place around them. Our case study shows that a successful way to lead requires a balance between control, personalization and ability to manage the team members emotions when problems arise. To fully engage their employees the leader's traits that emerge essential are empathy, emotional intelligence and listening. These traits humanize the solid technical knowledge that the leader needs to be authoritative and listened to. With our research we have just started to explore the WAS style trying to highlight the need to match technical skills to social and relational ones. Our aim is to overcome the limit of the present analysis having considered a single case study, by enlarging the number of interviews with megaproject leaders to meet new stories and new practices to support our model that we think could help to face challenges and take opportunities in situations of increasing complexity.

References

1. Al-Subaie AA, Faisal MN, Aouni B, Jabeen F (2021) Investigating the role of leadership styles and governance on project performance in megaprojects. Polish J Manag Stud 23(1):45–58. https://doi.org/10.17512/pjms.2021.23.1.03

2. Al-Subaie AA, Faisal MN, Aouni B, Talib F (2021) A strategic framework for transformational leadership development in megaprojects. Sustainability 13(6). https://doi.org/10.3390/su13063480
3. Anderson LL Jr, Polkinghorn B (2008) Managing conflict in construction megaprojects: Leadership and third-party principles. Conflict Resolut Q 26(2):167–198. https://doi.org/10.1002/crq.229
4. Barabaschi B, Barbieri L, Cantoni F, Platoni S, Virtuani R (2022) Remote working in Italian SMEs during COVID-19. learning challenges of a new work organization. J Workplace Learn 34(6):497–512. https://doi.org/10.1108/JWL-10-2021-0132
5. Caccialanza A, Cantoni F, Favari E, Mariani C (2022) Proposing a sustainability- centered approach to overcome critical issues in megaprojects. Sustain. Megaproject Dev 25–44. https://doi.org/10.4324/9781003305750-3
6. Cantoni F, Favari E (2022) Sustainability and megaproject development. Sustain Megaproject Dev 1–2.https://doi.org/10.4324/9781003305750
7. Cantoni F, Pagnone F (2020) Megaprojects. A special eye on sustainability to overcome the "Iron Law". In: Franca C, Cantoni F (eds) megaproject management: a multidisciplinary approach to embrace complexity and sustainability, Springer, Cham, pp 1–13. https://doi.org/10.1007/978-3-030-39354-0_1
8. Cao T, Locatelli G, Smith N, Zhang L (2021) A shared leadership framework based on boundary spanners in megaprojects. Int J Manag Proj Bus 14(5):1065–1092. https://doi.org/10.1108/IJMPB-03-2020-0090
9. Carte TA, Chidambaram L, Becker A (2006) Emergent leadership in self-managed virtual teams: a longitudinal study of concentrated and shared leadership behaviors. Group Decis Negot 15(4):323–343
10. Castellano S, Chandavimol K, Khelladi I, Orhan MA (2021) Impact of self-leadership and shared leadership on the performance of virtual R&D teams. J Bus Res 128:578–586
11. Coleman JS (1988) Social capital in the creation of human capital. Am J Sociol 94:95–120. https://doi.org/10.1086/228943
12. Damayanti R, Hartono B, Wijaya A (2021) Complexity, leadership, and megaproject performance: A configuration analysis. J Ind Eng Manag 14(3):570–603. https://doi.org/10.3926/jiem.3476
13. Di Maddaloni F, Davis K (2017) The influence of local community stakeholders in megaprojects. Int J Project Manage 35(8):1537–1556
14. Drouin N, Sankaran S, van Marrewijk A, Muller R (2021) Megaproject leaders: reflections on personal life stories. EE Elgar, Cheltenham
15. Faisal MN, Al Subaie AA, Sabir LB, Sharif KJ (2022) PMBOK, IPMA and fuzzy-AHP based novel framework for leadership competencies development in megaprojects. Benchmarking. https://doi.org/10.1108/BIJ-10-2021-0583
16. Fernandes CI, Veiga PM, Ferreira JJ, Rammal HG, Pereira V (2022) Assessing strategic leadership in organizations: using bibliometric data to develop a holistic model. J Bus Res 141:646–655. https://doi.org/10.1016/j.jbusres.2021.11.067
17. Fitzsimons D (2016) How shared leadership changes our relationships at work, HBR Digital edition, May 12
18. Gardner WL, Avolio BJ, Luthans F, May DR, Walumbwa F (2005) Can you see the real me? A self-based model of authentic leader and follower development. Leadersh Quart 16(3):343–372
19. Graham P, Nikolova N, Sankaran S, (2020) Tension between leadership archetypes: systematic review to inform construction research and practice. J Manag Eng 36(1)
20. Hamdy K (2010) the essential role of leadership in managing mega projects. paper presented at PMI® global congress 2010—North America, Washington, DC. Newtown Square, PA: Project Management Institute.
21. Han SJ, Hazard N (2022) Shared leadership in virtual teams at work: practical strategies and research suggestions for human resource development. Hum Resour Dev Rev 21(3):300–323
22. Havermans LA, Keegan A, Den Hartog DN (2015) Choosing your words carefully: Leaders' narratives of complex emergent problem resolution. Int J Project Manage 33(5):973–984

23. He Q, Luo L, Hu Y, Chan APC (2015) Measuring the complexity of mega construction projects in China—a fuzzy analytic network process analysis. Int J Proj Manag 33:549–63
24. Hoch JE, Dulebohn JH (2017) Team personality composition, emergent leadership and shared leadership in virtual teams: a theoretical framework. Hum Resour Manag Rev 27(4):678–693
25. Larson E, Gray C (2014) Project management: the managerial process, 6e. McGraw Hill, New York
26. Lee S, Sawang S (2016) Unpacking the impact of attachment to project teams on boundary-spanning behaviors. Int J Project Manage 34(3):444–451
27. Li Y, Lu Y, Cui Q, Han Y (2019) Organizational behavior in megaprojects: integrative review and directions for future research. J ManagEng 35(4)
28. Locatelli G, Mikic M, Kovacevic M et al (2017) The successful delivery of megaprojects: a novel research method. Proj Manag J 48(5):78–94
29. Locatelli G, Konstantinou E, Geraldi J, Sainati T (2022) The dark side of projects: dimensionality, research methods, and Agenda. Proj Manag J 53(4):367–381. https://doi.org/10.1177/87569728221103911
30. McCauley CD, Palus CJ (2021) Developing the theory and practice of leadership development: a relational view. Leadersh Quart 32(5)
31. Müller M (2014) After sochi 2014: costs and impacts of Russia's Olympic games. Eurasian Geogr Econ 55(6):628–655. https://doi.org/10.1080/15387216.2015.1040432
32. Ninan J, Clegg S, Burdon S, Clay J (2021) Overt obstacles and covert causes : An exploratory study of poor performance in megaprojects. Project Leadership and Society, March. https://doi.org/10.1016/j.plas.2021.100011
33. Packendorff J. (1995), Inquiring into the temporary organization: new directions for project management research, Scandinavian Journal of Management, 11(4), 319 333.
34. Pearce CL, Conger JA, Locke EA (2007) Shared leadership theory. Leadersh Q 18(3):281–288
35. Pitsis A, Clegg S, Freeder D, Sankaran S, Burdon S (2018) Megaprojects redefined – complexity vs cost and social imperatives. Int J Manag Proj Bus 11(1):7–34
36. Putnam RD (1994) Social Capital and Public Affairs. Bull Am Acad Arts Sci 47(8):5–19
37. Quinn L, D'amato A, (2008).: Globally responsible leadership a leading-edge conversation, center for creative leadership, CCL White Paper
38. Qiu Y, Chen H, Sheng Z, Cheng S (2019) Governance of institutional complexity in megaproject organizations. Int J Project Manage 37(3):425–443
39. Sankaran S (2018) Megaproject management and leadership: A narrative analysis of life stories–past and present. Int J Manag Proj Bus 11(1):53–79. https://doi.org/10.1108/IJMPB-07-2017-0081
40. Sweeney A (2022) Looking within: A longitudinal qualitative analysis of shared leadership behaviours in organisational teams. Team Perform Manag 28(7–8):441–460. https://doi.org/10.1108/TPM-02-2022-0013
41. Virtuani R, Barabaschi B, Cantoni F (2022) The human side of megaprojects: leadership style and traits to face growing levels of complexity and uncertainty. Sustain Megaproj Dev 3–24. Routledge. doi:https://doi.org/10.4324/9781003305750-2
42. Visser W, Courtice P, (2011) Sustainability Leadership: linking theory and practice SSRN. https://ssrn.com/abstract=1947221 or https://papers.ssrn.com/sol3/papers.cfm?abstract_id=1947221
43. Wan J, Le Y, Wang G, Xia N, Liu X (2020) Carrot or stick? the impact of paternalistic leadership on the behavioral integration of top management teams in megaprojects. Int J Manag Proj Bus 13(5):937–960. https://doi.org/10.1108/IJMPB-12-2019-0302
44. Wang G, Locatelli G, Zhang H, Wan J, Chen Y (2021) Playing the cards right: Exploring the way leadership influences organizational citizenship behavior for the environment in megaprojects. Eng Constr Archit Manag. https://doi.org/10.1108/ECAM-01-2021-0093
45. Whitty SJ, Schulz MF (2007) The impact of Puritan ideology on aspects of project management. Int J Project Manage 25:10–20
46. Woodward I, Shaffakat, S, Understanding Values for Insightfully Aware Leadership (February 17, 2016). INSEAD Working Paper No. 2016/05/OBH. SSRN: https://ssrn.com/abstract=2471492 or http://dx.doi.org/https://doi.org/10.2139/ssrn.2471492

47. Zhai L, Xin Y, Cheng C (2009) Understanding the value of project management from a stakeholder's perspective: case study of megaproject management. Proj Manag J 40(1):99–109
48. Zhu F, Wang L, Sun M, Sun X, Müller R (2019) Influencing factors of horizontal leaders' role identity in projects : a sequential mixed method approach. Int J Project Manage 37:582–598

Communications in Megaprojects: Two Projects, Two Approaches

Francesco Clemente

Abstract Communication is pivotal in the time, cost, risk and stakeholders' management of megaprojects. Communication is the backbone of engineering, but, since every megaproject is unique, the solutions adopted to ensure efficient communication must be therefore tailored case by case, in particular by adapting to the actual situation the general procedures suggested by the chosen management standard, regardless if they come from the best practices of PMBoK®, Agile or others. This article illustrates the difficulties that may be caused by communications in a megaproject in the Middle East, and compares the communication management choices for two megaprojects in the Arabic peninsula, thus exemplifying how two very different environments may and should require antithetic answers in order to reach efficient and effective management solutions in the field of communication.

Keywords Megaprojects · Communications · Design · Construction · DMIA · KAP4

1 Introduction

Carrying out a megaproject is a long and complex process. It starts with the design, which, simplifying the matter, is divided into two phases: first the design drawing phase, after which normally comes the selection of suppliers and the materials that will actually be installed; then comes the second phase of shop drawings, which integrate and replace the design drawings including the actual characteristics of the materials and equipment that will be installed. This phase brings major changes to the project's verification calculations, cost estimates and construction scheduling. But even after these two phases, the design in megaprojects is always constantly reviewed and updated during the construction according to the demands and needs of the site and the main stakeholders, the trends of the market, and new requests by the Client, resulting in the need for effective, fast and continuous communication

F. Clemente (✉)
Innovare Costruendo, Naples, Italy
e-mail: f.clemente1955@gmail.com

E. Favari and F. Cantoni (eds.), *Complexity and Sustainability in Megaprojects*, Lecture Notes in Civil Engineering 342,
https://doi.org/10.1007/978-3-031-30879-6_9

between the parties. In many ways it could even be said that the design work with its final product, the shop drawings, is nothing more than a long process of processing data, information, requirements which produces in the end a very complex and organized set of information. This information is transmitted to the construction site and allows the company to carry out the construction work.

After the processing of the shop drawings (or, in the case of Design – Build contracts[1], during their processing), comes the construction, which implies very complex communications between all the construction staff and between the construction site, the technical office, the management, the Client and the stakeholders. All the materials have to be approved, the Request for Information (RFI) by the building site have to be dealt with, changes requested either by the Client or other Stakeholders have to be implemented. And, of course, the Client and the Stakeholders have to be constantly updated about the phases of the project.

Bottom up communication with the management of the regional branch and the head office at home is also of great importance for engineering companies: only the continuous exchange of information in both directions can ensure that the central management is fully mastering and understanding the local professional and commercial environment of the staff on the ground, and that in the closing of the project the lessons learned are absorbed and processed also by the staff at home, thus becoming truly corporate assets.

It is therefore evident that, if communication fails, this will affect all the design and construction processes of the project, including the management of costs, time and quality, and will certainly compromise the successful outcome of the project.

Communication is therefore the backbone of engineering: indeed it can be said that much of engineering consists of the processing of information and its communication.

The problem is that the communication between stakeholders in international megaprojects suffers from an innumerable series of human, cultural, ethical, professional, even racial and religious filters that disturb it, generating what the PMBoK® calls noise that makes communications confusing.

On the other hand, anyone familiar with PMBoK® and with the management of engineering megaprojects in the Middle East and generally in the Third World will have experienced the sometimes enormous distance between their own working reality and the ideal situation presented in mostly American/British handbooks in which the best practices are set out. This distance is then aggravated by the particular communication difficulties that the technician and western management must face in transferring and adapting their professional, cultural and human skills in the Middle East, where communication and integration are especially difficult.

In the following, I will illustrate the communication management choices for two megaprojects in the Arabic Peninsula for which I have been Technical Manager, showing how two different environments may and should require anthitetic management solutions in order to get efficient and effective results. Indeed, every megaproject is unique, and the solutions adopted to ensure efficient communication should be

[1] In a design-build contract the contractor contemporarily carries on the design under the supervision of the Project Management Office and builds the works.

tailored case by case, particularly adapting the procedures suggested by the adopted method (PMBoK or whatever) to the case.

2 First Case: The DMIA - Development of Muscat (Oman) International Airport

The first megaproject to be analyzed is the DMIA, cost 4.3 billion USD, including: the new terminals for 12 million passengers over an area of 350,000 square meters, the new runway of 4,000 × 60 m with related taxiways, the new control tower, the air traffic control center, the training center, the new access routes, parking lots for a total of 6,000 parking spaces, and over 50 service buildings. The works were carried out with the employment of about twenty-two thousand workers on three shifts. All these constructions were performed keeping the existing airport working, with a very complex site occupation scheduling. The contract was a kind of mixed design-build one, since a new company was appointed to replace the original one during construction, due to the latter's unsatisfactory and late performances; the new company was in charge of managing the construction through a construction management office, and at the same time, through a technical office, updating the original design with the validation of new shop drawings, and the incorporation in such design of all upcoming requests by all the Stakeholders. These updates were generated mainly by the technical office, and sometimes by the contractor under the guidance of the technical office, with the final approval of the resident engineers.

The shared communication management procedure of the three offices involved (construction management, technical office and construction company personnel) created a particularly effective DMIA communications system, an example of a very effective approach to the question of communication in megaprojects in Middle East. This was the result of very well-established and tested organization assets, of a very favorable logistical situation, and of a communicative and collaborative attitude adopted by all the technical staff involved in the project. As a matter of fact, the international company in charge of the project was a very old and established one, with extensive experience in the field of construction inspection and management, and an asset of excellent and very well tested procedures. The on-site office included more than 400 engineers and surveyors, but its core consisted of seven resident engineers in charge of construction inspection, and material and new design approval. These very experienced resident engineers were supported by the technical office and by a staff including a number of teams, each one made up of experts in a particular aviation construction field (runway paving, luggage handling, data transmission, radar, lighting and signage, and so on). The procedures of the airport project identified the approval of the Inspection and Test Plan (ITP) as the starting point of the construction management and were based on three procedures with related forms:

1 - Material Sampling for Testing (MST), for taking a sample to be tested or for examining a particular item and equipment by the works management;

2 - Material Inspection Request (MIR), for quality assurance and acceptance for use in design or in the construction of a building component, machinery or other;
3 - Contractor Work Inspection Request (CWIR), for the acceptance by the construction management of a body of work, a lump sum or a finished and identifiable building or MEP item to be included in the cost estimates or the payment certificates.

According to the ITP, the other pivotal document during the construction works was the RFI (Request For Information), generated by the Contractor for any technical problem. The ordinary RFI were answered directly by the resident engineers, but the ones potentially involving a new design or a particular technical expertise were forwarded to the technical office. The requests for design variations by the Client or other Stakeholders were similarly forwarded to the technical office. The latter then produced a new or additional design, and sometimes supervised and reviewed a new design by the contractor's design office.

Oral and written communication between all parties took place exclusively in English; similarly, all project, contractual and economic documents were drawn up in that language. All the documents concerning the project – including drawings, reports, calculations, letters, Minutes of Meeting (MOM), Requests for Information (RFI), Non-Conformance Report (NCR), etc.—were uploaded to an ACONEX® platform, in my opinion the best data storage and project management program available on the market for megaprojects: it enables you to manage with ease huge amounts of information that normally distinguish these types of project and the latest versions also facilitate direct communication between construction sites, technical office and project control office.

In the data storage procedure of the airport project each document, once signed for validation and authorization by the person in charge, was delivered to the so-called *gatekeeper* (literally gatekeeper, a term currently in use in the multimedia industry), an office that dealt with the encoding of the file name by means of a string that contained all the information necessary for the identification of the object, the author, the discipline, and other matters.; the gatekeeper then proceeded to upload it onto the platform, officially turning it into project data.

Through the digitization of a partial code, each enabled user could then retrieve all the documents concerning a given topic and its development over time (RFI, first reports, correspondence between Customer and stakeholders, MOM, first designs, approvals of various kinds, economic evaluations, up to the executive design) and with a simple drag and drop transfer them to their desktop.

At the same time, the project had a transversal structure of communication, risk and issue analysis, and very effective work planning, borrowed from decades of tendering procedures at the highest level and facilitated by the logistical concentration of all the main actors and stakeholders within and near the airport area.

Onto this well-tested structure the various offices superimposed and adapted their communication plans and the related tools designed according to their need, their Enterprise Environment Factors Enterprise Environment Factors (EEF) and their Organization Process Assets (OPA). Communications were thus fed into a continuous cycle, sharing and controlling information throughout the time and managerial span of the project.

The process started with a meeting held every two weeks and called Liaisons Meeting (LM). It was organized in the contractor's offices in front of a 6-m long satellite photo. The meeting was led by the coordinator of the company, and was attended by representatives of the contractor himself, subcontractors, the customer, construction management, the project team, the risk management team, and the main stakeholders.

In the meeting each of the parties tabled what they hypothesized would be the problems—that is, the risks and issues, to use the language of the PMBoK® Guide—that would affect the progress of the megaproject in the months to come. The topics were discussed and if the coordinator felt that they were worthy of attention, at the end of the meeting he added them to an Excel sheet that served as MOM of the liaison meeting, with boxes for the description of the problem (with any photos or graphs deemed necessary), the actions to be taken, those responsible for these actions, and the expected times.

At the end of the Liaison Meeting the table/attendees was/were updated about both new entries and developments concerning old problems, which were used in the subsequent meeting as a basis for discussion and progress. Problems that persisted after a certain number of meetings formed the company's RFI or the Customer's Variation Order (VO).

The RFIs and VOs were discussed weekly in a technical meeting usually attended only by the chief resident engineer and the technical managers, and sometimes by other relevant engineers belonging to any of the two teams of staff. The meeting was led on the basis of an Excel sheet, symmetrical to that of the LM; the sheet reported the numbers of RFI or the extremes of VO, the design or in any case the clarification action began as a result, the person responsible for the action, and the status of the same. This table was also obviously updated at the end of the meeting and circulated among the participants.

The RFIs and VO that weekly called for new designs were termed Tasks in the internal system of the technical office and registered with a code and monitored monthly to keep control of the technical team's workload. The cycle finally ended in the weekly meetings with the Client conducted by the manager of the design team.

These meetings were attended by the Client, representatives of the works' management team divided up according to discipline, a representative of the PCM, and sometimes the stakeholders deemed necessary or the members of the risk management team. The meetings were based on a third Excel sheet similar to the previous ones, in which were set out in chronological order all the design actions and the technical studies carried out during the week for each RFI or VO.

The MOM of the meetings with the Client, countersigned by the parties in the following week's meeting, were valid as approval by the Client of the decisions made and the approaches proposed by the design team. Only sometimes—usually for particularly complex problems or those whose solution involved the involvement of many stakeholders or financially large VOs—were communications and approval carried out partly through official letters uploaded to Aconex®.

It is worth pointing out that in the Execution Plan of the airport project a large number of meetings and periodic reports were listed, but in actual working practice most of the problems soon became concentrated into the three aforementioned meetings, and special meetings called ad hoc by one of the parties were used only for the solution of particularly complex problems. On the other hand, thanks to the open plan structure of the offices, the staff of all parties had free and continuous access to the desks of the other parties, so that every problem was in practice analyzed by all interested parties with real-time communication, as the latter simply went to the workplace where the issue originated and discussed it freely, following design and construction developments.

It should also be noted that the Execution Plan contained an exhaustive list of Stakeholders, that the latter normally took part in liaison meetings and sometimes in meetings with the Client, and that in any case the relationship with them was straightforward and direct. The Client was advised that particular stakeholders would be contacted for a certain problem, and a direct approach was followed, using the aforementioned meetings and written communications to make the other parties aware of what was done, and to obtain their approval.

The stakeholders all had a good command of English and good analytical skills, mainly because those working in the airline industry are required to undergo years of preparation abroad and engage in international relations; it must be said, however, that the nation in which the project was located, Oman, is a nation with a Sunni Muslim majority that has an open and tolerant outlook and has great seafaring traditions, which historically has opened it to the outside world.

Unsatisfactory work was managed through NCRs, and all construction site problems not immediately solvable by inspectors on the basis of design documents and contract documents were sent back to senior resident engineers and managed through RFI. The less complex of the latter were solved by the resident engineer teams, otherwise the RFI was forwarded to the design team, as previously illustrated.

All in all, the resident engineers and their staff were pivotal in the communication and management plan, and therefore the balance of the management was tipped toward the building site.

Communication between the parties also plays a very important role in megaprojects in controlling time and costs, which is usually included in the management of the contract and is therefore the responsibility of the Contract Manager and staff dealing with scheduling and risk management. The data processed by the Contract Manager's office works comes from the technical office and the construction site, and, more importantly, once processed it must return to the latter two offices so they can take it into account in their management. Continuous and effective two-way communication is therefore necessary, ensuring the most immediate and controlled feedback possible. In the case of DMIA all new planning was forwarded to the cost evaluators and schedulers offices to draw up related costs estimates and time schedulings, which were promptly circulated among the parties and incorporated into the project documents.

As far as risk management is concerned, in my experience, in the case of the megaprojects it involves not so much risk management as understood in the best practices of the PMBoK®, but depends more on their hierarchy and their internal and external communication, since in the mental approach typical of engineers a project or construction risk often quickly becomes an issue or even a constraint on the work, and is almost automatically associated with its possible technical solution. On the other hand, the particular mentality and preparation of engineers makes it possible to identify a risk almost at the same time as providing its technical solution. In the case of DMIA, the risk management staff was coordinated by an interface manager and, at the beginning, they began to draw up tables including project risks, which were discussed in special meetings with interested parties. Soon, however, they turned their attention above all to the liaison meetings we have already talked about above. Actually, these meetings were structured and conducted in such a way as to highlight and develop possible future risks—understood as events or problems that would have affected the planned performance of the work—well before their manifestation, and focused the attention of the stakeholders concerned on their elimination or mitigation in advance.

Very soon, the risk and the interface manager became the main protagonist and contact person for what is the key risk in an airport project: the continuous updating of the phasing and the re-organization of the construction sites with related accesses in order to remain compatible with the simultaneous management of the airport. In fact, each change to processing or scheduling work in an airport in use makes it necessary to completely re-organize the fences of the spaces accessible to workers, the access routes of workers and construction vehicles, the circulation of airport service vehicles, but above all the ground and air movement of aircrafts, to be promptly communicated to international authorities, airlines and all the involved parties through NOTAMs (NOtice To AirMen).

3 Second Case: The KAP4 - King Abdulah Bin Abdul-Aziz Security Headquarters Development Project (Saudi Arabia)

The second megaproject under analysis is KAP4, a project leading to the construction of 24 building complexes for the Saudi Ministry of the Interior in 9 different locations over a total area of 3 and a half million square meters, with 386 buildings to be designed and built, for a total walkable area of 1.3 million square meters and 8.2 million mc. at a total cost of more than 2 billion USD. The contract was a typical design-build one, with a very highly empowered Project Control Management (PCM) contract awarded to an Italian Company.

Upon the arrival of the author in Saudi Arabia the technical office was still in training, and the communication situation was very difficult: the technical team could not see the construction company personnel, and neither the Client and the end users, and also communication with the central management of our own company was not easy. The difficulties were also enhanced by the profound differences between stakeholders which, if not addressed, could have a very strong negative impact. Moreover, there were initially huge problems of communication between our own colleagues, mostly from the Middle or Far East, even in the building of simple human relations. In the Middle East no one talks to anyone and everyone is jealous of their technical knowledge, which they do not share at all. So they do not know what technical coordination is: the architect does not ask the mechanical engineer where and how the air channels pass while the electrical engineer does not communicate to anyone the real size of the electric panels, and so on.

In the case of KAP4, much of the communication work was based on the initiative, the example and the good relations among the Italian management and staff, but much was also achieved through the preventive preparation of supporting materials such as check list, communication log and generally formats prepared for as many needs as possible. In a context in which people were not inclined nor able to communicate due to the lack of mastery of the language and of the necessary analytical skills, it is essential to provide staff with a series of guidelines and support about the topics to be analyzed and communicated.

Special care was taken to decide what and how much should be documented, based on experience and contractual documents, which are usually very important in the planning of communications: in fact, they usually indicate the owners of the information, the means of transmission, the frequency of communications, the conditions for considering a communication as official and, in the case of those of the Client, as a variant authorization. They also often contain precise instructions about communications with stakeholders. All this is normally included in the typical document management of an engineering program or project, namely the Execution Plan and Procedures, containing applicable policies and regulations, and all templates, forms and instructions deemed necessary.

The 5 procedures of KAP4 concerning communication were quite analytical, and included all forms, formats, templates and instructions for communications between the site, the technical office, the main office and the customer, and a large number of meetings and related agenda, including a weekly meeting with the participation of the Client, the construction company with its the designer and the PCM. They then provided for the so-called Steering Committee with the customer and the sponsor, and all the other regular meetings typical of engineering projects (risk control, quality control, etc.). They also included the Monthly Report, a voluminous monthly report for the Client and the sponsor on the current performance of the project, based in turn on numerous other internal reports prepared by the various departments and construction sites on a normally weekly basis, used internally to share information.

The procedures provided for weekly technical meetings with the contractor with normalized agenda for discussion relating to every technical aspect to be decided, but the technical office soon abandoned this system to move to informal daily meetings of all disciplines, recorded on Excel sheets®. The original design review procedure, planned in waterfall mode, was then greatly modified, leading to a cyclical and adaptive mode on the Agile model. The company's staff, site staff and Client staff had continuous access to the technical office, shared information, and discussed all the problems continuously and freely, which were practically analysed through real-time communication and implementation of the changes in the drawings or technical report being discussed without formal approval.

The first solution in difficult communication cases is in fact logistics, that is to bring all stakeholders together in open-plan offices: this simple fact makes communication easy and direct, and indeed encourages it. This was immediately done in KAP4: the technical office with its technical manager, construction manager and three location managers were all gathered in the same area of the main office, and this made it possible for them to be in continuous dialogue; on the contrary, the scattered location of construction sites made their communications difficult at the outset: many construction sites were thousands of kilometres away from each other and from the main office, with materially and instrumentally difficult communications (lack or inadequacy of data and telephone network, rudimentary road network, environmental and weather conditions that were often prohibitive). Moreover, the staff of single building sites often had diametrically opposed approaches to communication due to cultural reasons, professional and organizational incapacity, or because of lack of adequate equipment.

Apart from the informal communication with the headquarters of the main office described above, the official communication of the site with the technical office took place in English through the two normal types of communication used in all megaprojects: RFI and RTI (Request for Technical Information); the use of forms defined in the procedures made these communications easier. There was also official bilingual correspondence between the customer, stakeholders and contractors, whose filing system, however, was so convoluted that only secretaries and document controllers were able to locate them.

In KAP4, oral and written communication between all parties should have been made exclusively in English while contractual and economic documents, specifications and correspondence were drawn up in two languages, English and Arabic. For contractual documents, the Arabic version prevailed while design papers (drawings, reports, calculations) were only in English. Unfortunately, the lack of English proficiency among many of the PCM staff and businesses, and especially the Client, made communication difficult. The presence of Arabic-speaking personnel (Syrian, Egyptian, Jordanian, Lebanese, etc.) with a very poor command of English made communication even more complex, as the latter tended to communicate with each other and with the representatives of the Client and the stakeholders in their language, cutting out the Italians. It also happened that translations of project documents and correspondence confused ideas even more, as they were prepared by low-level secretarial staff and poor technical preparation. And even Arabic-speaking colleagues disagreed with the Arabic interpretation of the contract documents, as these were the result of three subsequent translations: the Arabic documents that counted as the official version were in fact based on standard international documents in English, then translated into Arabic, and finally conveyed in English, often becoming almost incomprehensible. The shift to agile techniques was crucial in solving these problems, as the frequent meetings, accelerated procedures, and the need to select the staff best suited to Agile inevitably led to a more continuous and effective communication[2].

Communication with stakeholders and end users was also quite difficult since it took place through the Client, who transmitted very convoluted and inaccurate communications in Arabic, the translation of which was always very difficult and low-level. Moreover, the Client only organized meetings or visits to existing facilities in the case of extreme need but such meetings were undermined by the very poor mastery of the English language by stakeholders and end users and by their far from proactive attitude. The solution was to follow up on each meeting through letters in which the PCM communicated to the customer the decisions to which it would push the end user, thus making them official.

Communications were further complicated by the fact that the nation in which the project was located was a nation of Sunni faith whose majority belonged to one of the most closed and intolerant confessions, and was traditionally linked to the customs of tribal groups. Every relationship was made difficult for us Westerners by very complex unwritten rules of precedence and formal courtesy. Moreover, the project procedures did not contain an exhaustive list of stakeholders, thus making work planning difficult.

Great problems arose for the storage of the files of the technical documents, which had to be shared instrumentally with both the construction sites and the customer, thus making this sharing a substantial aspect of the communication. Archiving and document sharing is generally a key aspect of communication in major engineering projects, but is unfortunately often overlooked. In a megaproject, the number of documents produced – and the relative difficulty in storing them – is considerable.

The KAP4 contract provided for the creation of an online archive to which the construction sites, the Client and stakeholders named for this could access remotely, to be created through Primavera, a program designed for management and not for the storage of data. This was totally unsuitable for the storage of such a huge mass of documents as those generated for the technical works of KAP4, which numbered more than 160,000 drawings, 50,000 material and equipment approval volumes, and 12,000 technical calculation reports.

Considering, however, the shortcomings of Primavera in fulfilling the required purposes, and not being able to use other programs available on the market, the technical office was forced to set up an independent system of filing and retrieving technical data, based on the naming of documents using a uniquely descriptive string, huge pivot tables in Excel® and the typical tree structure of the lists of design processing (site, building, technical discipline, planting, etc.), thus going against the basic rules of management and common sense, but managing to survive and make the project survive.

A system was implemented by the technical office and individual building sites to transfer data regularly to the Contract Manager's office, which processed them in

---

[2] On the use of Agile in KAP4 and generally in mega engineering projects see "Can Megaprojects be Agile?" by F. Clemente and P.L. Guida, in "A multidisciplinary approach to embrace complexity and sustainability", AA.VV., G. Giappichelli, 2019.

bar diagrams and, using the Earned Value technique, in SPI (Schedule Performance Index) and Cost Performance Index (CPI) in charts that were presented and discussed in weekly meetings and steering committees. In this way they were shared and could provide appropriate feedback on the individual activities of individual projects.

Unlike the DMIA airport project, the KAP4 procedures provided for the approval of both the shop drawings and the materials by the technical office. The shop drawings approved by the technical office were uploaded onto a centralized Primavera system, and the site staff could download them at any time, although in reality there were difficulties in some peripheral construction sites not well connected to the wi-fi network.

Samples of approved materials were sent by the main office to centralized material depots by area, so that site technicians could access them to compare them with the material delivered on site. The main reason for the centralization of approvals in the technical office was that if the approval of shop drawings and materials were left to the individual area managers of the 24 sites scattered throughout the nation, there would have been the risk of having a variety of different solutions approved in buildings of the same type, even if located in different geographical areas.

However, the decision to centralize the procedures was taken also because the site staff available in the project area was traditionally of a very low technical level. Thus throughout the works, the aforementioned centralization further increased the responsibilities and burden of the technical office: the number of RFIs and requests for clarification or opinions sent by the site staff to the main office was exceptionally high.

Similarly, the construction site staff requested that many mock-ups be inspected and approved by the technical office as well. Indeed, the checking of the documents to be prepared on site and their storage using the Primavera system was of a good level: every week a long list of reports was updated according to procedures, including the Status reports of the Main Activities, Labor and Equipment, PCM Personnel, Requests for Instructions (RFI), Compression and Traction Tests on the Specimens, Incidents and Safety Control, Weekly Progress, and Weekly Photo Documentation. At the same time, the registers of Inspection Requests, Material Inspection, Soil Compaction Test, Calibration of Equipment and Measuring Instruments, Summary of Tests on Site, Risks and Issues, Construction Site Instructions (SI), Non-Compliance Reports (NCR), and Long Lead Lead Items were also updated.

The risk specialist drew up the classical Excel® table for risks and one for issues, initially collecting and updating the data by means of one-on-one meetings with the three project managers. At some point, however, monthly meetings of all area managers were organized for updating and discussing the tables together, and these meetings were very effective from the point of view of communication, managing to overcome cultural gaps and resistance to communication between parties who sometimes even saw themselves as opposed to each other.

Finally, communication continued to be bottom up, from the Project's onsite managers to the management of the regional branch and the head office at home. For KAP4 the system was based on regular personal visits to the branch, many emails sharing problems, and regular reports to management in Italy.

4 Conclusions

As pointed out by the PMBoK® Guide, communication is the essence of management and project managers spend most of their time communicating with their staff members and project Stakeholders. In particular, the Guide to PMBoK® reads: "*Effective communication creates a bridge between the various stakeholders, who can have different cultural and organizational formations, different levels of experience, and different interests and perspectives*".

In the two cases under study, the situations were completely different: an excellent construction management office at DMIA, and a good and strong technical office with very poor building site personnel at KAP4. The two communication procedures were therefore wisely and efficiently designed in different ways: for DMIA this involved giving a pivotal role to the resident engineers, and for KAP4 it meant placing all responsibilities on the technical team, which was given wide margins for intervention.

But, apart from this macroscopic difference in the balance of responsibilities between technical office and building site, our analysis of the two cases also shows how a large range of tools may be applied to obtain the best communication at all levels, and how choosing the most appropriate ones can brilliantly solve the communication problems related to the local environment.

The best practices of PMBoK®, like those in any other management handbook, are very general, and can be applied as they are without any adaptation only in a very few cases, and definitely not in megaprojects. Unfortunately, managers without the necessary experience in large civil projects blindly and slavishly apply what is prescribed by management handbooks, thus forcing their projects into the straight-jacket of uncalled for and even damaging procedures. But a good manager must have the courage, the experience and the skills to adapt all the suggested practices to the real local situation in order to achieve the most effective communication, and to choose the appropriate tools.

The same considerations apply to risk management, which in megaprojects is very often considered to be a component of communication management, as in the two cases analyzed in this article. Risk management must therefore be essential, fast and adaptable, and use the best communication channels. Again a number of different tools may be applied instead of the suggested best practices without limiting management's sphere of action, while a blind use of such best practices may seriously undermine the correct running of the project.

All in all, management is a practical discipline, as underlined by the fact that handbooks are based on best practices reported by managers working in a number of different environments; therefore, experience is everything, and this is even truer in the case of megaprojects, which are unique and very large projects where a mistake in communication may cause very serious loss. Here, the experience of the manager and his capacity to devise new and better channels based on the local situation, is of the utmost importance.

Knowledge Discovery Framework for Decision Support Systems in Tendering, Cost Analysis and Construction Phases

Claudio Mirarchi, Davide Simeone, Luca Sivieri, and Alberto Pavan

Abstract Taking decisions in the tendering and construction phases of complex infrastructures requires the integration of several information sources. This activity is nowadays mainly defined trough the experience of the single experts that work on the project with a consequent limitation in the used information and knowledge (the one that the specific expert knows) and the high risk on taking decisions that are based on personal biases.

Hence the need to explore the development of a DSS that can support the experts in taking decisions during these phases. As the expert does, also the DSS should be able to integrate several data sources that may be of different nature and require different techniques to be used. The objective of this paper is presenting a DSS knowledge framework highlighting the key data sources and the techniques that may be applied to extract the knowledge contained in the specific source. This research aims at creating a reference base for the development of analytical processes to discover knowledge in the different data sources (drawings, reports, etc.) that constitute the basis of design and construction activities.

Keywords Decision Support System · Knowledge Management · Artificial Intelligence · Decision Making · Value creation · Project success factors

1 Introduction

The construction of infrastructures like bridges, tunnels, etc. has a high degree of uncertainty (and consequently risk) due to the limited amount of information available at the design and tendering phases. The geological composition, for example, is based on sample tests and then verified during the construction phase with frequent

C. Mirarchi (✉) · A. Pavan
Politecnico di Milano, Milan, Italy
e-mail: claudio.mirarchi@polimi.it

D. Simeone · L. Sivieri
Webuild SpA, Rozzano (MI), Italy

E. Favari and F. Cantoni (eds.), *Complexity and Sustainability in Megaprojects*, Lecture Notes in Civil Engineering 342,
https://doi.org/10.1007/978-3-031-30879-6_10

changes in the technologies used for the construction and in the design of the infrastructure itself.

The possibility of capturing the experience gained in previous projects comparing similar situations and providing support for the development of decisions that can rely not just on the experience of the single but on an "enterprise knowledge" represents nowadays one of the drivers in the research and industry developments. The increasing use and reliability of machine learning (ML) is currently transforming the possibilities related to the collection, analysis, and use of vast datasets. Decision-makers in enterprises can draw on ML capabilities gaining new insights into emerging phenomena, making predictions and extracting information from enormous quantities of data [1] hence reducing the intrinsic decision risks.

1.1 Knowledge-Driven Approaches to Construction Tendering

For construction companies, tendering is a delicate phase, dense of critical decisions to be rapidly taken by multidisciplinary teams, often without sufficient time and data for detailed analysis or simulations. In this challenging context, knowledge derived by analogies with similar projects is a relevant competitive value that, if carefully used, can drive the offer to a successful project awarding. In construction tenders, that competitive advantage is related to the ability to effectively manage knowledge [2]: relying on previous projects knowledge, when accurately managed, prevents the re-invention of the wheel, quickly providing reliable data, although to be critically used, and favoring innovation, quality of the technical proposal and reducing uncertainties [3]. At the same time, portfolio data and information reduce the habit of relying on tacit, personal knowledge from the different team members, allowing a data-driven comparison with each decision t In the past, technical knowledge of previous projects was essentially stored in the engineers' brains, with clear problems of accessibility and reliability. o avoid bias or subjective judgement, and providing additional information to avoid unfair, incomplete and poorly constructed result [4].

The progressive adoption of digital systems has improved the availability of projects' information, usually on documents' digital archives, but accessibility to documents does not necessarily ensure the correct exploitation of the data, information and knowledge stored in them. Nevertheless, the portfolio knowledge base represents a key asset for a construction company that, if well organized, can act as underlying component of a new generation of Decisions Support Systems specifically conceived for construction tendering.

1.2 Decision Support Systems for the Construction Tendering

In construction projects, decisions during tendering phase have a strong impact on both the results of the bid and on the actual execution of the project, if awarded. With the increasing complexity of technical decision-making processes and with the increasing number of stakeholders involved, each with its own objectives, competencies and bias, some research has focused on the study of Decision Support Systems (DSS) to enhance quality of decisions during this phase. The concept of DSS is not new [5] but the applications in industry are increasing only in recent years [6]. A DSS can be seen as a combination of computer-based decision support tools with the aims of supporting in decision-making activities [7, 8]. Accordingly, a DSS may combine several tools and data sources such as building information models, geographical information systems (GIS), artificial neural networks techniques, Machine Learning, model-driven approaches, etc. [9].

Within the different typologies of DSS considered, Knowledge-driven ones are the most effective in construction tendering and require that knowledge is stored in a reusable form in order to support solution of problems and more generally technical or economical decisions [10]. In a turbulent environment such as a tender development, in fact, traditional approaches such as quantitative models or data mining DSS are not likely to provide reliable support to strategic decisions [4], while qualitative information, often derived by previous project knowledge, can support decisions and assessment of different options.

Knowledge-based DSS for construction tendering mainly focuses on a decisions flow - guided by the proposed systems - where knowledge is essentially used at each sequential decision point to assess its weight in order to reach final indicators regarding the strategy to bid/not bid as well as evaluating different aspects such as risks, opportunities, profitability, etc. [11]. In competitive bidding, particular attention has also been paid on DSS to support decisions related to bid mark-up [12] to improve profitability and competitiveness of the proposal, while less interests have been made in supporting technical decisions that determine the correct estimation of the project.

Nevertheless, in these DSS knowledge is still in large part tacit or external to the process, and contributes to the bidding overall strategy but not to the analytic study of the project and, in case of technical offers within the tendering phase, to the development of project proposals and options. In the literature background of DSSs for tendering, it emerges the lack of systems that ensure effective knowledge sharing and management, as well as its capitalization, during the development of technical and strategical parts within the tendering organization, in particular when it is composed of multiple contractors, designers, consultants.

Considering the critical need of capturing and generating organization knowledge as input for the construction of a DSS, the objective of this paper is to present a framework for the knowledge preparation to pave the way for the development of DSS devoted to megaproject tendering and construction phases.

The rest of the paper is organized as follows. Section 2 presents the theorized DSS knowledge framework. Section 3 proposes a discussion about the possible technologies and processes that should be applied in the development of the framework components. Finally, Sect. 4 reports the conclusions and future works.

2 DSS knowledge Framework

Taking decisions in the tendering and construction phases of complex infrastructures requires the integration of several information sources. As explored in the introduction, this activity is nowadays mainly defined trough the experience of the single experts that work on the project with a consequent limitation in the used information and knowledge (the one that the specific expert knows) and the high risk on taking decisions that may be based on personal biases.

Hence the need to explore the development of a DSS that can support the experts in taking decisions during these phases. As the expert does, also the DSS should be able to integrate several data sources that may be of different nature and require different techniques to be used. In the following a DSS knowledge framework is presented highlighting the key data sources and the techniques that may be applied to extract the knowledge contained in the specific source (Fig. 1).

Similarly to other industries, knowledge sources in the construction sector can be classified in terms of how data are organized in information carriers such as documents, models, reports, drawings, etc. The first group, providing *structured data*, includes information models, tables, calculation data, applications' relational databases and models (i.e. a construction planning model or an quantity take-off model). In this area, information is already available, and a good use of data standardization can easily provide well-defined interrelated data to act as reference knowledge for a DSS.

The second group of sources – related to provision of semi-structured data – usually implies that data, although well-formed, consistent, and coherent, is not organized following a rigid schema, so it does not conform to a data table or relational database structure. This can be found in specific sources such as reports, drawings, emails and other textual documents that present data following a certain logic or, if generated by machines, images, videos.

Unstructured data are those that, for their nature, are not presented or presentable through databases. This group can include reports, drawings, Project' briefings, etc. Unstructured data are also the ones usually provided in extremely large amount as single, non-related information (for instance geographic data, topography, audio, video, images, etc.

In current construction processes, semi-structured and non-structured data sources are then the majority of the technical documents produced by different actors such as clients, designers, contractors, consultants, etc. At the same time, these technical documents usually represent relevant knowledge related to construction methodologies, design assumptions, logistics and contextual information currently

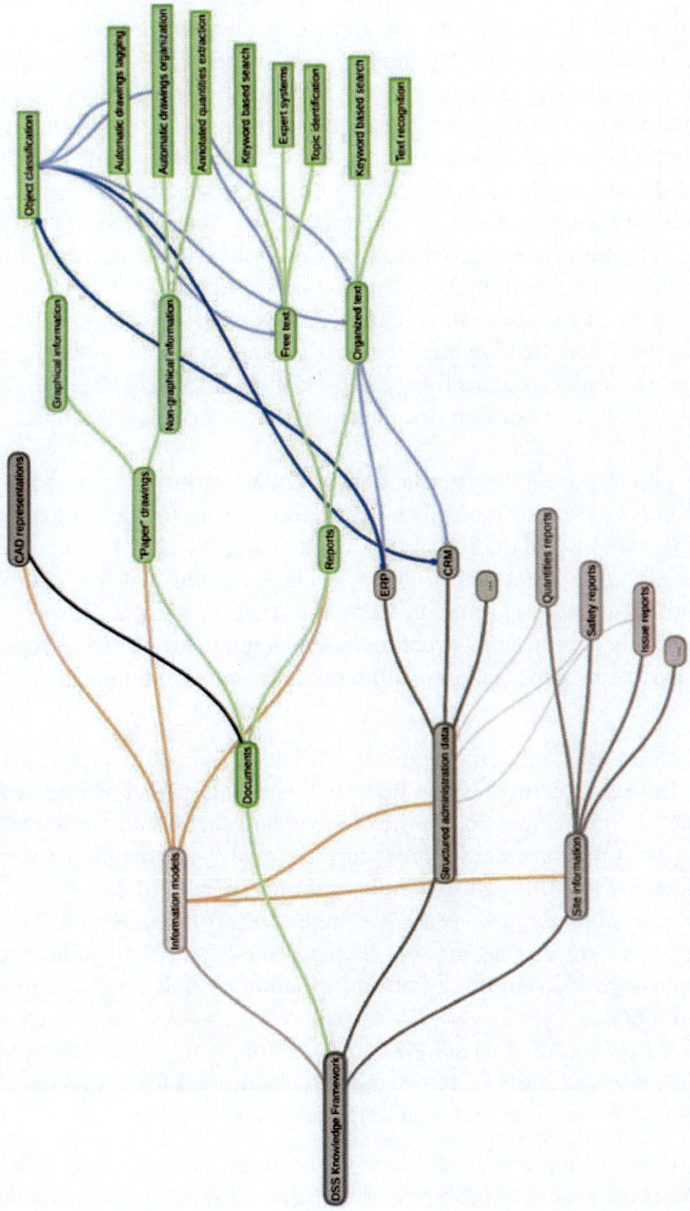

Fig. 1 Overview of the DSS knowledge framework

not formalized in information models that mainly focuses on the future artefact configuration.

Drawings are a particular sub-set of documents that falls into this large set of sources: they mix geometrical data and textual or numerical data and, although produced by cad and BIM applications, the knowledge they are providing is subject to the interpretation of the person. By looking at a shape, a line or a symbol in a drawing, taking into consideration the context in which those elements are placed, and referring to experience, an engineer is able to comprehend some specific features of a design, a methodology, a performance, a process. The interpretation of drawings represent one of the most difficult tasks in preparing a knowledge base for a DSS in construction, since it means mimicking in part the human interpretation process.

In the proposed framework the green area represents the core knowledge structure that should be used in the creation of the foundation for the DSS development. The grey elements represent the other data sources that should be related to the core knowledge framework and the blue links the possible interconnection highlighted in this phase. Of course, further connections can be identified like possible direct links between site information and design documents highlighting possible future works on this topic.

In the construction sector, the application of Building Information Modelling is nowadays shaping the innovation of the entire construction chain impacting on both design, construction, and operation and maintenance. Hence, together with the core knowledge structure identified in green, it can be noted that the information models are identified as central joint with all the information sources. However, even if information models can provide structured data as input for a DSS (simplifying the knowledge extraction process), can be identified some limitations as following detailed.

- The use of BIM is relatively new, and it is difficult to collect historical data that are based on information models and that can be used to extract past knowledge.
- BIM represents a core innovation in the construction sector and consequently is still evolving with new standards, procedures, etc. [13–17] creating a dynamic environment where it is difficult to identify common peculiarities.
- Even if BIM can potentially embrace all project information it is still difficult finding projects where all the process is based on BIM. The majority of the applications are hybrid combining both information models and traditional 2D drawings and reports.
- Even when information models are available, the frequent lack of data standardization hinders the possibility of retrieving and elaborate information, as well as transforming it in a meaning source of knowledge for a DSS.

According to these limitations, the focus of the framework is on the 2D documents considering both graphical representations (drawings) and reports as following described.

3 Discussion

As specified in the previous chapter, the core part of the DSS knowledge framework analyzed in this paper is the one based on documents divided in two main groups, i.e. "paper" drawings (to distinguish these from the CAD sources) and reports. It is worth mentioning that the term "paper" drawings does not means only drawings that are developed by hand (and then digitalized with scanning procedures). In this group are included also the drawings that have been developed with CAD software and/or BIM modelling tools and then directly printed in 2D pdf. As better explained in the following section in both cases (scanned documents and/or natively digital documents) the drawings will be analyzed using image recognition approaches.

3.1 Drawings

Figure 2 depicts the framework of the analysis that should be carried out on "paper" drawings. The schema highlights that there is the need to analyze the documents considering both the graphical information (the drawing itself) and the non-graphical information, i.e. all the annotations (tag, dimensions, text, etc.) as well as the information reported in the title block.

Graphical Information
Studying the different systems of drawing recognition for the classification of elements based on their characteristics (such as a type of pile or a specific type of tunnelling system) it emerged that the use of image recognition techniques can provide excellent results.

The use of machine learning techniques for the development of algorithms able to classify large set of images has been studied long since in the field of Artificial Intelligence. Many current approaches on image recognition are based on machine learning [18]. Between these techniques, ConvNets demonstrated high performances in image and object recognition tasks [19–21]. ConvNets are deep-learning methods,

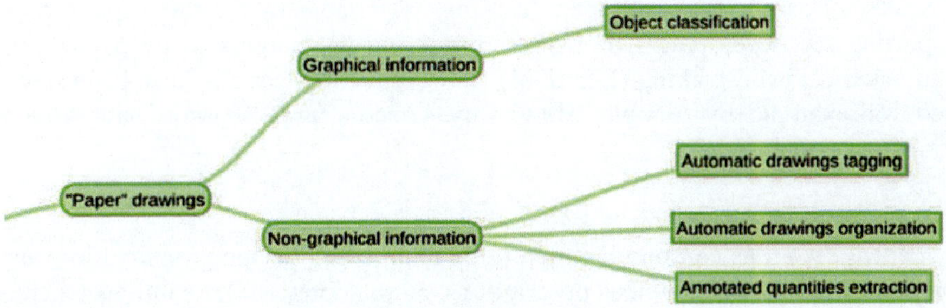

Fig. 2 "Paper" drawings knowledge analysis framework

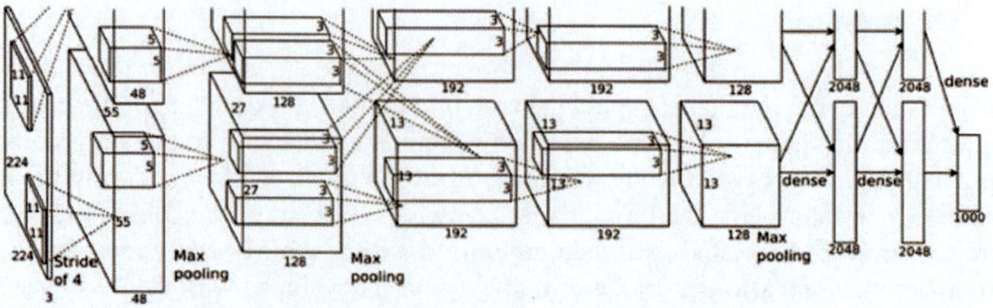

Fig. 3 Graphical representation of a ConvNet from Krizhevsky et al. [18]

i.e. *"representation-learning methods with multiple levels of representation, obtained by composing simple but non-linear modules that each transform the representation at one level (starting with the raw input) into a representation at a higher, slightly more abstract level"* [22]. Figure 3 shows a graphical representation of a ConvNet reported from Krizhevsky et al. [18].

ConvNets demonstrated excellent performance in the recognition of both similar objects with a great different in the image (for example two dogs in two complete different positions) and different objects that have similar features (for example a Samoyed dog and a white wolf) [22]. Hence, ConvNets can be used in the classification of objects but also in the identification of peculiar parameters such as the characteristics of tunnels, etc.

Commonly, machine learning methods (deep or not) are based on a supervised learning procedure. This means that a training set of data or example (e.g. a set of images) is labelled and processed by the machine to train itself. After the training procedure, the performance of the system is measured on a different set of examples called test set.

There are several open data sets of images that can be used for training and testing tasks, e.g. [23]. However, available images are real world images representing real world objects such as animals, furniture, or cars and cannot be used to recognize technical drawings. Hence, one of the first issue to solve it to create a common knowledge base that can be used for the training of the algorithm identifying the specific set of characteristics that can impact in the classification process.

Nevertheless, the use of deep learning is demonstrating the capability of supporting the development of DSSs considering the outputs as deep learning–augmented decision-making (DLADM) [24]. Some applications have been developed for example by Google, Best, Buy, Cisco, Zara, Amazon and Alibaba [25–27].

Non-graphical Information

In "paper" drawings can be identified three main types of non-graphical information namely comments (such as descriptions, annotations, etc.), quantities (such as dimensions, quantities annotations, etc.), and drawings description (that is usually the content of the title block). Each one of these types can be used for the enrichment of both graphical information analysis and the reports analysis.

In detail, comments can be used to support the development of automatic drawings tagging combining the text associated to a specific part of the drawing with the classification systems that can be developed according to the previous chapter indications. The ability to combine these two groups can reduce the time required in the creation of the knowledge base for the image recognition algorithm creating an automated tagging system based on both graphical and non-graphical information.

Quantities can be extracted and compared with comments information to identify elements that can be compared with the bill of quantities identifying possible discrepancies and issues between drawings and reports.

Title block information can be used to automatically organize the documents thanks to the use of document title, codification, date, etc. Considering the analysis of projects that are developed by big customers (e.g. important public authorities) it is possible developing automated systems based on the recognition of the title block schema able to populate the document with the relevant metadata.

Usually, drawings do not contain complex text structure and the analysis of the different information types can be based on defined schemas and rules (such as in the case of title blocks) or using basic natural language processing (NLP) approaches that will be better explored in the next chapter.

3.2 *Reports*

In the reports area are considered all the text-based documents that are related to one or more projects. For example, technical reports on asset performance requirements, bill of quantities, price list, etc. In these documents can be identified areas where the text is organized and structured considering industry-based standards (for example the identification of costs based on the specific tendering areas) and areas (usually the majority of the text) where the text is reported in natural language describing the different topics.

In the first case it is possible applying techniques like the one already described for the title blocks where the algorithm is able to research specific keywords and or text areas according to predefined rules then extracting in separated databases (and or associating the metadata to the documents) the identified information. On the other hand, natural language text (free text) represents an important source of knowledge that should be included in the knowledge framework for DSS also considering the need to create relations between the reports information and the one contained in drawings (Fig. 4).

Several studies have been developed in the analysis of text documents for the extraction of meaningful information and knowledge. Text mining (or text data mining) can be defined as the process of deriving information from text data which is not previously known and not easy to be revealed [28]. Together with text mining techniques it is fundamental considering natural language processing (NLP) as the

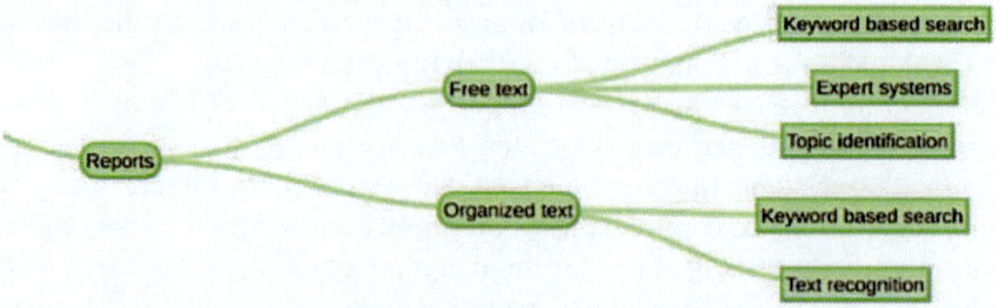

Fig. 4 Reports knowledge analysis framework

process of making computer understanding natural language and performing certain tasks [29]. NLP is used to analyze the semantic and grammatical structure of the text hence supporting the development of text mining applications [30].

In the literature can be found several examples of text mining applied to the construction sector in particular in the area of the safety management and of the maintenance prediction [31–33] as well as examples of applications in the project monitoring and control [34–36] that represent possible applications of one of the pieces of the proposed DSS knowledge framework.

3.3 Knowledge Interpretation and Combination

Even if not extensively described and analyzed in this paper, Fig. 1 highlights how the detailed analysis proposed in the previous chapter should be combined both among them and with the other data sources that may be used in the specific company (such as the Enterprise Resource Planning – ERP – information, the Customer Relationship Management – CRM – information, etc.). It is of crucial importance to integrate the discovered knowledge in the specific context of the project to guarantee a correct knowledge interpretation. Moreover, the ability to combine knowledge generated from different sources can increase the value of the identified knowledge [37].

4 Conclusions

This paper proposes a DSS knowledge framework that aims at identifying the different components that should be considered in the development of a DSS.

The results show how AI and more in detail ML should be combined considering different techniques and integrating these techniques with structured data and other branches of the knowledge representation and analysis (expert systems, ontologies, inference engines, etc.).

This paper represents a first analysis of the DSS knowledge framework and can be considered as an input for the development of further research around the topic. In particular the Authors are developing dedicated research activities on the different key points of the framework developing the relative procedural and software (prototype) components. The results of this step will impact on the refinement of the framework and on its extension toward the integration of all the components required for the effective development of a DSS useful both during the tendering and the construction phases.

References

1. Ghasemaghaei M (2020) The role of positive and negative valence factors on the impact of bigness of data on big data analytics usage. Int J Inf Manage 50(2018):395–404. https://doi.org/10.1016/j.ijinfomgt.2018.12.011
2. Kivrak S, Arslan G, Dikmen I, Birgonul MT (2008) Capturing Knowledge in construction projects: knowledge platform for contractors. J Manag Eng 24(2):87–95
3. Hari S, Egbu C, Kumar B (2004) Knowledge capture in small and medium enterprises in the construction industry: challenges and opportunities. In: Proceedings of 20th Annual Conference Association of Research in Construction Management (ARCOM), vol 2, pp 847–855
4. Sauter V, Free D (2005) Competitive intelligence systems: qualitative DSS for strategic decision making. ACM SIGMIS Database 36(2):43–57
5. Poch M, Comas J, Rodríguez-Roda I, Sànchez-Marrè M, Cortés U (2004) Designing and building real environmental decision support systems. Environ Model Softw 19(9):857–873. https://doi.org/10.1016/j.envsoft.2003.03.007
6. Walling E, Vaneeckhaute C (2020) Developing successful environmental decision support systems: challenges and best practices. J Environ Manage 264:110513. https://doi.org/10.1016/j.jenvman.2020.110513
7. Wong-Parodi G, Mach KJ, Jagannathan K, Sjostrom KD (2020) Insights for developing effective decision support tools for environmental sustainability. Curr Opin Environ Sustain 42:52–59. https://doi.org/10.1016/j.cosust.2020.01.005
8. Barzehkar M, Parnell KE, Soomere T, Dragovich D, Engström J (2021) Decision support tools, systems and indices for sustainable coastal planning and management: a review. Ocean Coast Manag 212:105813. https://doi.org/10.1016/j.ocecoaman.2021.105813
9. Yariyan P, Zabihi G, Wolf ID, Karami M, Amiriyan S (2020) Earthquake risk assessment using an integrated fuzzy analytic hierarchy process with artificial neural networks based on GIS: a case study of Sanandaj in Iran. Int J Disaster Risks Reduct. 50:101705
10. Mohemad R, Razak Hamdan A, Othman Z, Maizura N, Noor M (2010) Decision Support Systems (DSS) in construction tendering processes. Int J Comput Sci Issues 7(1): 35–45
11. Egemen M, Mohamed A (2008) SCBMD: a knowledge-based system software for strategically correct bid/no bid and mark-up size decisions. Autom Constr 17(7):864–872
12. Dikmen I, Birgonul M, Gur A (2007) A case-based decision support tool for bid mark-up estimation of international construction projects. Autom Constr 14(1):30–44
13. ISO, ISO 16739-1:2018 - Industry Foundation Classes (IFC) for data sharing in the construction and facility management industries - Part 1:Data schema (2018)
14. ISO, ISO 19650-1:2018 - Organization of information about construction works — Information management using building information modelling — Part 1: Concepts and Principles (2018)
15. ISO, ISO 19650-2:2018 - Organization of information about construction — Information management using building information modelling — Part 2: Delivery phase of the assets. (2018)
16. ISO, ISO 19650-3:2020 - Organization of information about construction works — Information management using building information modelling — Part 3: Operational phase of the assets (2020)

17. ISO, BS EN ISO 19650-5:2020 - Organization and digitization of information about buildings and civil engineering works , including building information modelling (BIM) - Information management using building information modelling - Part 5: Security-minded ap (2020)
18. Krizhevsky A, Sutskever I, Geoffrey HE (2012) ImageNet classification with deep convolutional neural networks. In: Advances in Neural Information Processing Systems 25 (NIPS2012), pp 1090–1098. https://doi.org/10.1109/5.726791.
19. Karpathy A, Toderici G, Shetty S, Leung T, Sukthankar, R, Fei-Fei L (2014) Large-scale video classification with convolutional neural networks. In: IEEE Conference on Computer Vision and Pattern Recognition (CVPR), pp 1725–1732. https://doi.org/10.1109/CVPR.2014.223
20. Lawrence S, Giles CL, Tsoi AC, Back AD (1997) Face recognition: a convolutional neural-network approach. IEEE Trans Neural Networks 8(1):98–113. https://doi.org/10.1109/72.554195
21. Liang M, Hu X (2015) Recurrent convolutional neural network for object recognition In: IEEE Conference on Computer Vision and Pattern Recognition, pp 3367–3375. https://doi.org/10.1109/CVPR.2015.7298958
22. Lecun Y, Bengio Y, Hinton G (2015) Deep learning. Nature 521(7553):436–444
23. Google, The open images dataset (2018). https://github.com/openimages/dataset. Accessed 20 Mar 2018
24. Shrestha YR, Krishna V, von Krogh G (2021) Augmenting organizational decision-making with deep learning algorithms: principles, promises, and challenges. J Bus Res 123:588–603. https://doi.org/10.1016/j.jbusres.2020.09.068.
25. Davenport TH, Harris J, Shapiro J (2010) Competing on talent analytics. Harv Bus Rev 88(10):52–58
26. Kaplan A, Haenlein M (2020) Rulers of the world, unite! The challenges and opportunities of artificial intelligence. J Bus Ethics 63(1):37–50
27. Gemawat P, Nueno JL, Dailey, M (2003) ZARA: Fast faschion (Vol. 1). Harvard Business School, Boston
28. Miner G, Elder J, Hill T, Delen D (2012) A Fast, Practical Text Mining and Statistical Analysis for Non-structured Text Data Applications, Academic Press, Cambridge
29. Chowdhury GG (2005) Natural language processing. Annu Rev Inf Sci Technol 37(1):51–89. https://doi.org/10.1002/aris.1440370103
30. Zhang F, Fleyeh H, Wang X, Lu M (2019) Construction site accident analysis using text mining and natural language processing techniques. Autom Constr 99:238–248. https://doi.org/10.1016/j.autcon.2018.12.016
31. Tixier AJP, Hallowell MR, Rajagopalan B, Bowman D (2016) Application of machine learning to construction injury prediction. Autom Constr 69:102–114. https://doi.org/10.1016/j.autcon.2016.05.016
32. Arif-Uz-Zaman K, Cholette ME, Ma L, Karim A (2017) Extracting failure time data from industrial maintenance records using text mining. Adv Eng Inf 33:388–396. https://doi.org/10.1016/j.aei.2016.11.004
33. Rivas T, Paz M, Martín JE, Matías JM, García JF, Taboada J (2011) Explaining and predicting workplace accidents using data-mining techniques. Reliab Eng Syst Saf 96:739–747
34. Choudhary AK, Oluikpe PI, Harding JA, Carrillo PM (2009) The needs and benefits of text mining applications on post-project reviews. Comput Ind 60:728–740. https://doi.org/10.1016/j.compind.2009.05.006
35. Marzouk M, Enaba M (2019) Text analytics to analyze and monitor construction project contract and correspondence. Autom Constr 98:265–274. https://doi.org/10.1016/j.autcon.2018.11.018
36. Williams TP, Gong J (2014) Predicting construction cost overruns using text mining, numerical data and ensemble classifiers. Autom Constr 43:23–29. https://doi.org/10.1016/j.autcon.2014.02.014
37. Moody D, Walsh P (1999) Measuring the value of information: an asset valuation approach. In: Seventh European Conference on Information Systems (ECIS 1999), pp 1–17

Practice Enterprise and MOOCs in the Design and Implementation of Megaprojects. Some Lessons from European Projects

Massimo Bianchi

Abstract One of the problems highlighted in the preparation of megaprojects, generally conducted by a consortium of participants from different countries and with a different organization, concerns the size of the resources, the number of partners and the sustainability of the project itself.

The experiences of some recent European projects can be considered to test the sustainability of organizational structures designed and of the potential performance of the partners, reducing the gap between expected and actual results that afflicts most of these projects.

A dilemma that can be posed to overcome these problems is whether it is necessary to reconsider the concepts of performance or if it is more convenient to review the working methods by replacing the growing amount of information currently required for the presentation of the projects with that realized through learning by doing.

In the paper, after a brief demonstration on the limits of the major types of performance indexes, will be described the use of didactics based on Practice Enterprise (PE) and Massive Open On line Courses (MOOCs) experimented in some Tempus and Erasmus projects and with the aim of increase the performance of organizations whose design is the basis of the projects themselves.

Keywords Learning by doing · Practice Enterprise · MOOCs · Project Management · New metrics for performance management · Change of paradigm · Change of approach

1 Introduction

The preliminary question connected to the improoving of performances of mega projects, represented firstly by the Adequacy of resources available to declared objectives, it concerns the limits itself of the Adequacy as representative of projects performances. Even with the prolipheration of indexes, once negatively considered by

M. Bianchi (✉)
Alma Mater Bologna, Bologna, Italy
e-mail: massimo.bianchi@UNIBO.IT

E. Favari and F. Cantoni (eds.), *Complexity and Sustainability in Megaprojects*, Lecture Notes in Civil Engineering 342,
https://doi.org/10.1007/978-3-031-30879-6_11

Cameron with his studies on organizational performances [6], the recent multiplying of project descriptions in Tempus, and after in Erasmus projects, didn't consider this criticism and, on the contrary, produced an increasing dimension of submission templates gone from tens to hundreds of pages.

To give an example, the Tempus project forms, initially used for the submission, consisted of a few tens of pages while today, in Erasmus projects, it amounts in a few hundred pages with the request to the extender of the project, among other things, of a SWOT analysis, an analytical description of the procedures for mitigating the risks of failure and the preparation of a Logical Matrix containing the criteria and activities aimed at facing the unforeseen difficulties that a project may have to face.

In last Erasmus calls, in addition to the increase of information required, the most relevant innovation was, in the Detailed Project Description (about 180 pages), the change of the order of subjects with the Relevance of the project (a description about project motivations) at the first place while the description of applicant and partners with their affiliates (once placed at the beginning of the template) was moved to the third place (F3 Organizations and activities).

Any case, the approach of the project description to be submitted for the funding remains mainly based on the SWOT analysis and on a Logical Matrix describing the Wider Objective, the Indicators of Progress, the Assumptions and Risk and how the Risks will be mitigated.

The basic question relating to the frequent disappointing results that the Literature denounces regarding the performance of mega projects is connected to the approach to the project design.

To be mentioned, in the period we will examine (2010—2022), is the change of the orientation of the evaluating commission of European projects passed from advising against the involvement of non-academic organizations to the positive evaluation of the participation of non-academic organizations and stakeholders associations, previously denied.

On the one hand, we can set ourselves the objective of reviewing the definition of performance and in concert that of the indexes used for the purpose or, if this path is not feasible, as we will try to demonstrate, it is appropriate to proceed with a radical change in the methods of work by focusing on the project management process and on the simulation of the behavior of the organizational structures that each project proposes to implement its objectives.

Now the question is related to a change of perspective owing to the appliance of an approach to project management based on the Practice Enterprise (PE), a kind of learning by doing based on the simulation, with the support of Massive Open On line Courses (MOOCs) to supply the scarcity of didactical staff prepared to this purpose.

In the paper, after a brief demonstration on the limits of the major types of performance indexes in the bettering of projects management adequacy, will be described the use of didactics based on PE and on MOOCs experimented in some Tempus and Erasmus projects with the aim of carrying out simulations concerning the functioning of organizational structures proposed by the projects.

The paper methodology can be sinthetized in four steps. After a summary on a critical point in the evaluation of projects performances, it will be described the PE methodology as a tool to simulate project management takes the third part, then it will be described the appliance of PE in some european projects and the experimental case AIPO-AIVO with consequent conclusions.

2 Some Topics on Megaprojects Performance Evaluation

Altough the variety in the definition of mega-projects concerning the parameters chosen as resources engaged, population involved, territory interested to, the most diffused criticism is concentrated on unsatisfactory performances due to many different reasons and mainly on: increasing costs, insufficient resources, negative reaction from the social environment [13]. Scholars didn't recognize that the dimension will introduce particular conditions connected to the magnitude of resources committed, to the cost overrun exposure, or to associated societal benefits or disbenefits and their lasting nature [Flyvbjerg, Bruzelius, Rothengatter 2003].

Two common traits to which we want to refer are: the complexity of mega projects and the involvement of numerous professionals, two traits to which the disappointing results of these projects are often attributed in relation to the expectations and resources used.

In these terms, the consideration of the adequacy of the projects applies to the design of the organizational structure of the project and to the simulation of the planned activities.

The very definition of performance, even in the attempt of a systemic quantification, is subject to ambiguities and contradictions that seem to be difficult to solve with some programming algorithm, but which rather require a simulation in the field.

Let's consider the case of Adequacy.

Performance dynamics make use of main parameters such as Objectives, Resources and Results which lead to the main domains of performance evaluation such as Effectiveness (Ez), Efficiency (Ec) and Adequacy (Ad) that synthetically could be expressed as shown at Fig. 1.

The nature of performances evaluation as a system of interconnected parameters shows some contradictions affecting in particular the Adequacy, cited as one of the most critics topics it regards Mega Projects also as it concerns the Feasibility and the Sustainability.

To this regard particularly interesting is the formula [1] of relationships among Indexes indicating Ad as directly proportional to Ec and indirectly proportional to Ec.

$$[1]\ \text{Adequacy} = \frac{\text{Resources}}{\text{Objectives}} = \frac{\text{Efficacy}}{\text{Efficiency}} = \frac{\dfrac{\cancel{\text{Results}}}{\text{Objectives}}}{\dfrac{\cancel{\text{Results}}}{\text{Resources}}}$$

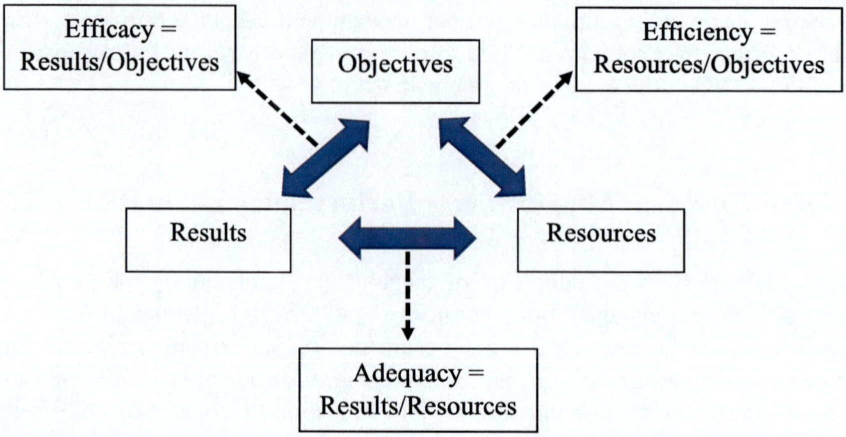

Fig. 1 The system of parameters and indexes typologies used in the performance evaluation

Finally, this problematic relationships among main performance indexes applied to Project Management told us that when the Project Efficiency increases, the Adequacy decreases. This contradiction undermines, in our opinion, the possibility of finding in these indexations the way to improve the performance of projects or to avoid their methodological limits which pushed Cameron to abandon his studies on Efficacy and Efficiency [5].

3 Business Simulation and Practice Enterprise

Once the field has been freed from indexing, as a tool for rationalizing project management, let's now examine some experiences of introducing learning by doing into design work. What we want to examine is the Business Simulation called PE coupled with MOOCs and applied in some International Projects.

It exist many kind of business simulations that are used in vocational education and training. In order to ensure a clear definition of the different simulation methods, the Practice Enterprise will be compared to other methods like Business Games, Learning Office and Junior Company. Table 1 provides an overview of the business simulations and of the main differences in terms of fictitious or real performances [8]:

Table 1 Comparison of four simulation methods Adapted from: [8], 19

	Business Games	Learning Office	Practice Enterprise	Junior Company
Flow of goods and services	fictitious	fictitious	**fictitious**	real
External contacts	fictitious	fictitious	**real**	real

– PE and Business Games

Unlike Practice Enterprises, Business and Simulation Games have a pre-designed game structure. Under the constraint of specific resources and information (e.g. market data) learners try to achieve the maximum success for a fictitious business [14]. The game is framed into time units or periods in which learners try to solve the underlying problem. Furthermore, Business Game didn't consider Social aspects or external contacts and no commercial tasks are undertaken. Briefly, the didactics is focused on making decisions under different situations.

– PE and Learning Offices

Literature [10] defines the Learning Office as a model of a real business. The learners are divided into smaller learning groups than in Business Games. The company's daily business (e.g. business correspondence, bookkeeping) is performed by the different departments and upcoming business activities are planned. There are no real external contacts and the flow of goods and money are fictitious as can be seen in Table 1. In contrast to the Business Game, the focus lies on daily office activities (e.g. correspondence) instead of sole decision making. The difference between the PE and the Learning Office is that the Learning Office is a closed system. There is neither a market where other Learning Offices co-exist, nor real outside contacts since all communication is simulated by the trainer. Therefore, the students cannot observe and react to any market changes [17].

– PE and Junior Companies

The Junior Company is a usual set for the duration of one school year, where students establish a company and offer self-developed products and services on the real market. In the beginning, the students develop independently their own business idea and go through all phases of a real business project. These phases include the idea generation, the team building as well as the planning and the production of the products or services. The students are also responsible for the marketing and selling of their products or services and, at the end of the school year, for the closing of their business.

As Table 1 indicates, in contrast to the Learning Office and the PE, a Junior Company trades with real products and money and has (just like PEs) real external contacts. The Junior Company participates in the real market. Real capital is used (although only limited amounts) and are established business relationships with customers and suppliers [18]. The Junior Company and the PE have a different degree of reality. A PE has no real economic risk [Fix 1989]. Learning and working takes place in a protected area. Junior Companies bear entrepreneurial risk and therefore often operate under the legal and financial protection of a mother company [20].

4 The PE Applied to International Projects

Business simulation is a teaching methodology that refers to constructivism [1] and connectivism theories through a learning by doing approach that embodies all the instances of blended learning.

Connectivism [7].

The main features of this teaching are related to the fact that it is:

1. authentic, contextualized and situated, as it is designed to reproduce authentic situations, representative of an organized entity, generally a company, in all its organizational s and therefore highly significant function;
2. active, as the participants are pushed to process the incoming information, connecting and organizing it in an efficient and meaningful way, in order to have a hologram view of the situation;
3. constructive, in the sense that the new learning, through reflection, integrates with the previous one, developing and enriching it, thus giving a meaning to the learning process;
4. intentional, as it is oriented towards a specific objective, the management of the business process;
5. cooperative, because one's knowledge is the result of a re-elaboration of the group that discusses and shares.

Here the learning cooperation dimension represents a valid tool for integrating cognitive, professional and social skills. In this didactic process knowledge is the result of a re-elaboration of the group that discusses and shares [9].

In 2001 the University of Bologna, Faculty of Economics, the Author with Daniele Gualdi as Coordinator introduced in a Business Management Course the PE, once named Simulimpresa. It was the first University in Italy to insert PE into the study plan of students, followed a few years later by the University of Parma [3].

The Simulimpresa course allows participants to immerse themselves in the life of a simulated company, applying the skills acquired during the years of study and test themselves in solving problems that may run into the management of a company.

Business Simulation is a participatory learning environment, in which the dimension of cooperative learning represents a valid tool for integrating cognitive, professional and social skills, as the members of the group, in order to achieve common goals, are called to work constructively and responsibly in the decision-making process with a positive interdependence, dialogic exchange and collective reflection. All these activities aimed at improving the problem solving and teamwork skills of participants, qualities that the modern working context strongly requires and seems to be fruitful of results in terms of Project Management and of their performance.

This teaching is based on Learning by doing and is aimed at applying the theoretical knowledge acquired during studies in the business context to achieve fundamental skills such as problem solving, decision making, working team, autonomy, responsibility, communication.

It also allows students to compete at an international level with students from all over the world, participating in the international fair of simulated companies. It should be noted that the teaching staff is made up of three profiles: the Teacher, the Tutor and the Mentor.

5 The Experimental Field

PE, managed by Europen Pen, an international association that groups together most of PEs in the world, has been applied in some recent Tempus Projects with the active participation of the Author as Project Manager or Partner Contact Person (Table 2). These projects are mainly aimed to modernize Eastern countries as it concerns public and educational structures with a particular focus on the teaching of entrepreneurship and pro-active attitudes in private and public organizations.

In these projects the PE was appied mainly to teach how to create and manage a start up in different sectors or to create new organizational structures in public bodies. All the initiatives foresee the preparation of Teachers, Tutors and Mentors in PE (Fig. 2) with the aim to support the sustainability of the project during and after its end.

The didactics of PE is supported by real enterprises that contribute to make this experience available through mentors, This engagement of external organizations, called "mother companies" introduces an updated practice in the activity of the learners.

6 The Case AIPO-AIVO

Among these experiences is the case of AIPO-AIVO managed with the collaboration of Astrakan State University (ASU), the Municipality of Astrakan and the Administration Authority of the Governator of the Astrakhan region for interaction with local authorities[1].

The Interregional Agency for the Po River (AIPO), is a public body that provides engineering and environmental services across the full spectrum of operations in support of the interests of the Italian regions crossed by the Po River: Piemonte, Lombardia, Emilia—Romagna and Veneto. Its catchments covers an area of more than 74.000 square kilometers - a quarter of the national territory. In 2010, AIPO and ARNI (Regional Agency for Inland Navigation for Emilia—Romagna) drawn up an agreement of fusion for project coordination and planning. AIPO performs its mission concerning the governance of the River Po Basin. The Agency operates

[1] Actuated within the Exploratory seminar on models and methodologies for entrepreneurial and tourist development in the lower Volga financed by Cooperazione Interuniversitaria Internazionale AF 2008, 27th September 2010.

Table. 2. Projects in which was experimented the methodology of PE and MOOCs

Project Acronym	Project Subject	Project Grant Holder	Countries involved	Period	Distance Learning
DOCSMES	TEMPUS - Regional Joint Doctoral Programme in Entrepreneurship and SME Management for Western Balkan Countries	University of Bologna	Albania, Italy, Kosovo, Republic of Macedonia, Poland, Spain,	2010-2014	Simulimpresa/Practice Enterprise
CHTMBAL -	TEMPUS - Network for Post Graduate Master in Cultural Heritage and Tourism Management in Balkan Countries	Gabriele D'Annunzio University of Chieti-Pescara (IT)	Italy, Poland, Kosovo, Albania, Spain	2014-2016	Simulimpresa/Practice Enterprise
RESINT	ERASMUS - Collaborative Reformation of Curricula on Resilience Management with Intelligent Systems in Open Source and Augmented Reality	Unibo Campus Forlì (IT)	Italy, Lithuania, Spain, United Kingdom,	2013-2015	Simulimpresa/Practice Enterprise
BECK	ERASMUS - BECK Integrating Education with Consumer Behavior Relevant to Energy Efficiency and Climate Change at the Universities of Russia, Sri Lanka and Bangladesh	Vilnius Gedeminias Technical University Vilnius (LT)	Bangladesh, Italy, Lithuania, Russia, Sri Lanka, United Kingdom.	2019-2022	Simulimpresa/Practice Enterprise - Moocs
EUROTECH	ERASMUS + - European Technologies for Cultural Heritage. Diagnostic, research, training, valorization and dissemination of archaeological heritage at risk	Gabriele D'Annunzio University of Chieti-Pescara (IT)	Cyprus, France, Italy, Poland, , Spain	2019-2022	Simulimpresa/Practice Enterprise - Moocs
HEIPNET	ERASMUS + - Inclusion of Innovative Work-Based-Learning and Business Partnerships in HEI Curricula Development	University of Pavia (IT)	Austria, Germany, Italy, Lithuania	2020-2022	Simulimpresa/Practice Enterprise - Moocs
PICASP	Pilot Courses and new didactics for teachers training in cultural tourism for the development of Caspian Area	University of Chieti (IT)	Azerbaijan, Italy, Kazakhstan, Lithuania, Poland, Russia	2021-2023	Simulimpresa/Practice Enterprise - Moocs

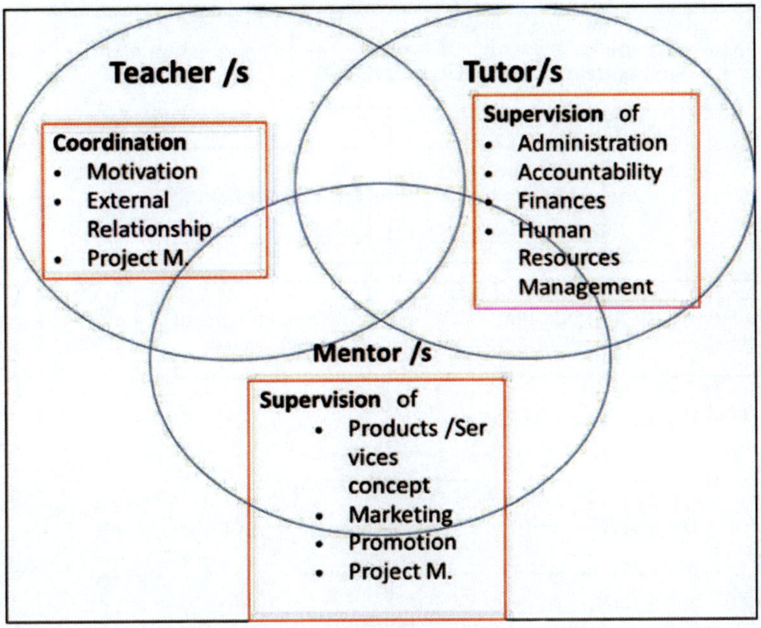

Fig. 2 Composition of didactical staff and roles in PE

through its main Headquarter in Parma, 12 divisions throughout the River Po Basin and a Research and Development District. It cooperates with the Po River Authority which has an institutional planning and programming competence for the elaboration of the "Piano di Bacino" (Basin Plan).

The Team engaged in the PE experimentation had the purpose to evaluate the feasibility of establishing a structure, similar to the AIPO, as it regards the low Volga River from Volgograd to the delta located in Astrakan and tributary of the Caspian Sea.

On the one hand the AIPO P.E. it was established with the organizational chart in Fig. 3 and agreed by AIPO Managers participating to PE.

At the other hand, on 27 th September 2010 the AIVO PE, Interregional Agency for Volga River, was created, by ASU and Local Public Managers, with the Organizational Chart at Fig. 4.

Both structures, AIPO and AIVO, with their respective assignments, were the result of the application of teaching by PE with a double preparation path regarding the mentors from AIPO and the learners selected among potential Managers and the Local stakeholders interested in developing a Volga Delta Authority.

The experience led to other proposals submitted in 2019- 2021 with three Erasmus Projects approoved: Beck, Heipnet and PICASP (Table 2), in which the MOOCs-PE combination was proposed or/and applied.

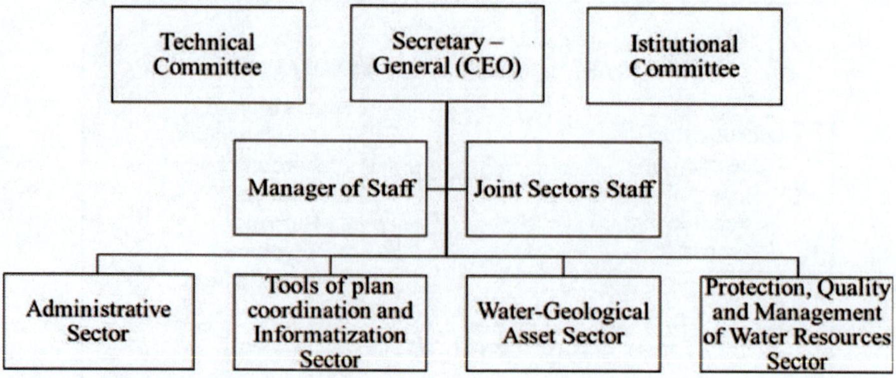

Fig. 3 AIPO P.E. Organizational Chart

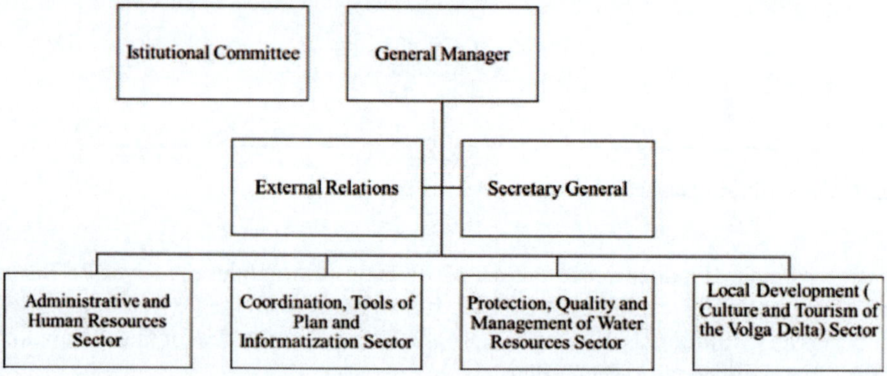

Fig. 4 AIVO P.E. Organizational Chart

Perticularly MOOCs were applied for the preparation of the Teaching Staff by the PE Unit of the University of Bologna managed by SERINAR[2] (https://www.serinar.unibo.it/m-o-o-cs-for-teaching-staff/) and PE were applied for the preparation of learners with the collaboration of the PE Unit in Forlì of the University of Bologna (http://perting.poloforli.unibo.it/).

7 PE and MOOCs

The dissemination of the experience, particularly as it concerns the on going PICASP Project[3], places PE and MOOCs in two different prespectives. On the one hand PE

---

[2] Servizi Integrati D'Area Ser. In. Ar. Forlì. Società Consortile.

[3] "Pilot courses and new didactics for teachers training in cultural tourism for the development of Caspian Area" Programme EPPKA2 - Cooperation for innovation and the exchange of good

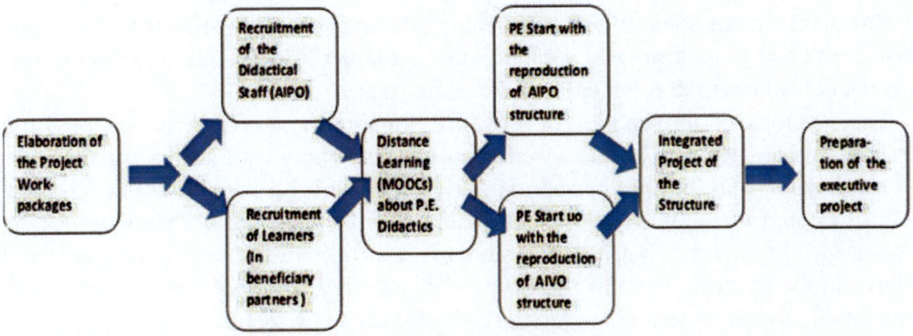

Fig. 5 Procedure for the Preparation of the Implementation Project with the use of PE and MOOCs

is submitted to the concentration of the Didactics in a classroom with direct and personal relationships among the teaching staff and the learners. At the other side, in Tempus and Erasmus projects in which the beneficiary partners, with no experience in PE have locations mostly distributed over long distances on the territory [21] the use of MOOCs can contribute to a fast preparation of the PE Teaching Staff.

This elects MOOCs as an adequate tool for the preparation of the teaching staff. This does not mean that the experience in the field of PE Classroom is not useful for the Teaching Staff. On the contrary, with the use of MOOCs, the preparation of the Teaching staff can be concentrated in the shortest time if the teaching staff previously received a preparatory education from the MOOCs.

The procedure followed in project activities is summarized at Fig. 5 with the steps of recruitment, didactical activity by MOOCs and PE.

8 Practical Appliance of MOOCs - PE Integration. The Case of PICASP Project

In recent Erasmus calls, the integration of MOOCs and PE was practically proposed in the Project PICASP, "Pilot courses in Practice Enterprise to implement the University-Enterprise Cooperation for the development of Caspian Area[4]". The project, still in progress, has the purpose to The main PICASP objectives are: Developing new teaching methodologies including Practice Management with Practice Enterprise; Developing courses in Entrepreneurship and SME Management; Enhancing the exchanging of best practices with EU partners through mobility of academic and technical staff; Enforcing stakeholders' involvement

practices Action CBHE-JP - Capacity Building in higher education - Joint Projects Proposal Call for Proposals 2020 - EAC-A02-2019-CBHE. Application No 617540-EPP-1-2020-1-IT-EPPKA2-CBHE-JP.

[4] A project for Capacity Building in the field of Higher Education, Action KA2, Cooperation and Innovation in the Exchange of Good Practices.

in curriculum development and graduates placement; Establishing Standards and Providing Quality assurance in didactics and sustainability. The main sector chosen was the entrepreneurship related to cultural heritage.

The network of universities of PICASP support the implementation of Pilot Courses in HEI Project Partners verifying the sustainability and the possibilities for implementation in other partner countries. This will be positive not only in the field of Economic Activities but more generally among the partners how to design higher education, according to UE Standards. Furthermore, it will have a positive impact in the perspectives, for the Caspian Area, trigged by policies to implement Sustainable Development related to the Silky Road. To be mentioned, with the position of Grant Holder represented by the University of Chieti-Pescara, the contribution of Europen-Pen, managing the international network of Practice Enterprises, enroled in the project as Associate Partner with the Italian ICA, Central Institute of Archaeology.

This will be induced by the education and up to date of teachers, tutors and mentors, coming from the entrepreneurial and managerial sector, to the applying of new didactical technolgies like Practice Enterprise and MOOCs. To develop the didactical methodology of courses (short, curricular or inserted in existing Masters) by multidisciplinary MOOC modules in Practice Management by Practice Enterprise for the preparation of teaching staff in condition to promote and manage courses oriented to the start up of SMEs and to the modernization of the existing ones.

Also in this case was essential, for the fulfillment of the project, the preparation of The Teaching Staff (Teachers, Tutors and Menthors) throughout MOOCs with a particular attention for the coaching activity having the purpose of creating a real team and an adequate network among the staff, an attitude that was replied in the project management applied in PE Courses.

9 Theoretical Approaches and Lessons from the Practice

The Literature [19] considers three main approaches to apply PE:

- PE as training of specific skills. With this experience learners should acquire and strengthen their professional knowledge and skills in a narrowly defined functional area.
- PE as a tool to aggregate different subjects through which the learners have acquired various business administration subjects and work together to understand the connection between all subjects. In this case the PE teaching is used to reinforce the traditional teaching methods and to strengthen acquired knowledge and skills.
- PE as a specific learning environment. The last approach is based on the the link between theory and practice. As a dynamic simulation model of an organization, the PE offers to learners the opportunity to carry out real activities, make business decisions and experience their effects [12] in a specific learning environment [15].

Table 3 Relevance of the three main PE approaches in the Case AIPO-AIVO

Orientation of PE approach	PE	AIPO/AIVO Case
training of specific skills	33%	10%
aggregation of different subjects	33%	60%
specific learning environment	33%	30%

In particular by comparing the traditional PE applied to the teaching of entrepreneurship in the case of AIPO-AIVO oriented to the design of an adequate organizational structure, the relevance of these three approaches can be estimated as at Table 3.

The design of a public organization such as an agency in which different subjects are involved from a managerial, administrative and technical point of view requires an interfunctional and interpersonal dynamics that cannot be taught without practical experience like the one experienced by PE.

This approach could also be considered a further sign of the recovery of the connectivist theories evoked by some Authors in relation to the growing use of online platforms in higher education [22].

To this purpose, the proposal to apply the MOOCs to the dissemination of the PE didactic approach seems fruitful, in a transdisciplinary perspective, for the design, preparation and management of mega and complex projects.

References

1. Amineh RJ, Asl HD (2015) Review of constructivism and social constructivism. J Soc Sci Lit Lang 1(1):9–16, 30 April. Retrieved from jssll.blue-ap.org
2. Bent F, Nils B, Werner R (2003) Megaprojects and risk: an anatomy of ambition. Cambridge University Press, Cambridge. ISBN 0–521–804205.
3. Bianchi M (2005) Dal progetto scientifico alla creazione d'impresa. Il ruolo della communities of practice nei paesi in transizione In: Bianchi M, Tampieri L (eds) Life long learning and managerial development in transition countries. Il Ponte Vecchio, Cesena. ISBN: 8883125541.
4. Bianchi M (1999) Cross border public centres of support and technical assistance for small business development" nella Riv. Ekonomska Istraživanja . Economic Research, Pula 12(1–2):21. ISSN: 1331–677X
5. Cameron K (1986) Effectiveness as paradox: consensus and conflict in conceptions of organizational effectiveness. Manage Sci 32(5):539–553
6. Cameron K, Whetten D (1983) Some conclusions about organizational effectiveness. In: Cameron K, Whetten D (eds) Organizational effectiveness: a comparison of multiple models, pp 261–278. Academic Press, New York
7. Corbett F, Spinello E (2020) Connectivism and leadership: harnessing a learning theory for the digital age to redefine leadership in the twenty-first century, Heliyon 6(1). https://doi.org/10.1016/j.heliyon.2020.e03250. ISSN: 2405–8440
8. Gramlinger F (2000) Die Übungsfirma auf dem Weg zur Lernfirma? Eine empirische Darstellung in Form zweier Portraits. Bergisch Gladbach, Hobein. ISBN: 3–924985–30–8
9. Gualdi D (2016) Manuale di simulazione d'impresa; Il Ponte Vecchio, Cesena; Gualdi D. 2016. Indagine sulle motivazioni e aspettative degli studenti di Forlì e Paper Univer, Parma
10. Kaiser FJ, Weitz BO (1992) Handlungsorientiertes Lernen in schulischen Modellunternehmen. In: Wirtschaft und Erziehung, vol 44, no 3, pp 89–92. Retreived from Research Gate (1992)

11. Kiatkoski KM, et al (2022) When to use transdisciplinary approaches for environmental research. Front Environ Sci 10:840569. https://doi.org/10.3389/fenvs.2022.840569
12. Linnenkohl OE, Ziermann H (1987) Schulische Übungsfirmenarbeit im Rahmen des Deutschen Übungsfirmenrings (DÜF). Erfahrungen und Perspektiven. In: Wirtschaft und Erziehung, vol 36, no 11, pp 357–361 (1987)
13. Mian A, Mehmood K, Hanan A (2019) Exploring factors behind project scope creep a stakeholders perspective. Int J Manag Proj Bus Res Gate. https://doi.org/10.1108/IJMPB-10-2018-0228
14. Prim R (1997) Planspiel. In: May H, May U (eds) Lexikon der ökonomischen Bildung, pp 380–382. De Gruyter, München
15. Reetz L (1984) Berufliches Lernen in der Übungsfirma unter den Bedingungen von Arbeitslosigkeit und Ausbildungsmangel. Zeitschrift für Berufs- und Wirtschaftspädagogik, Beiheft 5:16–28
16. Reetz L, Seyd W (1995) Curricularе Strukturen beruflicher Bildung. In: Rolf A, Antonius L (eds) Handbuch der Berufsbildung, pp 203–219. Opladen, Budrich (1995)
17. Riebenbauer E (2020) Agieren auf globalen Märkten – welche Anforderungen stellt dies an die Übungsfirma? In: bwp@ Berufs- und Wirtschaftspädagogik – online, Ausgabe Nr. 10. (2006). http://www.bwpat.de/ausgabe10/riebenbauer_bwpat10.pdf
18. Speth H (2002) Theorie und Praxis des Wirtschaftslehreunterrichts – Eine Fachdidaktik. 7. Auflage. Merkur, Rinteln
19. Tramm T (1996) Die Übungsfirma als Lernfirma oder Einzig ärgerlich an der Übungsfirma ist ihr Name. Vortrag gehalten auf dem Symposium Mobilität, Flexibilität, Sprachkompetenz. BMUKA, Wien
20. Trummer M, Berchtold S (2000) Practice Firms - Businesses without strategies: An approach to promote further development of Practice Firms through TQM. Creative Teaching ACT 4, Klein HE (ed) OMNI Press, Madison Wisconsin, USA
21. University of Graz Department of Business Education and Development: Syllabus of the course 'Practice Firm and Quality management' (2020). https://static.uni-graz.at/fileadmin/sowi-institute/Wirtschaftspaedagogik/Syllabi_Wip%C3%A4d/Syllabus-Uefa-abSJ1516.pdf
22. Utecht J, Keller D (2019) Becoming relevant again: applying connectivism learning theory to today's classrooms, Crit Quest Edu 10(2):107–119, September 2019. ISSN: EISSN-2327–3607

Extended Reality (XR) as an Interaction Tool for Digital Twins: Application to Safety Management in Construction Megaprojects

Felipe Muñoz La Rivera, Edison Atencio, and Javier Mora-Serrano

Abstract Construction megaprojects represent major challenges. Their time extension, high budgets, the multiplicity of stakeholders and multidisciplinary teams, together with their social, economic, political, and environmental impact, demand technical innovation and the implementation of trend management tools that seek to maximise the productivity of these projects. However, in a sector characterised by low levels of productivity and digitalisation, this is no easy task. In the context of Construction 4.0, the emergence of digital twins represents the possibility of centralising project success factors, focusing on the design and management of projects based on the monitoring, prediction, automation, and optimisation of processes, synchronising physical and virtual environments. While computer visualisation allows access to and control of the digital twin environments, achieving higher levels of immersion and interaction would increase its benefits and allow aspects not considered to be addressed. Extended reality (XR), understood as virtual, augmented, and mixed reality technologies, is proposed as the new paradigm for human–machine interaction with digital twins. Through them, the technologies of Construction 4.0 are linked to these environments. To this end, considering safety management in construction as an application case, this research integrates safety management processes and actions through XR interaction environments linked to the digital twins of construction. Thus, a process flow is presented with examples to demonstrate the application potential.

F. Muñoz La Rivera (✉) · J. Mora-Serrano
International Centre for Numerical Methods in Engineering (CIMNE), 08034 Barcelona, Spain
e-mail: felipe.munoz@pucv.cl

F. Muñoz La Rivera · E. Atencio
School of Civil Engineering, Universitat Politècnica de Catalunya, 08034 Barcelona, Spain

F. Muñoz La Rivera
School of Civil Engineering, Pontificia Universidad Católica de Valparaíso, 2340000 Valparaíso, Chile

F. Muñoz La Rivera · E. Atencio · J. Mora-Serrano
Department of Management, Economics and Industrial Engineering, Politecnico di Milano, 20156 Milan, Italy

E. Favari and F. Cantoni (eds.), *Complexity and Sustainability in Megaprojects*, Lecture Notes in Civil Engineering 342,
https://doi.org/10.1007/978-3-031-30879-6_12

Keywords Extended reality (XR) · digital building twins · construction megaprojects · interdisciplinary teams · construction safety management · Building information modeling

1 Introduction

1.1 Challenges of Construction Megaprojects

Construction megaprojects present a series of characteristics that make their development complex. Their large size and engineering challenges demand the implementation of new technological innovations for their resolution. Their long duration and large budgets, together with their social, economic, environmental and political impacts, demand higher levels of rigour in their management and execution [1]. Given these characteristics of megaprojects, the typical problems of the construction sector are maximised. The risks of time and cost deviations increase considerably, and therefore, the risk to the overall productivity of the projects [2]. Moreover, considering that the management of stakeholders and multidisciplinary teams is one of the main challenges of megaprojects in a sector characterised by fragmentation and low integration, the problem becomes even more relevant [3].

Thus, the scale of such projects requires highly advanced planning and control systems. Methodological best practices and technological trends are at the service of meeting these challenges. Thus, Lean Construction management practices and the Building Information Modelling (BIM) methodology meet these challenges. Likewise, Construction 4.0 promotes a technological framework for the deployment of automation and digitalisation of working environments [4].

Thus, different areas of action in the sector are incorporating these initiatives [5]. Safety in construction is a relevant aspect. Accident rates have not been reduced significantly, and good management practices and new technological tools have a relevant role to play in this area [6]. A change of paradigm and approach is needed to respond to construction safety challenges, especially in the context of mega-projects. For this, training, safety-focused design and site planning are relevant. A correct understanding of the workplace and its associated risks under a safety culture in the organisation allows more efficient safety plans to be established. However, awareness of protective measures is not an easy task, even more so in large projects with multidisciplinary teams, where transversal safety awareness is difficult to implement [7].

1.2 Digital Building Twins

Construction 4.0 promotes the automation, digitalisation and virtualisation of the construction sector, promoting the implementation of collaboration, integration and process efficiency methodologies, such as Lean Construction and Building Information Modelling (BIM), together with technologies that allow the interconnection of systems associated with the management of construction projects (sensors, additive manufacturing, big data, virtual reality, among others) to be deployed and supported.

In response to demand for the automation and digitisation of industries, the concept of digital twins (DT) has emerged in recent years. The term refers to a virtual representation of a physical system to provide a more precise and closer remote monitoring and control [8]. This term is associated with different modelling of the systems, simulation tools and synchronisation devices, which links elements of the real and virtual worlds by means of sensors. Comprising a link between these worlds, they facilitate seeing changes based on real-time data [9], what is a key feature of the DT. In construction (digital building twin – DBT), this ability is directly related to the real-time monitoring, diagnosis and forecasting of the infrastructure under construction and operation [10].

Although the last decade has seen technological advances in the industry, gaps in the technological-methodological coupling with the transition to new working methods concerning the traditional working methods make it difficult to know how to relate these elements and identify the role of DTs in construction [11]. Special attention must be paid to the human factor and multidisciplinary teams at the building site when considering the transition to more technological environments, given that construction workers are one of the least qualified, with one of the highest accident rates and with the increasing ageing of most of its workers. That is why the human–machine interfaces, such as those based on augmented, virtual and mixed reality, can be key to easing the transition to the new Construction 4.0 paradigm [12].

1.3 Extended Reality (XR) as New Paradigm for Human–Machine Interaction

Given the challenges of the construction sector, the great advantage of digitalisation is prediction, the ability to anticipate real events and schedules in controlled virtual environments, which makes it possible to model reality and not only to know what might happen but also to study why and prevent it [13]. In this regard, extended reality (XR), which includes virtual reality technologies (VR - virtual environment of total immersion for the user), augmented reality (AR - superimposing virtual elements on reality through viewers), and mixed reality (MR - the combination of virtual reality and augmented reality that creates environments in which real and virtual objects and people interact) have shown a series of potentialities for the integrated management of construction projects [14].

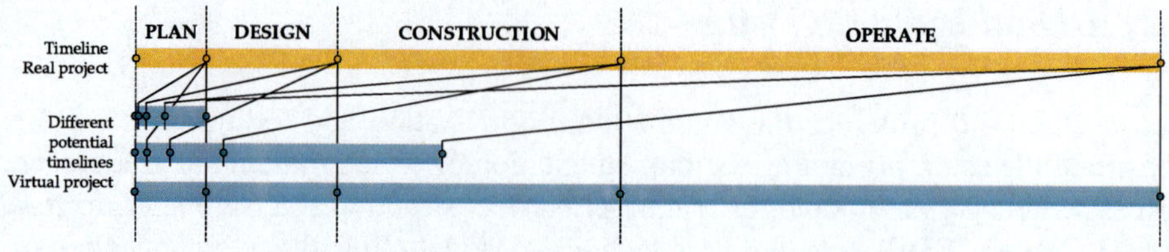

Fig. 1 Different timelines through extended reality

It is relevant to consider the channels through which engineers, builders, workers and users can interact with the new technologies of Construction 4.0 and the digital twins. Extended reality (XR) allows effective interactions in virtual, augmented or mixed environments, enabling new contexts and significantly expanding the possibilities of adaptation to multiple problems and, above all, to the various users of these contents [15]. Thus, it facilitates understanding models, processes and works dynamics in immersive environments. It also allows the interaction of different roles and from different perspectives, facilitating integration and work in integrated environments. Moreover, within a virtual environment, it is possible to interact with any part of the project without physical restriction [16]. Also, in the field, it allows physical-virtual interaction through augmented or mixed reality technologies to link both environments more efficiently [17].

Different uses are being tested from the XR perspective. In safety issues, its applications have focused on training to train work teams, designing safe environments and on-site safety inspections. In large construction projects, integrated safety management is key to addressing all project stages, teams and disciplines [18]. For a mega-project, where one of its main characteristics is the long time it takes to develop, build and operate, being able to anticipate all possible scenarios at an early stage is a great advantage. There, the extended reality has a relevant role. As shown in Fig. 1, in virtual environments, the time scales can be multiple, as well as the scenarios and simulations of the multiple factors of interest. This way, digital twins could be visualised and interacted with on reduced time scales or in parallel ("real-time") with the project's progress.

1.4 Research Objectives

Digital twins allow the monitoring and control of construction projects, achieving the synchronisation of physical and virtual environments. Their benefits are even more tangible in the context of mega-projects, where the physical and temporal extension of projects, added to the technical complexity, require more sophisticated mechanisms for their control and correct management and execution. However, the multiplicity of technologies and information that feed the models linked to digital twins requires new tools to facilitate their interaction. While computer visualisation allows access to and control of digital twin environments, achieving immersion in these environments and higher levels of interaction with the models and processes would increase the benefits and, furthermore, incorporate new aspects.

Thus, the extended reality is proposed as the new paradigm for human–machine interaction with digital twins, and through them, the technologies of Construction 4.0 are linked to these environments. To this end, considering safety management in construction as an application case, this research seeks to deploy safety management processes and actions through XR interaction environments linked to the digital building twins. Thus, a process flow is shown, together with application examples, to demonstrate the potential of its implementation.

2 Research Methodology

The Design Science Research Methodology (DSRM) was used to guide the entire research process [19]. In general, the DSRM involves four stages: (1) identifying problems and motivations; (2) defining the objectives of a potential solution; (3) design and development; and (4) demonstration. Figure 2 shows a diagram of the research methodology.

In the first stage, aspects associated with the challenges of construction megaprojects are identified, along with the role of digital twins in responding to those challenges and the opportunities that extended reality provides to enhance the ability to interact with digital twin environments. With this information, in the second stage, the objective of a potential solution is identified: XR enhances the ability to interact with digital twins, responding to the challenges of construction megaprojects. The third stage proposes a workflow for construction safety management by exploiting digital twins in XR environments. Finally, in the fourth stage, examples of the implementation of some functionalities are shown to exemplify the proposal's feasibility.

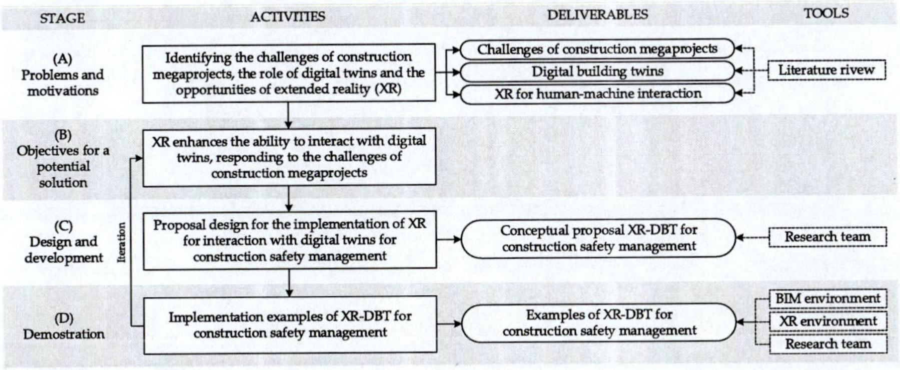

Fig. 2 Research methodology

3 Proposal

Figure 3 shows the proposal for implementing extended reality to support the interaction with the digital building twins, focusing on safety management in construction. The upper part of the figure shows the processes of the construction projects, with a focus on safety management. Then, based on the inputs from the processes at the top, specific safety sub-processes are broken down and developed in the XR environments. These, in turn, link to the digital twin of the building, which in turn links to the physical work site. Thus, the XR environments allow interaction with the digital twin, facilitating its understanding, design, and project review, among other multiple uses.

In the upper section of the figure, four stages associated with construction projects, with a focus on safety management, are considered: (a) Design development, (b) Construction planning, (c) Construction safety planning, and (d) Safety monitoring and control. These are based on BIM working environments, which consider 3D dimensions (development of parametric 3D models of the project), 4D/5D (integration of time and cost planning into 3D models, considering the construction stages), and 8D (integration of safety aspects into models, construction simulations and management plans in BIM environments).

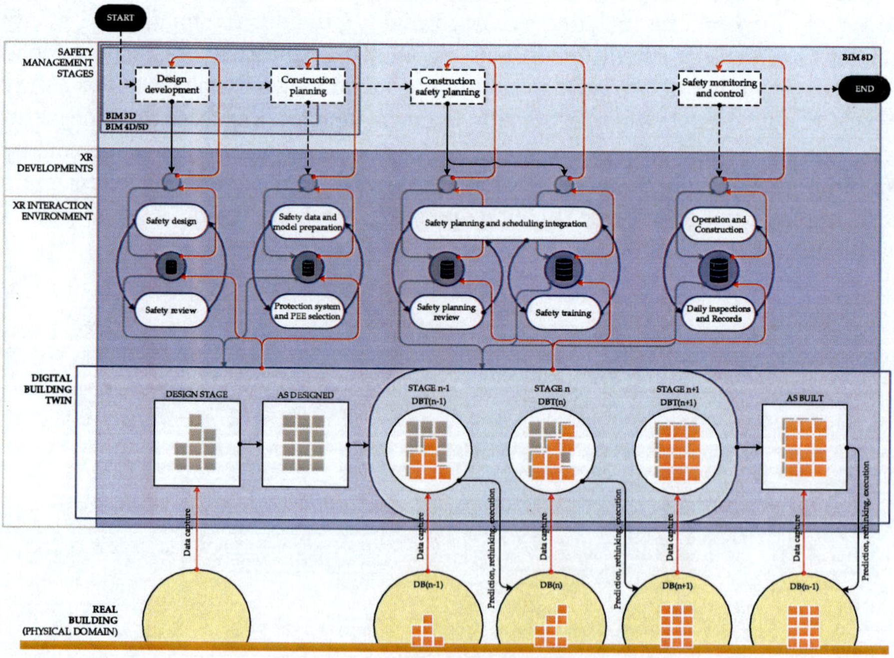

Fig. 3 Workflow proposal XR-DBT for safety management in construction

In the middle section of Fig. 2, the processes for safety management in X environments are broken down. The four processes are broken down into new sub-processes that address the entire construction safety management throughout the project lifecycle. These processes are developed in the extended reality environments, with the BIM models coming from the project's design phase. In turn, the information from the models and the actions performed in the XR interaction environments are synchronised with the digital twin of the building. Although at this level, only the interaction with the model itself is considered, the link with the digital twin allows incorporating all the information from the construction site with the temporal updates it offers. Therefore, the possibility of XR interaction is extended to all management aspects linked to sensors and updates of the worksite.

The digital twin now has greater potential for decision-making and project monitoring, as the immersion and new levels of interaction offered by XR technologies provide new possibilities for analysing the aspects of the project that need to be monitored. For example, as shown in the lower part of Fig. 2, in a DT of a building at the "n" stage, the DBT(n) will receive the real-time data inputs of the project that has been built based on the information provided by the DBT(n-1), which has ensured that what has been built in the RB(n) (physical building) is correct. Therefore, DBT(n) will have the original model and the changes made up to that point. With this information and the captured data, it is possible to process and correct the model according to its behaviour in real-time during project execution. This information allows the DBT(n) to deliver reliable information to continue with the RB(n + 1) construction. The process continues until the end of the project, where the digital building is equal to the constructed building and where the high representativeness as a constructed model is obtained since it was generated before and after each real-time instant considered.

4 Implementation Examples

Figure 4 shows examples of implementation of the proposed use of XR for interaction with digital twins in the case of construction safety management. For this case, a building project is used. It is possible to view the BIM model of the building with the collective protection elements arranged in the digital twin environment at a specific stage of the construction process (based on a 3D/4D/8D BIM model). On the other hand, an image of the real project is displayed. The interaction with a digital twin is limited to the visualisation of models (in theoretical terms, with all the data and simulations you wish to incorporate). However, by incorporating the XR layer and using these technologies to enhance the interaction, advantages are possible to see.

Thus, images (a), (b) and (f) in Fig. 4 show the incorporation of an immersed user in the building. This could be a first-person immersed user (e.g. through virtual reality goggles), a third-person immersed user (from non-immersive devices), or bots interacting with the user to incorporate other simulations. Also, multi-user considerations allow the incorporation of other users in the first person, allowing interaction between

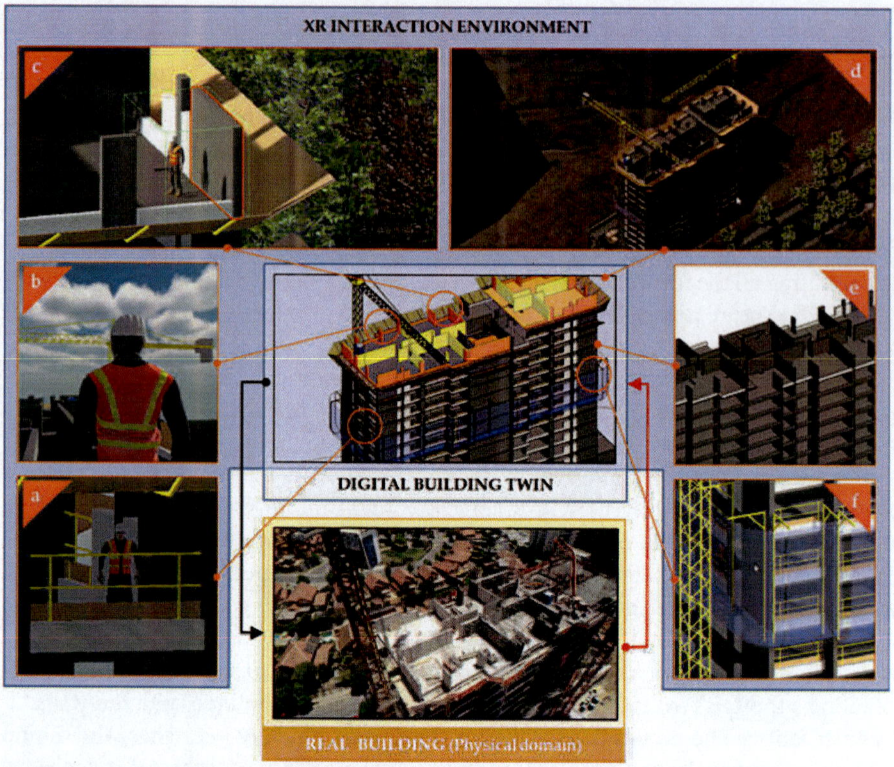

Fig. 4 Implementation examples XR-DBT for safety management in construction

different team members, disciplines or stakeholders in general, with different layers of information and access limitations, as appropriate. This functionality gains much relevance in mega-projects, where the management and, even more so, the integration of interdisciplinary teams are key. In these interaction environments, teams can co-design, monitor and manage the project more immersively, facilitating the integration of different perspectives.

Aligned with this interaction of users with the three-dimensional models for monitoring the project, image (c) in Fig. 4 shows the visualisation of an interaction zone (in green), which delimits that user's actions. Thus, for example, in areas where there is a danger of falls, visualising these zones allows the behaviour of workers, machinery, and transport of materials, among other elements, to be controlled. This linked to sensorisation of the actual work site would allow real-time interaction control, planned in a simpler (and safer) way in XR environments.

Images (e) and (d) show the BIM model of the building, both as a model itself and immersed in the context where it is being built. This application represents a great advantage when planning the project's construction in order to place the work dynamics in the real context and how the context could affect the execution of the project. In safety management, this is also a great advantage. For example, the brightness of the work site (changing with the time of day or nearby structures) could affect the workers, generating unforeseen risks. Simulating other environmental or noise conditions makes it possible to evaluate the effect on work dynamics in virtual environments and thus propose mitigation measures or change previously defined actions.

5 Discussion

While this research shows the potential of extended reality as a mechanism for interaction with the digital twins of buildings, it is important to clarify that its full development is not simple and its mass implementation on construction sites is not something for the medium or long term. The sector is in the process of evolution towards the generation of conditions for the methods and technologies of Construction 4.0 to be fully deployed.

From a methodological point of view, the construction sector is in the process of implementing BIM and Lean Construction, but they are not yet standard practice worldwide. These are the basis for the next stages of digital transformation; therefore, accelerating their mass implementation is key. On the other hand, the implementation of digital building twins is complex. At a theoretical level, several authors have proposed mechanisms and architectures to develop a DBT. However, their use in construction works is limited. Its application is limited to specific uses explored according to particular project needs, mainly driven by research projects. Therefore, full synchronisation between the physical and virtual environments cannot yet be considered achieved.

Several focuses of work are necessary to achieve this interconnection of the DBTs: site sense, the capture of the built environment, and collection and management of high volumes of information, among other aspects, are necessary to collect information from the construction site. After that, challenges in processing that information and linking it to the digital twins is another front of work, followed by how that information is used and enables better decision-making. Finally, another problem is how those decisions get back down to the workplace. Added to all of this is the consideration of "real-time" to reduce action times and the risks associated with those decision-making latencies.

Thus, although the extended reality is proposed as the form of interaction with DBTs, its complete implementation, with total interconnection between the different environments it considers, is a pending aspect. While models and functionalities can be linked between BIM, DBT and XR environments, these are limited to particular aspects and require complex development processes. However, early consideration of the advantages and uses of XR allows us to broaden the uses of DBT and to think of its design from the XR perspective, increasing its action potential. Thus, particularly in the area of construction safety, the broad vision of the project and of multiple variables that allows the extended reality to be addressed provides possibilities for its holistic management, representing an interesting opportunity to improve these management processes and, therefore, to reduce accident rate indicators in the sector.

6 Conclusions

Construction megaprojects represent major implementation and management challenges with a high social, economic, environmental and political impact. Their large dimensions represent challenges for managing time and budgets, which are no more than a reflection of the management behaviour of the work teams, the technical capabilities and the correct management tools used. Thus, Construction 4.0 trends are put at the service of these megaprojects to respond to these problems and help in the integrated management of the projects. In particular, the management of security in these megaprojects is a relevant aspect so that the protocols and protection measures are complied with throughout the physical and temporal extension of the project and by all the members of the different teams that make it up (a task that is not easy to achieve).

Thus, digital twins represent an opportunity to improve construction megaproject management, monitoring and control. Real-time synchronisation with what is happening on the job site allows for up-to-date information, which after being processed and analysed, enables better decisions to be made to execute the project. This way, new decisions and actions are taken based on real project information, not outdated planning or different contexts. Despite these benefits, one more layer is required to improve the interaction with DBTs. To this end, extended reality (XR) represents a new paradigm for human–machine interaction. Through it, interaction and immersion are increased, facilitating work with DBTs, and accessing new functions, perspectives, and other variables to be analysed.

Faced with this new paradigm, this research proposed a workflow for integrating extended reality as the visualisation layer of DBTs. This flow considered the main stages and sub-stages associated with safety management in construction projects to link these processes with their management from XR environments and their link with DBT. The implementation examples shown show the potential of XR for DBTs in safety but can also be extrapolated to other aspects of project management.

As future lines of research, it is proposed to continue implementing XR functionalities for interaction with digital twins, driving developments towards the automation of these processes, with a view to the massification of its implementation.

Acknowledgements This work has been supported by the CONICYT through its economic support of Felipe Muñoz, beneficiary of a pre-doctoral grant (CONICYT - PCHA/International Doctorate/2019-72200306) and "DI emergente PUCV 2022" project of the Vicerrectoría de Investigación y estudios avanzados - Pontificia Universidad Católica de Valparaíso (039.350/2022). The authors also acknowledge financial support from the Spanish Ministry of Economy and Competitiveness, through the "Severo Ochoa Programme for Centres of Excellence in R&D (CEX2018-000797-S)".

References

1. Damayanti RW, Hartono B, Wijaya AR (2021) Clarifying megaproject complexity in developing countries: a literature review and conceptual study. Int J Eng Bus Manage 13:1–25. https://doi.org/10.1177/18479790211027414
2. He Q, Xu J, Wang T, Chan APC (2021) Identifying the driving factors of successful megaproject construction management: findings from three Chinese cases. Front Eng Manage 8:5–16. https://doi.org/10.1007/s42524-019-0058-8
3. Wang T, Chan APC, He Q, Xu J (2022) Identifying the gaps in construction megaproject management research: a bibliographic analysis. Int J Constr Manage 22:1585–1596. https://doi.org/10.1080/15623599.2020.1735610
4. Muñoz-La Rivera F, Mora-Serrano J, Valero I, Oñate E (2021) Methodological - technological framework for construction 4.0. Arch Comput Methods Eng 28:689–711. https://doi.org/10.1007/s11831-020-09455-9.
5. Feng Y, Teo EAL, Ling FYY, Low SP (2014) Exploring the interactive effects of safety investments, safety culture and project hazard on safety performance: an empirical analysis. Int J Project Manage 32:932–943. https://doi.org/10.1016/j.ijproman.2013.10.016
6. Muñoz-La Rivera F, Mora-Serrano J, Oñate E (2021) Factors influencing safety on construction projects (fSCPs): descriptions and types. Int J Environ Res Public Health 18:1–11. https://doi.org/10.3390/ijerph182010884.
7. Mora-serrano J, Muñoz-La Rivera F, Valero I (2021) Factors for the automation of the creation of virtual reality ex- periences to raise awareness of occupational hazards on con- struction sites. Electronics 10:1–21
8. Mateev M (2020) Industry 4.0 and the digital twin for building industry. Int Sci J Industry 4.0. 5:29–32
9. Keenan M, Rostami A (2019) The impact of quality management systems on construction performance in the North West of England. Int J Constr Manage 1–13. https://doi.org/10.1080/15623599.2019.1590974
10. Adjei P, Montasari R (2019) A critical overview of digital twins. Int J Strateg Eng 3:51–61. https://doi.org/10.4018/ijose.2020010104
11. Lu Q, Xie X, Parlikad AK, Schooling JM, Konstantinou E (2020) Moving from building information models to digital twins for operation and maintenance. In: Proceedings of the Institution of Civil Engineers - Smart Infrastructure and Construction 1–9. https://doi.org/10.1680/jsmic.19.00011
12. Boje C, Guerriero A, Kubicki S, Rezgui Y (2020) Towards a semantic construction digital twin: directions for future research. Autom Constr 114:103179. https://doi.org/10.1016/j.autcon.2020.103179
13. Gao Y, González VA, Yiu TW, Cabrera-Guerrero G, Li N, Baghouz A, Rahouti A (2022) Immersive virtual reality as an empirical research tool: exploring the capability of a machine learning model for predicting construction workers' safety behaviour. Virtual Reality 26:361–383. https://doi.org/10.1007/s10055-021-00572-9
14. Bruttini A (2018) A project framework to introduce virtual reality in construction health and safety. New Front Constr Manage Workshop. 09:166–175
15. Truong P, Hölttä-Otto K, Becerril P, Turtiainen R, Siltanen S (2021) Multi-user virtual reality for remote collaboration in construction projects: a case study with high-rise elevator machine room planning. Electronics (Switzerland). 10(22):2806. https://doi.org/10.3390/electronics10222806.
16. Teizer J (2016) Right-time vs real-time pro-active construction safety and health system architecture. Constr Innov 16:253–280. https://doi.org/10.1108/CI-10-2015-0049
17. Wolf M, Teizer J, Wolf B, Bükrü S, Solberg A (2022) Investigating hazard recognition in augmented virtuality for personalized feedback in construction safety education and training. Adv Eng Inform 51:101469. https://doi.org/10.1016/j.aei.2021.101469

18. Ramos-hurtado J, Muñoz-La Rivera F, Mora-Serrano J, Deraemaeker A (2021) Proposal and for the deployment of an augmented reality tool for construction safety inspection. 08034:1–35
19. Peffers K, Tuunanen T, Gengle, C (2006) The design science research process: a model for producing and presenting information systems research. J Manage Inf Syst 83–106

Towards the Integration Between Construction Projects and the Organization: The Connections Between BIM and Enterprise Architecture

Edison Atencio, Felipe Muñoz-La Rivera, Mauro Mancini, and Guillermo Bustos

Abstract Construction projects' complexity increases with their size, stakeholders' environment and engineering specialties. This condition is accentuated in the context of megaprojects.

On the one hand, BIM has helped to manage the complexity of construction projects, transforming all project components into a digital system. On the other hand, enterprise architecture (EA) allows managing the organization as a system and decomposing interrelated components into a model. BIM and EA aspire to interoperability and collaboration as their application maturity increases beyond project boundaries. The joint work between BIM and EA could manage the project's complexity and accelerate BIM maturity in project-business management.

This paper seeks to identify the connections between BIM and EA through a literature review showing the advantages and synergies of their combined application. Advances and challenges are discussed for future studies to achieve stronger exploitation of BIM and EA working together.

Keywords Building information modeling (BIM) · enterprise architecture (EA) · business process management (BPM) · interoperability · maturity

E. Atencio (✉) · M. Mancini
Department of Management, Economics and Industrial Engineering, Politecnico di Milano, Via Lambruschini 4b, Bovisa, 20156 Milan, Italy
e-mail: edisonpatricio.atencio@polimi.it

E. Atencio · F. M.-L. Rivera
Department of Civil Engineering, Pontificia Universidad Católica de Valparaíso, Av. Brasil 2147, 2340000 Valparaíso, Chile

E. Atencio · G. Bustos
Department of Industrial Engineering, Pontificia Universidad Católica de Valparaíso, Av. Brasil 2241, 2340000 Valparaíso, Chile

E. Favari and F. Cantoni (eds.), *Complexity and Sustainability in Megaprojects*, Lecture Notes in Civil Engineering 342,
https://doi.org/10.1007/978-3-031-30879-6_13

1 Introduction

The Architecture, Engineering and Construction (AEC) industry is one of the main relevant industries in worldwide economies [1]. However, this sector currently faces many challenges in terms of productivity and digitalization and currently, the AEC industry occupies the last position of productivity and digital maturity [2]. This situation turns into a significant concern in front of a megaproject: size, budget, complexity, environmental impact and a large number of stakeholders. In these cases, governance is needed to ensure the goals' achievement [3].

Facing these challenges, Building Information Modeling (BIM) appears as an alternative solution [4]. Moreover, BIM is considered an effective construction project governance tool when its complexity increases, as well as projects or megaprojects [5].

The first BIM approaches can be identified with the work of Eastman in the seventies, establishing the concept of the building model, and the plants and sections can be obtained from a 3D model assisted by computers [6]. Through time, several developments have led to the concept of BIM, such as the creation of Graphisoft Archicad in the eighties, the Industry Foundation Classes (IFC) as a data exchanging standard, the agreement between organizations towards adopting digital construction practices, for instance, the UK Cabinet Office BIM Plan or the Chilean Plan BIM [7].

BIM is structured in four pillars: people, processes, policies and tools. These pillars define a governance environment that constantly matures through collaboration and interoperability improvement [8, 9]. The above inescapably connects the BIM domain with the organizational domain. The expected high-maturity level of BIM encompasses integration and interoperability in data management and process management, considering different stakeholders around the project working collaboratively [9]. However, achieving maturity is still challenging for the AEC industry [10].

From the organizational point of view, one of the main characteristics that differentiate AEC companies from others is the project-based orientation that conducts its processes. So, within this field, a similar BIM integration challenge is identified: the integration between the project, the organization and the stakeholders' environment [11].

Projects in an organizational context entail complexity that has to be managed. Underperformance in megaprojects, for instance, has been attributed to the inherent complexity and the gaps because of misalignment between project management (PM) processes and PM teams [12]. This complexity could be understood in three dimensions (1) technical complexity, considering the technological and practical concerns, (2) human complexity, encompassing the formal and informal relationships and leadership and (3) political complexity, related to negotiations and managing stakeholders [13].

Systems thinking appears as an effective viewpoint for understanding, analyzing and managing the complexity in an organization, projects and the projects' organization [14]. Within this perspective, Enterprise Architecture (EA) approach is a systemic management and governance tool for managing the complexity of organizations and projects, aligning it to achieve a common goal [11, 14, 15]. EA is defined as a group of principles, methods and models used to design and realize the following organizational components: organizational structure, business processes, information systems and infrastructure [16].

Considering that, for construction projects, BIM is an enabler for managing the project as a whole interrelated system composed of people, processes, policies and tools and EA is a systemic approach focused on the same components, *BIM and EA could work together* facing the organizational and project complexity, as shown in Fig. 1.

Thus, considering the BIM-EA collaboration possibilities, the following research questions are proposed for this research:

- *RQ1: What is the relationship between BIM and EA in the construction projects context in the literature?*
- *RQ2: How BIM and EA could work together to manage the complexity of construction projects and achieve desired maturity goals such as interoperability and collaboration?*

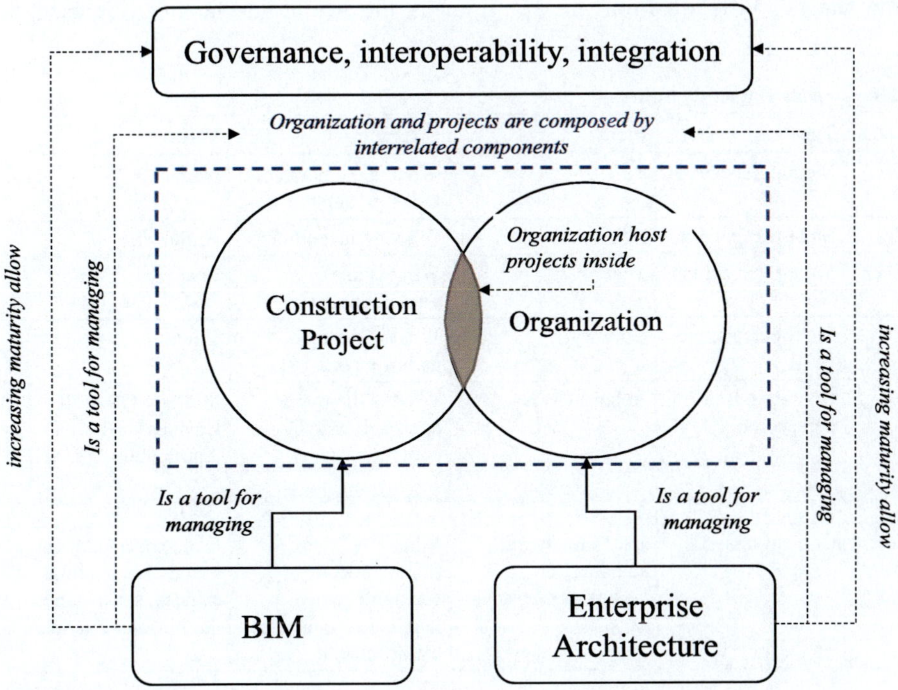

Fig. 1 BIM and EA collaboration

- *RQ3: How BIM and EA could work together to manage megaprojects?*

To answer these questions, a literature review is developed through the analysis of bibliometric and content data obtained from the 43 selected articles.

The remaining document is organized as follows: Section two presents the method that conducts this research and the article selection process. Section 3 contains the results and discussion; finally, section four describes the conclusion and future research.

2 Research Method and Article Selection Process

This literature review follows the analysis approach proposed by [17] and adapted by [18]. This proposal considers extracting and processing the meta-content perspective data (bibliometric data) and the content-perspective data (data obtained from the successive lecture of the articles). The following data-extraction form for the articles' content obtaining, as shown in Table 1.

The bibliometric data is available as metadata from the article and queries to databases obtaining structured data. This data is composed of the fields D1 to D4 of Table 1. For content data extraction and managing its variability, it is recommended to use a theoretical framework from the studied domains for better structuring and data analysis [18–20]. From the EA domain, the ArchiMate language is used as

Table 1 Data extraction form

ID	Type	Field	Question	Value
D1	Meta-perspective	Title	What is the name of the approach?	Name
D2	Meta-perspective	Authors	Who are its authors?	Author list
D3	Meta-perspective	Year	What is its publication year?	Year
D4	Meta-perspective	Country	What is the first author's country?	Country
D5	Content-based perspective	Relation to EA	What is the main approach with EA components?	Strategy OR/AND Business OR/AND Application OR/AND Technology OR/AND All EA layers
D6	Content-based perspective	EA Framework, standard, or language reference	What EA frameworks, standards, or language are used in the article?	Data extracted from successive readings of articles, for instance, ArchiMate, Zachman

(continued)

Table 1 (continued)

ID	Type	Field	Question	Value
D7	Content-based perspective	Relation to BIM	What is the BIM use(s) referenced in the article	(OR/AND) Capture Existing Conditions; Author Cost Estimate; Author 4D Model; Analyze Program Requirements; Analyze Site Selection Criteria; Author Design; Review Design Model(s); Analyze Structural Performance; Analyze Lighting Performance; Analyze Energy Performance; Analyze Engineering Performance; Analyze Sustainability Performance; Coordinate Design Models; Author Construction Site Logistics Model; Author Temporary Construction Systems Model; Fabricate Products; Layout Construction Work; Compile Record Model; Monitor Maintenance; Monitor System Performance; Monitor Assets; Monitor Space Utilization; Analyze Emergency Management
D8	Content-based perspective	Needs addressed	What are the needs of the EA-BIM application?	Data extracted and generalized from successive readings of articles, for instance: interoperability, data exchange
D9	Content-based perspective	Challenges identified for future research	What challenges are proposed for future research?	Data extracted and generalized from successive readings of articles (case study, architect–project manager integration, adding complexity)

(continued)

Table 1 (continued)

ID	Type	Field	Question	Value
D10	Content-based perspective	Type of project or activity	What is the type of construction project or activity addressed in the selected paper?	For instance, building, highways, facility management

a basis. ArchiMate is a widely used EA language that considers the EA components in four interrelated layers: (1) The Strategy layer encompasses the capabilities, resources and the proper course of action for modelling the organizations' strategy. (2) Business layer is composed of the different business units' processes, services, functions and events. (3) The Application layer considers the software supporting business components through application services. (4) The technology layer encloses the hardware and communication infrastructure required to service the applications. Thus, the relationship between articles and EA is labelled through these four layers and a fifth layer called "all layers" is considered when the articles refer to EA (See D5 of Table 1). The EA framework, standard or language reference is identified within the EA domain, as shown in D6 in Table 1.

The relationship with BIM domain in selected articles is structured using two labels (1) the connection with one or more BIM uses, according to the Penn State definitions [21] and is detailed in D7 in Table 1. (2) the BIM dimension considered such as conceptual design (2D), modelling and visualization (3D), planning (4D), cost estimating (5D), sustainable development (6D) and operation-maintenance (7D) [21, 22].

The main needs addressed, challenges identified, and type of project or activity considered in selected papers are extracted as shown in D8 to D10 of Table 1.

Regarding the article selection process, Web of Science and Scopus were considered as databases and the keywords used for searching were BIM (Building Information Modeling), Enterprise Architecture, Business Process Modeling Notation (BPMN), Business Process Management (BPM) and Business Process Architecture (BPA). The last three keywords are considered strongly related concepts with EA following the work of González [23]. The selection process considers two exclusion criteria: (EC1) the duplicated papers and (EC2) when the paper does not show a relationship between BIM and EA. The processes finished obtaining 43 articles, as shown in Fig. 2.

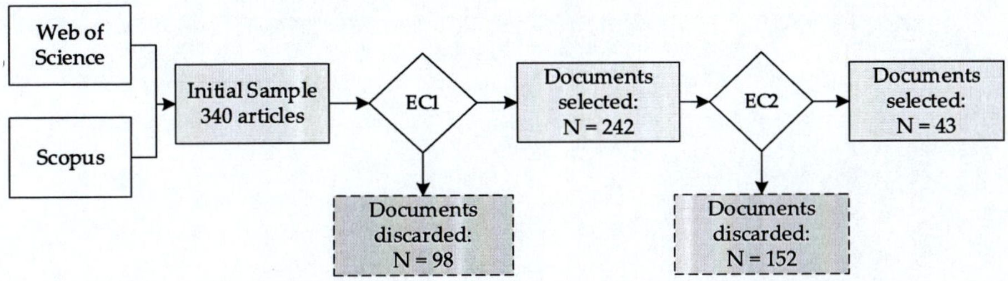

Fig. 2 Document selection flowchart

3 Results and Discussion

3.1 Bibliometric Analysis

Regarding the bibliometric data analysis, Fig. 2 shows that the selected articles were distributed between 2010 to 2022. It is observed that increasing productivity between 2017 to 2019. However, this trend has not been maintained during the last years (Fig. 3).

Regarding the country distribution, the top three countries that contribute to the fields' research are UK, Germany, and USA, concentrating about 50% of the selected papers, as shown in Fig. 4.

Keywords are concepts highlighted by researchers and give an idea of the focus of the articles. Several reviews have used this bibliometric data for finding relationships between articles using co-occurrence mapping [11, 24]. A co-occurrence appears when a concept (or keyword) is addressed in two or more articles. Figure 5 shows the clusters configured through the co-occurrence of the 43 selected articles.

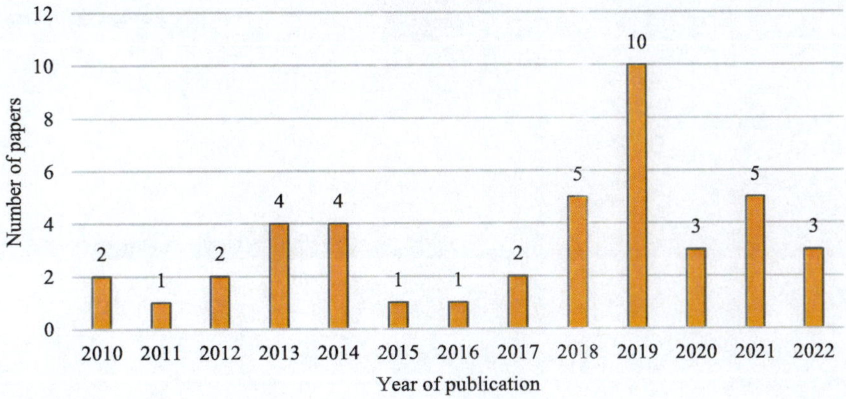

Fig. 3 Number of relevant articles published from 2010 to 2022

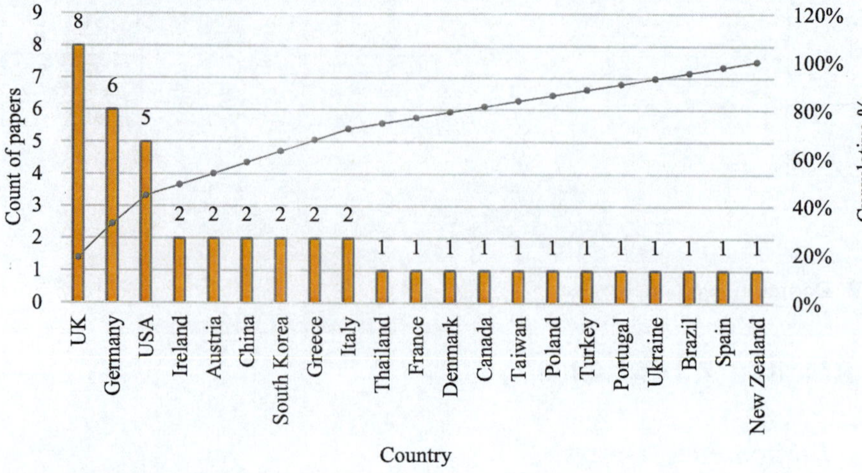

Fig. 4 Productivity by country

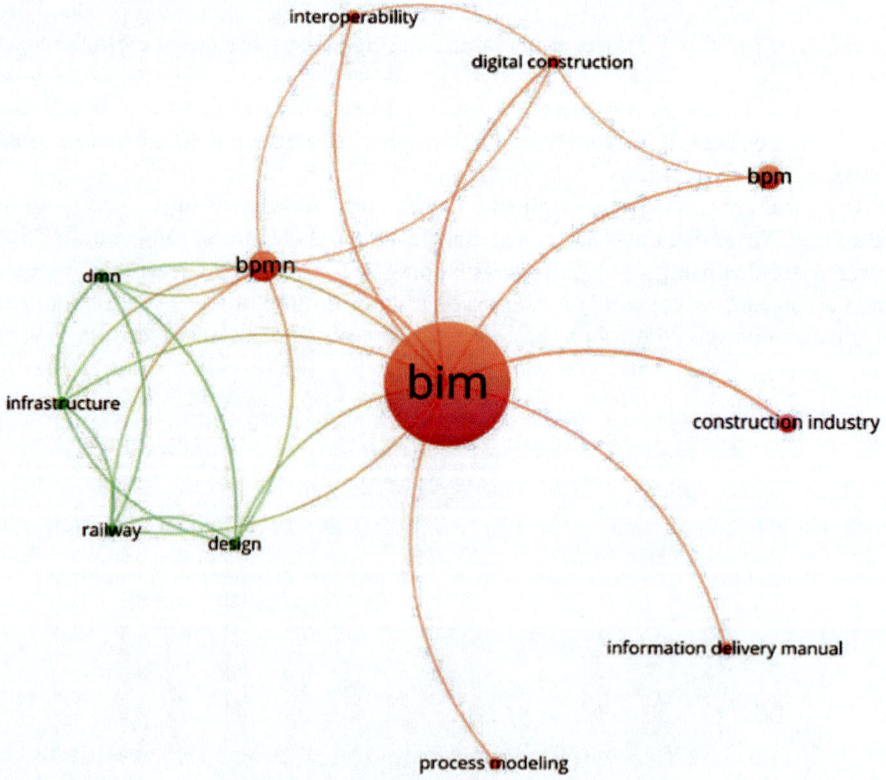

Fig. 5 Co-occurrence map for keywords of selected papers

As shown in Fig. 5, the most centric concept is BIM, and two clusters are formed. The red cluster connects BIM with the business process domain, such as business process modelling notation (BPMN) and business process management (BPM). Also, the process modelling keyword is related to these concepts. Digital construction and interoperability are connected with BIM through BPMN. The green cluster encompasses the design, railway, infrastructure and dmn keywords. This cluster refers to a group of papers which BIM has been applied in infrastructure projects and the design guidelines has been displayed in models' form for better understanding and connecting with BIM models and activities. DMN (decision model notation) is a modeling language for the specification of business decision and rules [25].

3.2 Content Data Analysis

Concerning the data content perspective analysis, the relationship between BIM and EA of the selected articles is mainly with the Business layer, as shown in Fig. 6. The application and technology layer appears in a similar level of connection. Relationships with the strategy layer were not identified.

The connection of BIM with the Business EA layer is mainly generated from the following concerns:

- *Managing roles and responsibilities:* Construction projects consider a group of different roles and responsibilities displayed inside and outside the project borders. Researchers have highlighted EA approaches for managing this environment when it is more complex. BPM approach and process mapping have been widely adopted in the selected papers, mainly the Business Process Model and Notation (BPMN). This language describes roles, activities, relationships and interactions with artefacts such as documents, procedures and software. Italian researchers highlighted the usefulness of business process mapping in heritage BIM projects for better managing and describing the different BIM-supported activities for restoration

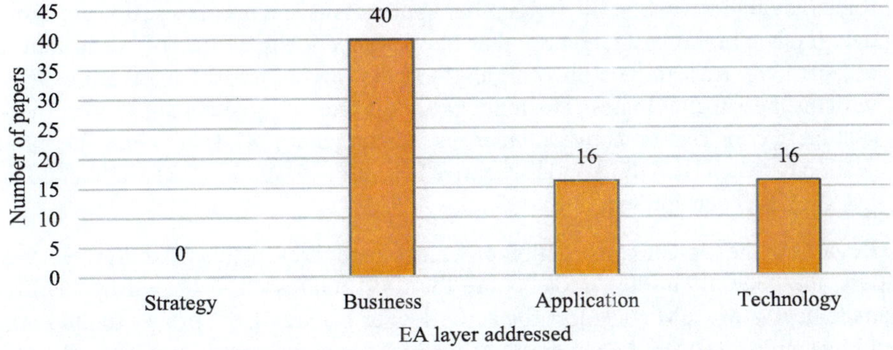

Fig. 6 Relationship with EA

projects, considering the complex bureaucracy in their local context [26, 27]. The fragmentation and collaboration in the whole project's lifecycle is a concern [24] and the BPM approach has been noted for covering this issue [28–30].

In some cases, the processes orchestrating construction projects are unknown. Process mining has been integrated with BIM tools for discovering the roles and activities involved and reengineering these processes [31, 32]. Process mining is a technology that draws a process based on log data and is recognized as a tool for improving business work and achieving customer satisfaction and cost reductions [33].

- *Guidelines representation and code compliance:* In the AEC industry, several standards, such as norms and guidelines, are available in written form that humans can read, but not by machines [34]. These elements that regulate the construction business are required to integrate with the business processes, in the level of human and machine-assisted activities, in this case, the BIM technologies. Through the BPM approach, railways projects have achieved higher levels of efficiency through a complete integration of the design guidelines with BIM. Häußler and Borrmann [34, 35] have shown that more than 50% of design guidelines' rules (applied in BIM environment) can be automated using a BPMN approach. Moreover, 90% of man-hours savings could be obtained.

  For sustainability code compliance for Author design/Analyzing sustainability BIM uses (within the 6D dimension of BIM), the BPM approach has shown advantages, allowing a better structure of the guidelines (from text to models) and, using this model, some activities have been able to be automated through workflows [36].

The connections between BIM and the technological layers appear together (with application and technology layers). The main concerns identified in this connection are:

- *Data exchanging:* From the technological viewpoint and its components such as software, hardware, infrastructure and the data and its architecture, interoperability and data exchange are the most need addressed in the selected papers. From BIM context, the Industry Foundation Classes (IFC) are a central resource for achieving interoperability with other applications and processes. IFC are open data models in the AEC industry that facilitate the identification of construction objects [37]. Researchers have highlighted the possibilities of integrating BIM with business technologies. The main gap is the lack of project management integration in construction between business management [38]. In this line, the integration between BIM tools and Enterprise Resources Planning (ERP) applications [31, 38] has been studied.

Regarding the type of construction projects or activities addressed in the selected papers, the general building projects are the most addressed, followed by facility management (FM). FM considers the activities for managing the proper function of buildings and infrastructures integrating people, space, processes and technologies [39].

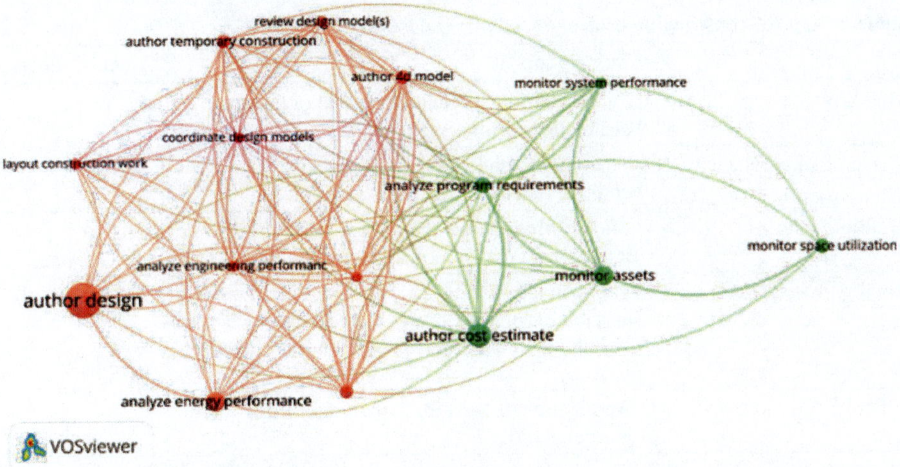

Fig. 7 BIM uses co-occurrence map

Regarding the BIM uses, mapping the co-occurrence between BIM uses addressed in the selected papers, two clusters are identified, as presented in Fig. 7.

According to Fig. 7, the most weighted BIM – and from the red cluster - use is author design, which corresponds to the use of BIM software to 3D model developing and processing attribute information for configuring a library of parametric design elements [21]. The green cluster considers the author's cost estimate the most weighted BIM use. This use considers the activities supported by BIM for cost estimation and the economic effects of additions or modifications in the project.

Parameters in the BIM model are used for integrating with components outside the boundaries of the model. These connections could be modelled using different architectural EA-related languages such as BPMN, UML (Unified Modeling Language) or IDEF[1] (integrated definition). For example, the work of Zanni [53] considers the integration between the BIM project model with business applications such as PowerBI (for metrics dashboard displaying) and Excel spreadsheets. The "bridge" for achieving these connections is parametric programming tools such as Dynamo©, which allows interoperability with the BIM model data and other software.

Regarding the relationship with the BIM dimensions, the EA connections are related mainly to 3D and 4D, as shown in Fig. 8.

[1] https://www.iso.org/obp/ui/#iso:std:iso-iec-ieee:31320:-1:ed-1:v1:en.

Table 2 Type of construction project or activities addressed

Type	Description	Count of papers
Building	Typical building with commercial or housing destination, for instance [40, 41]	19
Design build projects	The one-entity delivery of design and construction as a whole service [42–45]	4
Building retrofit	Projects that change the system or structure after its original construction and use [28, 46, 47]	3
Facility management	The management of facilities, using BIM models as a container [48], energy management [49] and basing the facility management in BIM models [50]	3
HBIM	Application of BIM for managing historical buildings [26, 27]	2
Railways	Development in German railways projects [34, 35]	2
Major construction projects	Projects characterized by their complexity, level of investment and stakeholders involved [51]	1
Mining	For discovering processes in mining construction projects [32]	1
Highways	Using BPMN for managing the complexity of standards and requirements [52]	1
Built to rent (BTR)	Whole life cost into BIM processes assisting decision making [53]	1
MEP	Describing the required interactions between project stakeholders during the design phase of a geothermal heat pump system [54]	1
Prefabrication	Processes: the initial focus is developing Process Data Models connected with Product Data Models (BIM models) and exchanging data in almost real-time [55]	1
Sustainable building	Energy analysis integrated with the BIM model using BPMN [56]	1
Airport asset management	BIM-based modularized platform architecture using systems modelling language (SysML) [57]	1
Asset Management	Studying the activity systems that drive BIM business value in asset management [58]	1
Green building	Proposing a framework for BIM project execution planning guidelines [36]	1
Total of papers		**43**

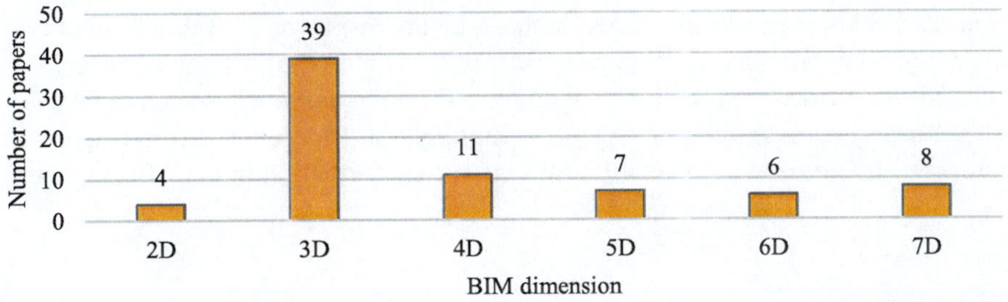

Fig. 8 Relation with BIM dimensions

4 Conclusions

In this research, connections between BIM and EA are showed. EA is presented as a system thinking alternative approach for managing construction projects and could collaborate with BIM together. Explicit mentions of EA have not been identified in the literature. However, other components of EA were jointly sought with BIM, such as BPM, BPMN and BPA. Regarding *RQ1*, the main finding is the relevance of the process approach for describing the interrelationship between BIM, the organization and stakeholders. Processes are part of the EA business layer, and this perspective has been widely accepted and applied in project management [11]. Processes can be reached on two levels, firstly for descriptive purposes, as shown in different articles for managing the roles and responsibilities and transcribing into model norms and guidelines that are difficult to apply in text form. The second level considers automatizing activities using processes and developing workflows that could cross the project's boundaries, organization, and applications [59]. BPMN has been widely used in the selected papers (30 of 43 articles). This modelling language has the advantage of containing a wealth of semantics and being easy to understand for users, even those who are not system experts. It facilitates meetings between AEC and IT (information technologies) professionals.

Regarding *RQ2*, the literature shows that BIM and EA could effectively work together, but the current focus is on the business and a second level, application and technology layers. BIM seems to be a hub where interactions between different actors and tools, mainly those directly linked to the project, are concentrated. EA provides a transversal view of the organization, its components and the external environment. As presented, relationships at the organizational-strategic level were not identified. It seems that the BIM and EA domains need to become more intertwined. The relationships are in the lower layers of an EA, but the strategic implications (strategy layer) can be considered an open field to explore.

Finally, answering *RQ3*, only one paper mentions the concept of megaprojects as a focus of application [51]. This work again highlights the usability of the business process approach for connecting with BIM and how BPM-BIM can scale up as the project becomes a megaproject. When the project-business landscape becomes more complex, transparency and the "lean" characteristic of the business approach are needed. However and indirectly, some projects displayed in Table 2, could be considered megaprojects, for instance, railways [34], mining [32] and airports [57]. In these three works, the scalability of systems thinking approaches such as BPM or architecture systems modelling SysML is highlighted. Moreover, this management approach is recognized as an enabler for digital transformation in the AEC industry.

References

1. Atencio E, Araya P, Oyarce F, Herrera RF, Rivera FM, Lozano-Galant F (2022) Towards the integration and automation of the design process for domestic drinking-water and sewerage systems with BIM. Appl Sci 12:9063
2. CNP (2020) Productividad en el sector de la construcción, Santiago de Chile
3. Ruijter H, van Marrewijk A, Veenswijk M, Merkus S (2021) 'Filling the mattress': trust development in the governance of infrastructure megaprojects. Int J Proj Manag 39:351–364
4. Collao J, Lozano-Galant F, Lozano-Galant JA (2021) BIM visual programming tools applications in infrastructure projects: a state-of-the-art review. Appl Sc 11:8343
5. Alreshidi E, Mourshed M, Rezgui Y (2017) Factors for effective BIM governance. J Build Eng 10:89–101
6. Eastman CM (1975) The use of computers instead of drawings in building design. AIA J 63(3):46–50
7. Loyola M (2019) Encuesta Nacional BIM 2019, Santiago, Chile
8. Smits W, van Buiten M, Hartmann T (2017) Yield-to-BIM: impacts of BIM maturity on project performance. Build Res Inf 45(3):336–346
9. Liang C, Lu W, Rowlinson S, Zhang X (2016) Development of a multifunctional BIM maturity model. J Constr Eng Manag 142(11):06016003
10. Rashidian S, Drogemuller R, Omrani S (2022) The compatibility of existing bim maturity models with lean construction and integrated project delivery. J Inf Technol Constr 27:496–511
11. Atencio E, Bustos G, Mancini M (2022) Enterprise architecture approach for project management and project-based organizations: a review. Sustainability 14:9801
12. Nyarirangwe M, Babatunde OK (2019) Megaproject complexity attributes and competences: lessons from it and construction projects. Int J Inf Syst Proj Manag 7:77–99
13. Mikkelsen MF, Venable J (2020) Researching navigation of project complexity using action design research. Int J Manag Proj Bus 14(1):108–130
14. Niemi E, Pekkola S (2020) The benefits of enterprise architecture in organizational transformation. Bus Inf Syst Eng 62:585–597
15. Hladik M (2012) Application of enterprise architecture supported by organization complexity and resource-based view. In: Soliman KS (ed) Innovation and sustainable competitive advantage: from regional development to world economies, Norristown, USA, vol 1–5. International Business Information Management Association, pp 254–1261
16. Lankhorst M (2009) Enterprise architecture at work. Springer, Berlin
17. Houy C, Fettke P, Loos P, Houy C, Fettke, P, Loos P (2011) Empirical research in business process management – analysis of an emerging field of research. Bus Process Manag J 16(4):619–661
18. Gonzalez-Lopez F, Bustos G (2019) Business process architecture design methodologies – a literature review. Bus Process Manag J 25:1317–1334
19. Miterev M, Turner JR, Mancini M (2017) The organization design perspective on the project-based organization : a structured review. Int J Manag Proj Bus 10:527–549
20. Atencio E, Mancini M, Bustos G (2022) An ontology for project-based organization design: the star model case. In: 5th IEOM European conference on industrial engineering and operations management. IEOM, Rome

21. State P (2022) Penn state BIM uses. https://bim.psu.edu/uses/. Accessed 07 Oct 2022
22. Muñoz-La Rivera F, Vielma JC, Herrera RF, Carvallo J (2019) Methodology for building information modeling (BIM) implementation in structural engineering companies (SECs). Adv Civ Eng
23. Gonzalez-Lopez F, Bustos G (2019) Integration of business process architectures within enterprise architecture approaches: a literature review. EMJ - Eng Manag J 31:127–140
24. Galaz-Delgado EI, Herrera RF, Atencio E, Rivera FML, Biotto CN (2021) Problems and challenges in the interactions of design teams of construction projects: a bibliometric study. Buildings 11(10):461
25. OMG (2022) Decision, model and notation specification. https://www.omg.org/dmn/. Accessed 10 Oct 2022
26. Malagnino A, Mangialardi G, Zavarise G, Corallo A (2018) Process modeling for historical buildings restoration: an innovation in the management of cultural heritage. Acta IMEKO 7:95–103
27. Malagnino A, Mangialardi G, Zavarise G, Corallo A (2019) Business process management and building information modeling for the innovation of cultural heritage restoration process. In: IMEKO international conference on metrology for archaeology and cultural heritage, MetroArchaeo 2017. IMEKO, Lecce, pp 184–189
28. D'Angelo L, Hajdukiewicz M, Seri F, Keane MM (2022) A novel BIM-based process workflow for building retrofit. J Build Eng 50:104163
29. Likhitruangsilp V, Ioannou P, Leeladejkul S (2014) Mapping work process and information exchange of construction entities for BIM implementation: case study of an academic institute. In: Computing in civil and building engineering, pp 955–1865
30. Ajam M, Alshawi M, Mezher T (2010) Augmented process model for e-tendering: towards integrating object models with document management systems. Autom Constr 19:762–778
31. Kouhestani S, Nik-Bakht M (2020) IFC-based process mining for design authoring. Autom Constr 112:103069
32. Wang J, Bi L, Wang L, Jia M, Mao D (2019) A mining technology collaboration platform theory and its product development and application to support China's digital mine construction. Appl Sci 9:1–33
33. Kipping G et al (2022) How to leverage process mining in organizations - towards process mining capabilities. In: Di Ciccio C, Dijkman R, del Río Ortega A, Rinderle-Ma S (eds) Business process management. BPM 2022. Lecture notes in computer science, vol 13420. Springer, Cham, pp 40–46
34. Häußler M, Esser S, Borrmann A (2021) Code compliance checking of railway designs by integrating BIM BPMN and DMN. Autom Constr 121:103427
35. Häußler M, Borrmann A (2021) Knowledge-based engineering in the context of railway design by integrating BIM, BPMN, DMN and the methodology for knowledge-based engineering applications (MOKA). J Inf Technol Constr 26:193–226
36. Wu W, Issa RRA (2015) BIM execution planning in green building projects: LEED as a use case. J Manag Eng 31:1–18
37. Tsay GS, Staub-french S, Poirier É (2022) BIM for facilities management : an investigation into the asset information delivery process and the associated challenges. Appl Sci 12(19):9542
38. You Z, Wu C (2019) A framework for data-driven informatization of the construction company. Adv Eng Inform 39:269–277
39. Jofre C, Muñoz La Rivera F, Atencio E, Herrera RF (2021) Implementation of facility management for port infrastructure through the use of UAVs, photogrammetry and BIM. Sensors 21:6686
40. Gigante-Barrera Á, Ruikar D, Crunden M, Ruikar K (2017) Lod object content specification for manufacturers within the UK using the idm standard. J Inf Technol Constr 22:80–103
41. Redmond AM, Smith B (2013) Designing a cloud BIM business process model case study. AACE Int Trans 1:1–18
42. Ali AK (2019) Mapping a resource-based design workflow to activate a circular economy in building design and construction. In: IOP conference series: earth and environmental science, vol 225

43. Owerko T, Owerko P (2018) Methodology for linking business processes undertaken during bridge load testing with the specification of required information according to OpenBIM standards. In: VI international conference of science and technology INFRAEKO 2018 modern cities. Infrastructure and environment. E3S web of conferences, Krakow, Poland, p 61
44. Tsai MH, Md AM, Kang SC, Hsieh SH (2014) Workflow re-engineering of design-build projects using a BIM tool. J Chin Inst Eng Ser A 37:88–102
45. Berard O, Karlshoej J (2012) Information delivery manuals to integrate building product information into design. Electr J Inf Technol Constr 17:64–74
46. Armijo A, Elguezabal P, Lasarte N, Weise M (2021) A methodology for the digitalization of the residential building renovation process through openbim-based workflows. Appl Sci 11:10429
47. D'Angelo L, Piccinini A, Seri F, Sterling R, Costa A, Keane MM (2019) BIM-based business process model to support systematic deep renovation of buildings. In: 16th IBPSA international conference and exhibition. International Building Performance Simulation Association, Rome, pp 137–144
48. Chung SW, Kwon SW, Moon DY, Lee KH, Shin JH (2019) Information exchange process for AR based smart facility maintenance system using BIM model. In: Proceedings of the 36th international symposium on automation and robotics in construction, ISARC 2019. The International Association for Automation and Robotics in Construction, Banff, pp 595–602
49. Chassiakos AP, Karatzas SK (2019) Process centric building energy management-as-a-service. In: International conference on smart infrastructure and construction 2019, ICSIC 2019: driving data-informed decision-making, Cambridge, UK, pp 11–20
50. Tian B, Liu K (2015) Semiotic approach to building information modelling based service model. In: Raidén A, Aboagye-Nimo E (eds) Proceedings of the 31st annual association of researchers in construction management conference, ARCOM 2015. Association of Researchers in Construction Management, Lincoln, pp 721–730
51. Wilczynski K (2018) Process modeling. In: Rheology in polymer processing. Springer, Cham, pp 171–221
52. Tsou JY, Li X, Tsou K, He J, Pan D (2018) Advanced computing strategies for engineering. In: 25th EG-ICE international workshop 2018, 2018 proceedings, part II, Lausanne, Switzerland, 10–13 June 2018, pp 535–553
53. Zanni M, Sharpe T, Lammers P, Arnold L, Pickard J (2019) Developing a methodology for integration of whole life costs into BIM processes to assist design decision making. Buildings 9:114
54. Parti R, Hauer S, Monsberger M (2019) Process model for BIM-based MEP design. In: IOP conference series: earth and environmental science, vol 323
55. Correa FR, Maciel AR (2018) A methodology for the development of interoperable bim-based cyber-physical systems. In: ISARC 2018 - 35th international symposium on automation and robotics in construction and international AEC/FM hackathon: the future of building things. The international association for automation and robotics in construction, Berlin, Germany
56. Asare KAB, Ruikar KD, Zanni M, Soetanto R (2020) BIM-based LCA and energy analysis for optimised sustainable building design in Ghana. SN Appl Sci 2:1–20
57. Keskin B, Salman B, Koseoglu O (2022) Architecting a BIM-based digital twin platform for airport asset management: a model-based system engineering with SysML approach. J Constr Eng Manag 148:1–15
58. Munir M, Kiviniemi A, Jones SW, Finnegan S (2020) The business value of BIM for asset owners: a cross-case analysis. J Facil Manag 18:469–486
59. Bider I, Johannesson P, Perjons E (2013) Do workflow-based systems satisfy the demands of the agile enterprise of the future? In: La Rosa M, Soffer P (eds) Business process management workshops. BPM 2012. Springer, Heidelberg, pp 59–64

Pillar Two and Tax Competition: What is the Future of Tax Incentives?

Paolo Arginelli and Francesco Reboli

Abstract This paper analyzes, from a theoretical perspective, the evolution of the use of tax incentive and considers the foreseeable implications that the Pillar Two Proposal could have in their respect. During the last two decades, the fight against (harmful) tax competition became indeed an overriding challenge for both the OECD and the EU institutions, which aimed at contrasting profit shifting to low or no tax jurisdictions. Such concerns were also fostered by the rapidly digitalizing economy that prompted governments to reach an international agreement on a "Two-pillars" solution. However, both the speed of implementation and some political compromises inherent in the Solution cast some shadows on its possible developments and interactions with the international tax framework.

Keywords GloBE Proposal · Pillar Two · Tax Incentives · Harmful Tax Competition · BEPS Project

1 Introduction

Tax incentives represent an important policy tool, which is widely used by the governments all over the world in pursuit of a variety of goals, ranging from stimulating the economic growth and development, to encourage certain behaviors or even to address market failures (like for the R&D incentives or those addressing Start-ups). Moreover, especially in the last few years, many governments have introduced tax incentives to support businesses during unfavorable conjunctural moments, for instance through the COVID-19 pandemic. At the same time, for many developing and emerging economies, tax incentives play a central role as part of their industrial and economic development strategy, in order to promote certain activities, sectors, investments in

P. Arginelli (✉)
Università Cattolica del Sacro Cuore, Piacenza, Italy
e-mail: Paolo.Arginelli@unicatt.it

F. Reboli
Pirola Pennuto Zei e Associati, Milan, Italy
e-mail: Francesco.Reboli@studiopirola.com

E. Favari and F. Cantoni (eds.), *Complexity and Sustainability in Megaprojects*, Lecture Notes in Civil Engineering 342,
https://doi.org/10.1007/978-3-031-30879-6_14

certain goods or in specific locations and indeed to attract foreign direct investment (hereinafter "FDI"). Nevertheless, some incentives have proven to be less effective, failing to achieve their intended goals, and becoming subject of an intense debate.[1]

Against this background, many scholars and public institutions have underlined how the proliferation of tax incentives, together with an increasingly globalized and interconnected world, have fostered a though tax competition, leading governments to reduce over-time the effective tax rates (the so-called "*race to the bottom*").[2] Such a phenomenon - and more specifically the harmful tax competition - has been addressed by both the OECD and the EU since the late nineties,[3] whose commitment spanned for more than 20 years, placing important milestones in the international tax framework such as the BEPS project and the current "Two-Pillars" proposal. Under the assumption that both digitalization and tax competition (fostered by the globalization) have created opportunities for the multinational enterprises (hereinafter "MNEs") to arrange their businesses to minimize their global taxes by shifting profits to low tax jurisdictions, in October 2021, a political landmark agreement was reached among 137 countries part of the OECD/G20 Inclusive Framework (hereinafter "IF") on BEPS to address the "twin challenges" of globalization and digitalization. Such political agreement is based on a set of rules (hereinafter "GloBE Rules", which are part of the wider "Pillar Two" set of rules) that introduces the so-called "*global minimum tax*" aimed at mitigating tax competition and discouraging profit shifting purported to achieve low or zero taxes.

Evaluating the compatibility and the interaction of tax incentives with Pillar Two rules will represent a key challenge for many countries. Although those Rules do not lead to harmonized tax rates, they set a floor on tax competition (which of course will not be eliminated) mitigating its effects and potentially have an impact on the effectiveness of tax incentives.

---

[1] *See* R. Gropp, K. Kostial, *The Disappearing Tax Base: Is Foreign Direct Investment Eroding Corporate Income Taxes?*, IMF Woking Paper WP/00/173, Washington, D.C., 2000; James, S. (2013), "*Effectiveness of Tax and Non-Tax Incentives and Investments: Evidence and Policy Implications*", SSRN Electronic Journal, https://doi.org/10.2139/SSRN.2401905.

[2] *See* T. Piketty, Il capitale nel XXI secolo, Milano, 2014, p. 776; OECD (2013), Action Plan on Base Erosion and Profit Shifting, OECD Publishing. http://dx.doi.org/10.1787/9789264202719-en, p. 17 ff.; M.P. Devereux, R. Griffith, *Evaluating tax policy for location decisions*, 10 International Tax and Public Finance 3(2003); OECD (2022), Tax Incentives and the Global Minimum Corporate Tax: Reconsidering Tax Incentives after the GloBE Rules, OECD Publishing, Paris. Available at: https://doi.org/10.1787/25d30b96-en.

[3] OECD, *Harmful Tax Competition – An Emerging Global Issue*, OECD, Paris (1998). Available at: https://www.oecd.org/ctp/harmful/1904176.pdf; Commission of the European Communities, *Communication From the Commission to the Council towards tax co-ordination in the European Union*, COM(97)495final, Brussels, 01.10.1997; European Union Council of Economics and Finance Ministers (ECOFIN), conclusions of 1 December 1997. *See* also in literature, M. Keen & K. Konrad, *The Theory of International Tax Competition and Coordination*, Max Planck Institute for Tax Law and Public Finance, Working Paper 06 (2012); A. Contrino, *La concorrenza fiscal tra gli Stati nel contesto della governance economica europea*, Riv. Dir. Trib. (Suppl. On.), 24 novembre 2022.

Thus, the vast landscape of tax incentives will probably undergo significant changes in the next future and governments all over the world will implement a reassessment of their tax policies in this respect.

It is well known that tax incentives, especially those related to capital investments, can play a relevant role as a stimulus also for Mega-Projects. Innovation, advanced skills and cutting-edge technologies are key components for all the entities involved in such Projects, shaping their effectiveness and efficiency.[4] Indeed, as depicted in the following section, tax incentives may lead to positive externalities and serve as a spur for the firms by easing their burden for such investments. An analysis of current changing tax framework addressing the possible interaction between the "Pillar Two" proposal and tax incentives may, thus, help to understand to what extent the latter can be still considered a valid stimulus tool to promote and support Mega-Projects.

Against this background, the chapter is divided into six main sections and is aimed at describing the current state of play of tax competition and the possible impacts triggered by the Pillar Two proposal in respect of tax incentives. Section 2 summarizes the wide range of tax incentives currently granted by jurisdictions all over the worlds. Section 3is dedicated to the analysis of the evolution of the last 20 years OECD's works on harmful tax competition, focusing on the current state of play. Then Sect. 4 describes the main rules and the general functioning of GloBE rules, while Sect. 5 focuses on their interaction with tax incentives in order to understand which tax incentives and to what extent are likely to be more affected by these new provisions. Section 6 concludes.

2 The Tax Incentives Landscape: A Brief Overview

Lacking a consensual view on the definition for "tax incentives", such term can be generally referred to a favorable tax treatment or a reduction of the overall tax burden for a taxpayer/investor,[5] as compared to a general reference framework. More in detail, the design of tax incentives could vary significantly across regions and countries.[6] As recently highlighted by the OECD, reduced tax rates and tax holidays

---

[4] *See* P. Arginelli, *The Relevance and Limits of Taxation as an Aid to Foster Megaprojects,* in *Sustainability and Megaproject Development* (F. Cantoni, E. Favari Ed., Giappichelli, 2022) p. 89 ff.

[5] Despite their spread, there is no consensus between scholars on the definition of the term "tax incentive". In this respect, examples of definitions that scholars have provided are the following: "Special provisions that allow for exclusions, credits, preferential tax rates or deferral of tax liability"; "Tax incentives can be defined in terms of their effect on reducing the effective tax burden for a specific project"; "They are exceptions to the general tax regime"; "All measures that provide for a more favourable tax treatment of certain activities or sectors compared to what is granted to the general industry". See L. G. Ogazón Juárez, D. Calderón Manrique, in M. Cotrut, K. Munyandi, *Tax Incentives in the BEPS Era* - Part One: Introduction - Chapter 1: Introduction to Tax Incentives in the BEPS Era, IBFD Tax Research Series, March 2018. *See* also A. Dagnino, *Agevolazioni fiscali e potestà normativa*, Padova, 2008, pp. 1–41.

[6] *See* OECD (2022), *Tax Incentives and the Global Minimum Corporate Tax*, Supra, 4.

are more common among developing countries, while developed countries appear to rely more frequently on "expenditure-based" tax incentives, such as accelerated depreciation, investment tax allowances or credits, than on "broad-based" income tax exemptions.[7] Nonetheless, according to the OECD, foreign affiliates of MNEs can face "*ETRs well below 15%, even in jurisdictions with high statutory tax rates*".[8]

In this respect, literature has – broadly speaking – identified the general categories of tax incentives set out below:

- "*Tax holidays*" regimes are generally shaped as a time-limited exemption from taxation (or a reduced tax rate), with the objective of attracting foreign direct investment (FDI) to sustain and stimulate the economic growth and the development.[9] The specific time limitation represents their distinctive element from. Tax holidays are generally targeted at certain types of investments, activities, or companies desirable for a (in general, developing) country. Despite their spread, they are often criticized by many international bodies by reason of their alleged ineffectiveness and inefficiency and because of their potentially distortive effects.[10]
- "*Capital investment incentives*", can take the form both of financial incentives (e.g., cash grants, loans at below-market interest rates, loan guarantees or rebates) and tax-based incentives aiming at decreasing the cost of certain investments like those in capital assets or the expenditure in research and development. Some examples are investment allowances (e.g., tax depreciation, accelerated depreciation and enhanced deductions), front-end incentives (e.g., R&D tax credits) and back-end incentives (e.g., IP box regimes). The rationale behind this policy considers that the economic benefits of those investments and, in particular, investments in innovation can spill over without charge. According to classical economic theory, markets will tend to underinvest in R&D, as they will not factor in all the benefits of the positive externalities caused by the R&D spillovers and, as a result, they will not generate the social optimal level of R&D. In addition, other market failures connected to innovation derive from the market inefficiencies caused by the

---

[7] *See* OECD (2022), Ibid.;also in literature: Celani A., Dressler L. and Hanappi T., *Corporate Effective Tax Rates in the presence of Investment Tax Incentives*, OECD Publishing, Paris. González Cabral, A., S. Appelt and T. Hanappi (2021), *Corporate Effectives Tax Rates for R&D: The case of expenditure-based tax incentives*, OECD Taxation Working Papers, No. 54, OECD Publishing, Paris, https://dx.doi.org/10.1787/ff9a104f-en; B. Ferreira Liotti, J. Waruguru Ndubai, R. Wamuyu, I. Lazarov, J. Owens, *The treatment of tax incentives under Pillar Two*, UNCTAD, World Investment Report, June 2022, p. 136 ff. Available at https://papers.ssrn.com/sol3/papers.cfm?abstract_id=4132515.

[8] *See* OECD (2022), Tax Incentives and the Global Minimum Corporate Tax, Supra, 11 et seq.

[9] *See* C.E. McLure Jr., *Tax Holidays and Investment Incentives a comparative analysis*, Bulletin of International Taxation, 1999; H.M. Bjerkestuen, H.G. Wille, *Tax holidays in a BEPS-perspective*, Intertax, Vol. 43, Issue 1, p. 106 ff..

[10] G. Mascagni, M. Moore, R. McCluskey, *Tax Revenue Mobilisation in Developing Countries: Issues and Challenges, Report to the European Parliament's Committee on Development*, 2014; A. Klemm, S. Van Parys, *Empirical Evidence on the Effects of Tax Incentives*, IMF Working Paper WP/09/136, 2009.

existence of intellectual property rights, and from the asymmetric information in the capital market, which may lead small and innovative companies to pay higher interest rates and to have a narrower access to the capital market. These potential market failures may justify the policy of subsidizing private investments in innovation, for the purpose of increasing the national welfare.[11] Furthermore, innovation is considered a prerequisite for strong economic growth, global competitiveness and social welfare[12] and the importance of innovative activities is accentuated during an economic crisis, as persistent R&D performers seem to survive crisis better than their competitors. Hence, investments by businesses in R&D are "*pro-cyclical on aggregate, and thus, apt to contracting in crisis*". Therefore, it is of primary importance for governments to provide a nourishing environment to R&D performing firms in general and hold tools ready to strengthen R&D in times of crisis, such as the COVID-19 pandemic, or to sustain growth in developing areas.

– "*Reduced tax rates*" are referred to as incentives granting a lower tax rate than the one generally applicable, to specific types of companies, sectors, regions or income sources. If a country grants a low tax rate to all companies, that low tax rate does not qualify as a reduced tax rate incentive but is rather a low general corporate income tax rate regime. Thus, such incentives constitute an "*exception to the general tax regime*" or policy of a state (and not a general reduction in the tax rate)[13]; examples are the regimes encouraging investments in new companies (e.g., venture capital companies) by way of reduced tax rates.
– "*Special economic zones*" regimes in general refer to specific areas (separated zones) where a jurisdiction, aimed at developing the economy therein or to attract

[11] P. Arginelli, *Innovation through R&D Tax Incentives: Some Ideas for a Fair and Transparent Tax Policy*, 7 World Tax J. 1 (2015), Journal Articles & Opinion Pieces IBFD; P. Arginelli, Chapter 5: The Interaction between IP Box Regimes and Compensatory Tax Measures: A Plea for a Coherent and Balanced Approach in EU Law and the Building of Global Supranational Tax Law: EU BEPS and State Aid (D. Weber ed., IBFD 2017).

[12] As highlighted by C. Spengel, B. Stage and D. Steinbrenner "*Empirical research finds that increasing research and development (R&D) activities by 1% of GDP increase economic growth by 0.32% to 1.18%*" (*See* C. Spengel, B. Stage, D. Steinbrenner, R&D Tax Incentive Regimes – A Comparison and Evaluation of Current Country Practices, World Tax Journal, 2022 (Volume 14), No. 2). Contrarily, according to an OECD Working Paper of 2008 (Å. Johansson et al., Tax and Economic Growth, Economics Department Working Paper No. 620, ECO/WKP(2008)28 (OECD 2008), p. 38), empirical studies using industry-level data support previous findings in that, although tax incentives for R&D appear to enhance productivity, their effect relative to best practice-level seems to be rather small. *See* also European Commission Taxation Papers, Working Paper no. 52, A Study on R&D Tax Incentives, Final Report (2014); P. Palazzi, Taxation and Innovation, OECD Taxation Working Papers, no. 9, OECD Publishing (2011), p. 3. http://dx.doi.org/10.1787/5kg3h0sf1336-en; OECD (2013), Supporting Investment in Knowledge Capital, Growth and Innovation, OECD Publishing, p. 17, available at http://dx.doi.org/10.1787/9789264193307-en.

[13] L. Gerzova & M. Olejnicka, *Chapter 4: Reduced Tax Rates at a Crossroads: Before and After the OECD BEPS Project in Tax Incentives in the BEPS Era* (M. Cotrut et al. eds., IBFD 2018); United Nations Conference on Trade and Development, Tax Incentives and Foreign Direct Investment: A Global Survey, ASIT Advisory Studies No. 16 (2000).

FDI, grants tax incentives and customs alleviations.[14] In general, *"[t]he term special tax zone applies to areas where tax regulations are more beneficial than in the generally applicable tax system of the surrounding jurisdiction or country"*.[15] SEZs are often accompanied by specific eligibility conditions aimed at targeting taxpayers with certain characteristics, certain types of income or expenditure or even specific economic activities. In general, SEZs involve real business activities (such as manufacturing, logistics and ports, as well as trade and business service companies) with local employees, being focalized on the development of a certain area. The development is fostered by specific provision of tax incentives, subsidies, a corporate-friendly administration, or a plentiful supply of labor or natural resources. Examples of SEZs are the zones that restrict their operations to foreign companies and activities, whereas entities in other SEZs may purchase their supplies and services from domestic providers and sell their products to domestic markets. Moreover, VAT or customs duties may not be charged in these zones and there is free movement of capital and goods for export and import purposes.

Considering the above and the aims of the analysis set out below, hereinafter the term "incentive" assumes and is used in a broad sense.

3 OECD Countering Harmful Tax Competition

More than two decades have passed since the OECD published the 1998 report on *Harmful Tax Competition: An Emerging Global Issue* (the "1998 Report"), seeking to identify the "harmful" elements of tax competition and pursuing a minimum "level playing field".[16] Indeed, tax competition was not a new phenomenon in 1998, but the

---

[14] Literature also identify SZTs as "economic and development zones", "enterprise zones", "free economic zones", "free zones", "tax-free zones", "export processing zones" and "industrial development zones".

[15] *See* A. Laukkanen, *Chapter 5: Special Economic Zones: The Acceptance of Tax Incentives in the BEPS World,* in *Tax Incentives in the BEPS Era* (M. Cotrut et al. eds., IBFD 2018), Books IBFD; A. Laukkanen, *The Development Aspects of Special Tax Zones,* 70 Bull. Intl. Taxn. 3, p. 152 (2016); A. Laukkanen, P. Pistone, J. de Goede, *Special Tax Zones in the Era of International Tax Coordination* (IBFD ed. 2019); R. Biçer, *An Assessment of Free Trade Zones from a Transfer Pricing Perspective,* 15 Intl. Transfer Pricing J. 5, at sec. 2 (2008); C. Cipollini, *The Concept of Special Tax Zones in EU Tax Law,* EC Tax Review, Vol. 28, Issue 6 (2019) pp. 307 – 321; R. S. Avi-Yonah & M. Vallespinos, *Special Tax Zones and the WTO,* University of Michigan Public Law Research Paper no. 545 (2017), https://ssrn.com/abstract=2928644; W. De Jong, *Establishing Free Zones for Regional Development,* Library of the European Parliament (2013), http://www.europarl.europa.eu/RegData/bibliotheque/briefing/2013/130481/LDM_BRI(2013)130481_REV1_EN.pdf.

[16] P. Schoueri: *Conflicts of International Legal Frameworks in the Area of Harmful Tax Competition,* chapter 1 – introduction, WU - Tax Law and Policy Series, 2019.

issue of the 1998 Report was rather a key turning point,[17] representing the seminal work of the OECD in the area of harmful tax practices and laid down the basis for the subsequent actions of the ad hoc created Forum on Harmful Tax Practices.[18]

The OECD action was indeed a response to the concerns raised by governments of its Member States with regard to the harmful tax practices put in place by member and non-member countries in respect of geographically mobile activities (e.g., financial activities, provision of services, licensing of intangibles), which could be easily shifted from one country to another, thus leading to a corresponding substantial shift of taxable profits.[19] Indeed, the 1998 report was limited in scope to preferential tax regimes aimed to attract geographically mobile activities – fostered by globalization and technological innovation - such as financial and other service activities (e.g., insurance, leasing, banking), but it did leave out the analysis of regimes intended to attract active investment in plant, or manufacturing.[20]

At the same time, the European Union started to move its first steps towards a more defined policy in respect of tax competition with the 1997 Communication from the Commission to the Council "*A package to tackle harmful tax competition in the European Unio*n". Furthermore, the EU Code of Conduct for business taxation was adopted in the conclusions of the Council of Economics and Finance Ministers (ECOFIN) of 1 December 1997.[21]

Since then, many further steps have been taken and new progress reports have been issued.[22] Furthermore, the Global Forum on Transparency and Exchange of

---

[17] R. Saw, *Tax Competition: And the Winner Is...*, in Cotrut M., Munyandi K., Tax Incentives in the BEPS Era, IBFD Tax Research Series, march 2018; T. Katsushima: *Harmhul tax competition*, Intertax, Vol. 27, Issue 11, 1999, p. 396.

[18] P. Arginelli, Innovation through R&D Tax Incentives, Supra, p. 3 ff.

[19] OECD (2014), Countering Harmful Tax Practices More Effectively, Taking into Account Transparency and Substance, OECD/G20 Base Erosion and Profit Shifting Project, OECD Publishing. *See* also in literature: R.S. Avi-Yonah, *The OECD Harmful Tax Competition Report: A Tenth Anniversary Retrospective*, Brook. J. Int'l L. 34, no. 3 (2009), 783–95; Lampreave Márquez, Fiscal Competitiveness versus Harmful Tax Competition in the European Union, 65 Bull. Intl. Taxn. 6, sec. 2. (2011); P. Piantavigna, Tax Competition and Tax Coordination in Aggressive Tax Planning: A False Dichotomy, 9 World Tax J. 4, sec. 2. (2017).

[20] In particular, the OECD stated that "*[t]he accelerating process of globalistion of trade and investment has fundamentally changed the relationship among domestic tax systems. [...] the removal of non-tax barriers to international commerce and investment and the resulting integration of national economies have greatly increased the potential impact that domestic tax policies can have on othereconomies. Globalisation has also been one of the driving forces behind tax reforms, which have focused on base broadening and rate reductions, thereby minimising tax induced distortions*". *See* J. Owens, *Curbing Harmful Tax Competition – Recommendations by the Committee of Fiscal Affairs*, Intertax, Vol. 26, Issue 8–9, p. 230 ff.

[21] Conclusions of the ECOFIN Council Meeting on 1 December 1997 concerning taxation policy - Resolution of the Council and the Representatives of the Governments of the Member States, meeting within the Council of 1 December 1997 on a Code of Conduct for Business Taxation - Taxation of saving, 98/C 2/01. Available at: https://taxation-customs.ec.europa.eu/system/files/2016-09/coc_en.pdf; *See* also, C. M. Radaelli, Harmful Tax Competition in the EU: Policy Narratives and Advocacy Coalitions, Journal of Common Market Studies 1999, 37, no. 4, 661–682.

[22] *See* OECD, *Towards Global Tax Co-operation: Progress in Identifying and Eliminating Harmful Tax Practices* (OECD 2001); OECD, *The OECD's Project on Harmful Tax Practices: The 2001*

Information for Tax Purposes (the Global Forum) was formed in 2000 and the Final Report on Action 5 of the OECD BEPS Project was issued in 2015, revamping the OECD's work in the area of harmful tax practices.[23] In the BEPS Action Plan, the OECD explicitly recognized that preferential regimes continue to be a key pressure area, prompting significant base erosion and profit shifting, and that the underlying policy concerns over geographically mobile income expressed in the 1998 Report are still relevant today. However, as mentioned by literature "*[...] despite the impressive convergence of norms and standardization manifested in the international tax regime, it is a competitive, beggar-thy-neighbor approach that has been guiding the norms themselves*".[24] At the same time, the OECD acknowledged that harmful preferential tax regimes less often took the form of traditional ring-fencing measures and more often took the form of across-the-board effective tax rate reductions on particular types of income, in particular income from the provision of intangibles benefiting from IP box regimes. To counteract such forms of harmful tax practices, the OECD highlighted the need to develop more effective solutions, taking into account transparency and substance.[25]

Such concern has evolved over the years and became both more complex and increasingly intertwined with the challenges arising from the digitization of the economy.[26] Digital and dematerialized economies have fostered harmful tax competition among states and the latter has in turn incentivized aggressive tax planning. MNEs deriving substantial income from countries, often through digital activities, without any substantial taxable presence therein and thereby avoiding the payment of substantial taxes to those countries soon became one of the main concerns of the OECD, implicitly putting the blame – "in a paternalistic way" as suggested by

Progress Report (OECD 2002); OECD, *Harmful Tax Practices: The 2004 Progress Report* (OECD 2004); and OECD, *The OECD's Project on Harmful Tax Practices: 2006 Update on Progress in Member Countries* (OECD 2006).

[23] OECD, *Countering Harmful Tax Practices More Effectively, Taking into Account Transparency and Substance, Action 5* – Final Report (OECD 2015); E. Traversa, A. Flamini, *The impact of BEPS on the fight against harmful tax practices: risks... and opportunities for the EU*, British Tax Review 3, pp. 396–407 (2015).

[24] *See* Y. Brauner, *What the BEPS,* 16(2) Fla. Tax Rev. 55 (2014), 76 et seq.

[25] *See* P. Arginelli, Supra, p. 50.

[26] *See* A.P. Dourado, *Digital Taxation Opens the Pandora Box: The OECD Interim Report and the European Commission Proposals*, Intertax, Volume 46, Issue 6 & 7, p. 565 ff.; A.P. Dourado, *Taxes and Competitiveness: How Much Competitive Is European Tax Competition?*, Intertax, Volume 46, Issue 12, 2018; A.P. Dourado: *Taxing the Digital Economy*, Intertax, Volume 47, Issue 2, 2019; A. Perrone, *Tax competition e giustizia sociale nell'unione europea, Padova*, 2019; S. Cipollina, *I redditi "nomadi" delle società multinazionali nell'economia globalizzata*, in Riv. dir. fin. sc. fin., 2014, I, 21 ss., par. 4; L. Carpentieri, *La crisi del binomio diritto-territorio e la tassazione delle imprese multinazionali*, in Riv. dir. trib., 2018, I, 351 ss.; T. Gasparri, *Stati sovrani ed imprese multinazionali alla sfida del Fisco, tra sostanza e trasparenza*, in Note e Studi Assonime, n. 15/2017; S. Micossi, *La fiscalità d'impresa nel nuovo mondo globalizzato*, in Note e Studi Assonime, n. 1/2017; Avi-Yonah R.S., *Globalization, Tax Competition, and the Fiscal Crisis of the Welfare State*, Harv. L. Rev. 113, no. 7 (2000), p. 1573–676.

Arnold[27] - on such countries for the "race to the bottom" phenomenon. BEPS project made important improvements in addressing low and zero taxation, increased tax transparency, ensuring that multinational enterprises (MNEs) would need substance in profit-earning countries, and put in place multilateral mechanisms for dispute resolution. However, according to the OECD, the international tax framework has remained under stress due to increasing globalization and the digitalization of the economy, soon becoming known as the "residual BEPS challenges".[28]

In this regard, since the very beginning of the BEPS Project in 2012, the OECD/G20 has been working on digital transformation, leading to the 2015 BEPS Action 1 Report.[29] As remarked by the OECD this phenomenon "*[…] spurs innovation, generates efficiencies, and improves services while boosting more inclusive and sustainable growth and enhancing well-being, but at the same time, the breadth and speed of this change introduces challenges in many policy areas, including taxation*".[30] However, such challenges remained unaddressed until March 2018, when the Inclusive Framework on BEPS (hereafter, also "IF") released an Interim Report on Tax Challenges Arising from Digitalization which recognized the need for a global solution,[31] based on the fact that the corporate tax systems still not fully reflect the realities of economy and global developments as they are still mainly based on the principles of local brick-and-mortar production.[32]

---

[27] B.J Arnold, *The evolution of controlled foreign corporations rules and beyond*, in Bull. Int. Tax., 2019, 73, 644.

[28] OECD (2018), *Tax Challenges Arising from Digitalisation – Interim Report 2018: Inclusive Framework on BEPS, OECD/G20 Base Erosion and Profit Shifting Project*, OECD Publishing, Paris http://dx.doi.org/10.1787/9789264293083-en; World Bank: *The Global Minimum Tax: from agreement to implementation. Overview booklet. Information and Communications for Development.* Washington, D.C., 2022.

[29] OECD (2015), Addressing the Tax Challenges of the Digital Economy, Action 1 - 2015 Final Report, OECD/G20. Base Erosion and Profit Shifting Project, OECD Publishing, Paris. http://dx.doi.org/10.1787/9789264241046-en.

[30] OECD (2020), Tax Challenges Arising from Digitalisation – Report on Pillar Two Blueprint: Inclusive Framework on BEPS, OECD/G20 Base Erosion and Profit Shifting Project, OECD Publishing, Paris, https://doi.org/10.1787/abb4c3d1-en.

[31] OECD (2018), Tax Challenges Arising from Digitalisation, Supra.

[32] In particular, the OECD remarked the fact that "*[w]hile the measures set out in the BEPS package have further aligned taxation with value creation and closed gaps in the international tax architecture that allowed for double non-taxation, certain members of the Inclusive Framework consider that these measures do not yet provide a comprehensive solution to the risk that continues to arise from structures that shift profit to entities subject to no or very low taxation […]*". (*See* OECD (2019), Addressing the Tax Challenges of the Digitalisation of the Economy – Policy Note, as approved by the Inclusive Framework on BEPS on 23 January 2019, OECD, Paris, www.oecd.org/tax/beps/policy-note-beps-inclusive-framework-addressing-tax-challengesdigitalisation.pdf; OECD, *Programme of Work to Develop a Consensus Solution to the Tax Challenges Arising from the Digitalisation of the Economy, Inclusive Framework on BEPS (2019)*, https://www.oecd.org/tax/beps/programme-of-work-to-develop-a-consensus-solution-to-the-tax-challenges-arising-from-the-digitalisation-of-the-economy.pdf).

The International Monetary Fund stated that: "*The G-20/OECD project on Base Erosion and Profit Shifting (BEPS) has made significant progress in international tax cooperation, addressing*

The works on a solution continued through the OECD/G20 Inclusive Framework on BEPS. However, the lack of progress, pushed many countries to choose unilateral action, often adopting gross-based digital services taxes (hereinafter "DSTs").[33] The spread of unilateral measures spurred the IF to intensify its work, and in 2019 it announced that agreement had been reached on a "Two-pillars" solution. While Pillar One is aimed at reallocating tax revenues to the country of the consumer, Pillar Two introduces a global minimum effective tax for MNEs. In this regard the OECD identifies the Pillar Two proposal's rationale in addressing the "*remaining BEPS risk of profit shifting to entities subject to no or very low taxation*",[34] based on the axioms that all counties have the right to "*tax back profits where other countries have not sufficiently exercised their primary taxing rights*".[35] Indeed, Pillar Two puts forward a comprehensive and coordinated solution to ensure that multinational enterprises are subject to a minimum tax rate[36]. Throughout 2019 and 2020, IF works had continued, culminating in the publication of detailed "Blueprints" for both pillars in October 2020 (the "2020 Blueprints)".[37] These goals were also fully embraced by the European Union, which at the same time published a Directive Proposal for a European digital service tax addressing the "*problem that the current corporate tax rules are inadequate for the digital economy*".[38] In addition, with the Communication on Business Taxation for the 21st Century, the European Union acknowledged that

---

some major weak points in the century-old architecture. But vulnerabilities remain. [...] Tax competition remains largely unaddressed. And concerns with the allocation of taxing rights across countries continue. Recent unilateral measures, moreover, jeopardize such cooperation as has been achieved" (*See* International Monetary Fund, *Corporate Taxation in the Global Economy, Policy Paper No 19/007*, Washington D.C., 2019, Available at:: https://www.imf.org/en/Publications/Policy-Papers/Issues/2019/03/08/Corporate-Taxation-in-theGlobal-Economy-46650).

See also, A.P. Dourado: *In Search of an International Tax System in a Post-BEPS Tax Competition Setting.* Intertax, 2019, 47, 2 ss.; M. Devereux, J. Vella: *Taxing the Digitalised Economy: Targeted or System-Wide Reform*, Brit. Tax Rev., 2018, 4, 381.

33 *See* P. Pistone, J.F. Pinto Nogueira, A. Turina, *Digital Services Tax: Assessing the Policy Reasons for its Introduction in the European Union,* International Tax Studies, 2021 (Volume 4), No. 7.

34 OECD, Programme of Work to Develop a Consensus Solution to the Tax Challenges Arising from the Digitalisation of the Economy, 28 May 2019, p. 5 ff. Available at: https://www.oecd.org/tax/beps/programme-of-work-to-develop-a-consensus-solution-to-the-tax-challenges-arising-from-the-digitalisation-of-the-economy.pdf.

35 OECD, Addressing the Tax Challenges of the Digitalisation of the Economy, Public Consultation Document, Paris, February 2019, par. 90.

36 Id., relating to the need for a coordinated solution, OECD stated that "*[...] global action is needed to stop a harmful race to the bottom, which otherwise risks shifting taxes to fund public goods onto less mobile bases including labour and consumption, effectively undermining the tax sovereignty of nations and their elected legislators*".

37 OECD (2020), Tax Challenges Arising from Digitalisation – Report on Pillar Two Blueprint: Inclusive Framework on BEPS, OECD/G20 Base Erosion and Profit Shifting Project, OECD Publishing, Paris, https://doi.org/10.1787/abb4c3d1-en.

38 European Commission, *Proposal for a Council Directive on the common system of a digital services tax on revenues resulting from the provision of certain digital services, COM(2018) 148 final,* Brussels, 21.3.2018. Available at: https://eur-lex.europa.eu/legal-content/IT/ALL/?uri=CELEX%3A52018PC0148.

the context for EU business taxation policy has changed radically in the past year and claimed the need for ambitious reforms at international level, pushing to reach an agreement towards the "Two-pillars" solution.[39]

Finally, in October 2021, an historical political agreement on the "Two-pillars" solution was reached among 137 countries within the IF to address the twin challenges of globalization and digitalization and, on December 2021, the OECD issued model legislation - the "GloBE Model Rules" (hereinafter also "MR") - in respect of Pillar Two, whose implementation is envisaged for 2023[40–41].

The rationale behind the Pillar Two project appears to be twofold.[42] Indeed, albeit being initially conceived as an anti-BEPS instrument to curb profit shifting (complementary to the BEPS Project),[43] it is characterized by the further objective to reduce the international tax competition, not limited to abusive or harmful practices. In the view of these authors, this latter goal overwhelms the former. This second rationale, moreover,[44] reveals some inconsistencies in respect of the previous OECD's works. As highlighted by scholars,[45] the idea of introducing a global minimum taxation, setting a floor to tax competition, seems to betray both the very promise of the

[39] European Commission, Communication from the Commission to the European Parliament and the Council, Business Taxation for the 21st Century, COM(2021) 251 final, Brussels, 18.5.2021. https://taxation-customs.ec.europa.eu/system/files/2021-05/communication_on_business_taxation_for_the_21st_century.pdf. *See* also Communication from the Commission to the European Parliament and the Council, A Fair and Efficient Tax System in the European Union for the Digital Single Market, Brussels, 21 Sept. 2017, COM (2017) 547 Final, at 6.

[40] The Commentary on the Model Rules and a separate document with examples were released on 14 March 2022.

[41] Few days later, on 22 December 2021, the European Commission published a proposal to transpose the OECD's GloBE MR into EU secondary law (the "Directive"). Such proposal has been amended during the first half of 2022 (*See* the amended draft compromise text of 12 March 2022, 6975/22, Available at: https://data.consilium.europa.eu/doc/document/ST-6975-2022-INIT/en/pdf), but currently EU is still struggling to have this Directive proposal adopted. The current expectation is to reach an agreement within the end of the 2022. Otherwise, some states expressed their willingness to implement such provisions by any possible legal means within 2023 (*See Joint Statement* by France, Germany, Italy, Netherlands and Spain, 9 September 2022).

[42] P. Pistone, J. F. Pinto Nogueira, B. Andrade, A. Turina: *The OECD Public Consultation Document "Global Anti-Base Erosion (GloBE) Proposal – Pillar Two": An Assessment.* Bulletin for International Taxation, 2020 (Volume 74), No. 2.

[43] A.P. Dourado: *The Global Anti-Base Erosion Proposal (GloBE) in Pillar II*, Intertax, Volume 48, Issue 2, 2020, p. 152 ff.

[44] B. Ferreira Liotti: *Limits of International Cooperation: The Concept of "Jurisdiction Not to Tax" from the BEPS Project to GloBE*, Bulletin for International Taxation, 2022 (Volume 76), No. 2, pp. 77 ff.

[45] L.E. Schoueri: *Some Considerations on the Limitation of Substance-Based Carve-Out in the Income Inclusion Rule of Pillar Two.* 75 Bull. Intl. Tax 11 (2021); J. Englisch: *GloBE Rules and Tax Competition*, Intertax, Volume 50, Issue 12 [pre-publication]; B.J. Arnold, Supra, 631 ff.; S. Tandon: *The Need for Global Minimum Tax: Assessing Pillar Two Reform*, Intertax, Volume 50, Issue 5; J. Englisch, J. Becker: *International Effective Minimum Taxation – The GLOBE Proposal*, 11 World Tax J. 4, sec. 2.1. (2019), Journal Articles & Papers IBFD; Y. Brauner: *Agreement? What Agreement? The 8 October 2021, OECD Statement in Perspective*, Intertax, Volume 50, Issue 1, 2022; B. Ferreira Liotti, Supra, p. 79 ff.; J. Becker, J. Englisch, *Implementing an international*

OECD, apparently willing to leave the states "*free to determine their own tax system, including whether they have a corporate income tax and the level of their tax rates*"[46] and the previous OECD works on harmful tax competition, when it was supported the idea that each country is free to set its own tax rates and that low tax rates cannot automatically be a synonym of harmful tax practice.[47]

As a side effect, and relevant for the present study, the GloBE proposal - being focused on cases of no or low taxation – neutralizes the effects of tax incentives any time the level of taxation falls below the 15% minimum rate, irrespective of the existence of substantial activities.

Thus, it is evident Pillar Two takes a substantial departure from the OECD's traditional focus on harmful tax practices,[48] involving a significant change in the approach to tax competition.

4 The Pillar Two Proposal: Functioning and Architecture

As previously mentioned, GloBE has a broad scope and is aimed at reducing tax competition between jurisdictions regardless of abusive or aggressive tax practices. In this regard, the OECD has implemented, through the GloBE rules, a system of top-up taxes that raises the taxes paid on a MNE profits in a jurisdiction up to the minimum effective rate of 15%.[49]

The computation of the top-up tax liability is based upon the multi-step process set out below[50]:

---

effective minimum tax in the EU (June 23, 2021). Available at SSRN: https://ssrn.com/abstract=3892160 or http://dx.doi.org/10.2139/ssrn.3892160.

[46] *See* also OECD, *Cover Statement by the OECD/G20 Inclusive Framework on BEPS on the Reports on the Blueprints of Pillar One and Pillar Two*, OECD/G20 Inclusive Framework on BEPS (OECD 2020).

[47] According to the Action 5 Final Report (2015): "*the work on harmful tax practices is not intended to promote the harmonisation of income taxes or tax structures generally within or outside the OECD, nor is it about dictating to any country what should be the appropriate level of tax rates. Rather, the work is about reducing the distortionary influence of taxation on the location of mobile financial and service activities, thereby encouraging an environment in which free and fair tax competition can take place*" (para. 3, p. 11 ff.).

[48] C. Cipollini, *GloBE Proposal and Possible Carve-Outs: Is There a Future for Preferential Tax Regimes?*, World Tax Journal, 2020 (Volume 12), No. 2; M.P. Devereux et al., The OECD Global Anti-Base Erosion Proposal, Oxford University Centre for Business Taxation, 2020. Available at: https://www.sbs.ox.ac.uk/sites/default/files/2020-02/OECD_GloBE_proposal_report.pdf.

[49] *See* in literature L. Parada, *Taxing somewhere, no matter where: what is the GloBE proposal really about, Multinational Group Tax & Transfer Pricing News (2 September 2020)*, available at https://mnetax.com/taxing-somewhere-no-matter-where-what-is-the-globe-proposal-really-about39996.

[50] OECD (2022), Tax Incentives and the Global Minimum Corporate Tax: Reconsidering Tax Incentives after the GloBE Rules, OECD Publishing, Paris. Available at: https://doi.org/10.1787/25d30b96-en; M. Dietrich, C. Golden, *Consistency versus "Gold Plating": The EU Approach to Implementing the OECD Pillar Two*, 76 Bull. Intl. Taxn. 4 (2022).

- *Step 1*: the MNE Group determines whether it is within the scope of the GloBE rules and identifies the constituent entities (hereinafter "CE") within the Group and their location
- *Step 2*: computation of the profits of each CE
- *Step 3*: computation of the relevant taxes of each CE
- *Step 4*: calculation the Effective Tax Rate of the aggregate of all CE located in the same jurisdiction and determination of top-up tax in the event an MNE is subject to an effective tax rate below 15% in any jurisdiction
- *Step 5*: charging of a top-up tax under IIR or UTPR in accordance with agreed rule order

In a nutshell, once determined if an MNE group falls within scope of the GloBE rules (Step 1), the core of the analysis is represented by the computation of the ETR for each jurisdiction in which the MNE group operates and its comparison with the 15% threshold (Step 2, 3 and 4). The results of ETR computation may trigger a top-up tax to be allocated to the group entities as per the IIR and the UTPR (Step 5).

Step 1 – In scope MNEs (Articles 1.1.-1.5. of the GloBE MR)

The GloBE rules applies to entities and permanent establishments (hereinafter "PEs"), being Constituent Entities (hereinafter "CE"), that are part of a multinational group (MNEs Group) with a consolidated group revenue that exceeds EUR 750 million in at least two of the four fiscal years preceding the fiscal year that is being tested. However, certain types of entities are excluded from the scope of the GloBE Rules.[51]

Nonetheless, considering the possible Pillar Two effects on tax incentives, such a limited scope may result interesting and could give rise to some concern in terms of equity as further discussed below.

Step 2 – GloBE income determination (Articles 3.1.-3.5. of the GloBE MR)

The GloBE income/loss computation relies on the ultimate parent entity's consolidated financial statements before any consolidation adjustments.[52] Indeed, the use of financial accounting rules provides a more consistent outcome among jurisdictions to determine the GloBE tax base if compared to the use of local tax rules.

The financial accounting net income or loss is then subject to certain adjustments to better align financial accounts and tax accounting.[53] The financial accounting net

[51] Under Article 1.5. of the MR are considered excluded entities Governmental Entities, International Organizations, Non-profit Organizations, and Pension Funds. Investment Funds and Real Estate Investment Vehicles are also excluded if they are the UPE of the MNE Group. Entities owned by these excluded entities can also be considered as excluded entities under certain circumstances.

[52] Id., Article 3.1. MR.

[53] These adjustments reflect common permanent differences like additions for net tax expenses, disallowed expenses (for example, bribes and certain fines and penalties) and reductions for dividends and capital gains in connection to non-portfolio shareholdings with regard to participation exemption regimes (*See* Article 3.2. MR).

income (or loss) may further be re-allocated to account for the tax treatment of flow-through entities, hybrid entities and PEs.[54] Further adjustments may be also required for intragroup transactions to comply with arm's length principle.

Once determined, the GloBE income for each jurisdiction serve as denominator for the Jurisdictional ETR computation.[55]

Considering the potential impact of tax incentives two provisions must be highlighted about Step 2:

- GloBE MR states that certain tax credits granted by governments to engage in certain activities whenever meet the definition of Qualified Refundable Tax Credits (hereinafter "QRTC") shall be treated as income in the computation of globe income or loss.[56] On the other hand, Non-Qualified Refundable Tax Credits (hereinafter "NQRTC") shall not, resulting in a reduction in the ETR.
- Certain types of income are excluded from the GloBE calculation, like international shipping income and ancillary international shipping income.[57] Moreover, GloBE MR sets out an industry-specific adjustment in relation to insurance companies.[58]

Step 3 - Determination of taxes attributable to income of a CE (Articles 4.1.-4.6. of the GloBE MR)

Under Article 4.2 MR, the term Covered taxes is broadly defined to include taxes recorded in respect of a Constituent Entity's net income (potentially including withholding taxes on interest and royalties), taxes imposed in lieu of a corporate income tax, taxes imposed under eligible distribution tax systems and taxes on retained earnings and corporate equity. Also, taxes accrued under Pillar One and taxes arising from the Subject to Tax Rule (STTR) should both meet such definition. On the other hand, consumption taxes (sales taxes or value-added taxes), excise taxes, digital services taxes and stamp duties would fall outside the scope of the covered taxes definition.[59]

---

[54] Id., Articles 3.4, 3.5. MR.

[55] More specifically, according to Article 5.1. MR, the denominator is represented by the Net GloBE Income of the jurisdiction for the Fiscal Year which is computed as the sum (difference) of the GloBE Income (loss) of all Constituent Entities located in the jurisdiction.

[56] Id., Article 3.2.4. MR.

[57] Id., Article 3.2. MR.

[58] Id., Article 3.2.3. MR.

[59] Under Article 4.2.1–4.2.1 "Covered Taxes" means: *"(a) Taxes recorded in the financial accounts of a Constituent Entity with respect to its income or profits or its share of the income or profits of a Constituent Entity in which it owns an Ownership Interest; (b) Taxes on distributed profits, deemed profit distributions, and non-business expenses imposed under an Eligible Distribution Tax System; (c) Taxes imposed in lieu of a generally applicable corporate income tax; and (d) Taxes levied by reference to retained earnings and corporate equity, including a Tax on multiple components based on income and equity*". However, Article 4.2.1 provides that: "*Covered Taxes does not include any amount of: (a) Top-up Tax accrued by a Parent Entity under a Qualified IIR; (b) Top-up Tax accrued by a Constituent Entity under a Qualified Domestic Minimum Top-Up Tax; (c) Taxes attributable to an adjustment made by a Constituent Entity as a result of the application of a Qualified UTPR; (d) A Disqualified Refundable Imputation Tax; (e) Taxes paid by an insurance company in respect of returns to policyholders*". *See* also the Commentary on Article 4 MR, p. 85 ff.

Once computed the covered taxes amount, some further adjustments are required to find the Adjusted covered taxes amount, which serves as ETR numerator. These includes reductions for tax expenses relating to excluded income (as non-portfolio dividends) and uncertain tax positions, as well as reductions for certain tax credits and accrued taxes that are not paid within three years.[60]

The Model Rules also requires adjustments to allocate Covered taxes from one Constituent Entity to another, according to the nature of the taxpayer (for example, flow-through entities, hybrid entities or PEs) and to the cross-border character of the tax (for instance CFC rules and withholding taxes).

Article 4.4 sets out a mechanism to address certain timing differences arising when income (or a loss) is recognized in a different year for financial accounting and tax purposes. Such a provision becomes relevant in respect of certain tax incentive that defers tax payments. As discussed further below (*see* Sect. 5), some adjustments are required where certain deferred tax liabilities are not unwound within five years to avoid long-term deferral. However, under Article 4.4.5. of MR, the above-mentioned five-year recapture does not apply to an agreed list of "recapture exception accruals".

Step 4 - Calculation the Effective Tax Rate (Articles 5.1.-5.6. of the GloBE MR)

The ETR calculation determines whether in a fiscal year the top-up tax is required by triggering a top-up tax. The GloBE ETR of a jurisdiction is equal to the sum of Adjusted Covered Taxes of all CEs located in the jurisdiction divided by the Net GloBE Income of the jurisdiction.[61] If the jurisdictional GloBE ETR is below the 15% minimum rate, then a top-up tax is triggered.

The jurisdiction top-up tax is calculated by multiplying the top-up tax percentage (calculated as a difference between the 15% minimum rate and the jurisdictional ETR) by the excess profit for that jurisdiction.

Further, the excess profit is the positive amount resulting as a difference between the GloBE income (and the losses of all the Constituent Entities) in a jurisdiction and the substance-based income carve-out (hereinafter the "SBIE").[62]

In general terms, the SBIE consists of a reduction to the jurisdictional tax base and is computed as a percentage mark-up on tangible assets[63] and payroll costs.[64]

---

60 Id., Article 4.1. MR.

61 In case of a Net GloBE loss, no ETR is calculated for the jurisdiction.

62 Id. Article 5.3 MR.

63 As per Article 5.3.4., are considered "eligible tangible assets": "*(a) property, plant, and equipment located in that jurisdiction; (b) natural resources located in that jurisdiction; (c) a lessee's right of use of tangible assets located in that jurisdiction; and (d) a license or similar arrangement from the government for the use of immovable property or exploitation of natural resources that entails significant investment in tangible assets*". The GloBE MR further excludes from the tangible asset carve-out both "*the carrying value of property (including land or buildings) that is held for sale, lease or investment*" and "*the carrying value of tangible assets used in the generation of a Constituent Entity's International Shipping Income and Qualified Ancillary International Shipping Income*".

64 Under Article 10.1.1., the term "eligible payroll costs" means "*employee compensation expenditures (including salaries, wages, and other expenditures that provide a direct and separate personal benefit to the employee, such as health insurance and pension contributions), payroll and employment taxes, and employer social security contributions*". However, as per Article 5.3.3, are not

The exclusion is equal to 5% of the carrying value of the eligible tangible assets and payroll costs in the jurisdiction. A transition rule provides a phased introduction of the SBIE over the first ten years.[65]

Finally, the Model Rules also provide for an option for low-taxed jurisdiction to impose themselves a "qualified" domestic top-up tax (hereinafter "QDMTT").[66] The aim of such provision is to avoid/reduce the revenue transfer effect of tax incentives by reducing the general top-up tax liability of the parent jurisdictions.[67] Indeed, if a jurisdiction opts for the QDMTT, the parent entity applying the charging provision (the "IIR") is required to offset the QDMTT amount when calculating the top-up tax payable for that jurisdiction.

Step 5 – Charging provisions (Articles 2.1.-2.6. of the GloBE MR)

Unless low-tax profits are taxed under a QDMTT, top-up taxes are collected under a charging provisions' system which primarily involves an Income Inclusion rule (IIR) and, secondly, its backstop, the Undertaxed Payments Rule (UTPR). In general, the top-up tax charged through the IIR is based upon a top-down approach, which gives priority to the parent entity at the highest point in the ownership chain (Ultimate Parent Entity, hereinafter "UPE") and charges the top-up tax in proportion to the UPE's allocable share of the low-taxed Constituent Entity's income.[68]

On the other hand, the UTPR acts as a "backstop rule", aimed at ensuring that the top-up tax is actually collected in cases where the IIR does not apply. The UTPR's allocation mechanism distributes the top-up tax among jurisdictions where the MNE Group operates. These jurisdictions will then apply the UTPR and collect the top-up tax by denying a deduction or requiring a similar adjustment to the CEs located in their jurisdiction.[69]

considered "eligible" those payroll costs "*[…] that are: (a) capitalised and included in the carrying value of Eligible Tangible Assets; (b) attributable to a Constituent Entity's International Shipping Income and Qualified Ancillary International Shipping Income under Article 3.3.5 that is excluded from the computation of GloBE Income or Loss for the Fiscal Year*".

[65] In 2023, the exclusion would be equal to 8% of the carrying value of tangible assets and 10% of the payroll costs, which would be reduced gradually (in ten years) up to 5%.

[66] Article 10.1.1. defines QDMTT as "*a minimum tax that is included in the domestic law of a jurisdiction and that: (a) determines the Excess Profits of the Constituent Entities located in the jurisdiction (domestic Excess Profits) in a manner that is equivalent to the GloBE Rules; (b) operates to increase domestic tax liability with respect to domestic Excess Profits to the Minimum Rate for the jurisdiction and Constituent Entities for a Fiscal Year; and (c) is implemented and administered in a way that is consistent with the outcomes provided for under the GloBE Rules and the Commentary, provided that such jurisdiction does not provide any benefits that are related to such rules*".

[67] Article 5.2.3. MR.

[68] *See* Articles 2.1 - 2.3 MR.

[69] *See* Articles 2.4 - 2.6 MR.

5 The Globe Rules' Interaction With the Tax Incentives

As anticipated by the analysis of the architecture of the MR, the GloBE Rules do not directly (and expressly) prohibit jurisdictions to adopt neither tax incentives nor reduced tax rates within their tax systems, rather, they set a floor (on tax competition), which will likely reduce the incentives for profit shifting to achieve low or zero taxes. Nevertheless, the compatibility of tax incentives with GloBE Rules has been one of the key discussion points since the agreement was reached in October 2021 and the Model Rules were finalized. In effect, the provisions contained therein will probably have profound implications on the tax incentives policy for many jurisdictions. Evaluating the compatibility of tax incentives offered by a certain tax system will likely become a key issue for many countries. In general terms:

- GloBE Rules do not distinguish between different tax incentives and apply in the same way regardless of the policy goal, the type of tax instrument and whether it is income or expenditure based.
- The extent to which tax incentives will be impacted by the GloBE Rules will also depend on other design elements that affect the generosity of tax support.[70]
- For developing countries, the "switch" in terms of tax incentives policy could prove to be challenging.[71]

However, the impact won't be the same for every tax incentive, some of them will be affected to a lesser extent because of GloBE rule functioning. As highlighted by the OECD, it is possible to identify three principal provisions that could directly affect the impact on tax incentives[72]:

- The specific scope of application of the same GloBE rules.
- The calculation of the GloBE ETR and its components.
- The SBIE.

[70] According to the OECD: *"[...] tax incentives that apply only to new investment would apply to a lower incremental base than volume-based incentives applying to old and new investment. Similarly, targeted provisions to certain types of income such as export income or IP income will have a lower impact on firms' ETRs than permanent reduced rates or tax holidays applying to the full income of the firm. Ceilings that bound the maximum amount of relief or investments that can qualify for support or thresholds that establish minimum investment requirements, act to affect the ultimate tax benefits firms can obtain. This means that similar tax instruments can deliver very different ETR outcomes depending on the details of their design features*". (OECD (2022), Tax Incentives and the Global Minimum Corporate Tax: Reconsidering Tax Incentives after the GloBE Rules, OECD Publishing, Paris, https://doi.org/10.1787/25d30b96-en, p. 33 ff.).

[71] *See* in literature L. Parada, *Global Minimum Taxation: A Strategic Approach for Developing Countries* (November 18, 2022). Available at SSRN: https://ssrn.com/abstract=4280669 or http://dx.doi.org/10.2139/ssrn.4280669; A. Oguttu, *Preventing International Tax Competition and the Race to the Bottom: A Critique of the OECD Pillar Two Model Rules for Taxing the Digital Economy - A Developing Country Perspective*, 76 Bull. Intl. Taxn. 11 (2022), Journal Articles & Opinion Pieces IBFD; A. Titus, *Pillar Two and African Countries: What Should Their Response Be? The Case for a Regional One*, Intertax 50(10) (2022).

[72] Id., p. 33 ff.

Out of Scope Entities and Incomes

First of all, tax incentives aimed at enterprises outside the scope of application will not be impacted by the GloBE rules. In other words, tax competition could continue unchallenged and standalone investors that are not part of a group (purely domestic firms) as well as entities belonging to groups below GloBE's threshold will not be subject to the rules. To a similar extent, entities earning out-of-scope income (e.g., shipping income), are also less likely to be affected.

ETR Calculation

Secondly, the GloBE Rules impacts will depend on the interaction between a certain tax incentive and the functioning of the GloBE ETR. In other words, it must be evaluated if a tax incentive: (i) reduces the numerator (the amount of "Covered Taxes"); (ii) increases the denominator (the amount of "GloBE Income"); or (iii) is adjusted to ensure a neutral effect on the GloBE ETR.

Accordingly, the majority of "income-based" and "expenditure-based" tax instruments are the tax incentives most likely to be affected by GloBE Rules provided that they are treated as reductions in covered taxes for the GloBE ETR computation.[73] Examples of such incentives are preferential corporate income taxes rates (through either reduced rates or exemptions), investment tax allowances, or credits aimed at mitigating the taxable income or the tax liability on certain investments.[74]

On the other hand, certain "*(qualified) refundable tax credits*"[75] are expected to be less affected by the GloBE Rules. Indeed, as a safeguard provision, certain tax credits are treated as income for GloBE purposes whenever they are deemed "refundable" under Article 3.2.4. MR. In other words, these refundable tax credits – such as research and development credits - are granted a treatment similar to that of government grants that form part of income, being in effect government support for a certain type of activity that can ultimately be received in cash or cash equivalent.[76]

---

[73] OECD (2022), Tax Incentives and the Global Minimum Corporate Tax, Supra, p. 35 ff.

[74] OECD (2022), Tax Incentives and the Global Minimum Corporate Tax, Supra, p. 36.

[75] According to the Commentary on Article 3.2.4. MR (par. 110) "*[…] The refundable tax credits referred to in Article 3.2.4 are government incentives delivered* via *the tax system. They are not ordinary refunds of tax paid in a prior period due to an error in the computation of tax liability or pursuant to an imputation system. Instead, they are incentives to engage in certain activities, such as research and development, whereby the government allows the company to offset its taxes dollar-for-dollar for engaging in specified activities or incurring specified expenditures or the government will refund the amount of the unused credit if the company doesn't have any tax liability. In this way, the government effectively pays for the activity or expenditure in a similar manner to a grant. The basic idea is that the incentive or grant is delivered by a tax reduction to the extent possible because it is more efficient than having checks from the government and taxpayer crossing in the mail*". (OECD (2022), Commentary to the Global Anti Base Erosion Model Rules, Supra p. 119 ff.).

[76] Article 10.1.1. MR further states that: "*A tax credit that is refundable in part is a Qualified Refundable Tax Credit to the extent it must be paid as cash or available as cash equivalents within four years from when a Constituent Entity satisfies the conditions for receiving the credit under the laws of the jurisdiction granting the credit*".

Similarly, "*tax payments deferment provisions*" (e.g., accelerated depreciation and immediate expensing) are generally unlikely to generate top-up taxes under the GloBE Rules. Despite such tax incentives lead to a deferral of taxation and thus to a timing benefit, the GloBE Rules incorporate certain deferred tax adjustments into the calculation of Covered Taxes allowing to consider the differences between book and tax values.[77] The effect of these adjustments is that, under a moderate tax rate and assuming no recapture is required, tax incentives such as accelerated depreciation, will generally on their own not give rise to additional tax liability under the GloBE Rules absent any other base narrowing provisions.

However, where these differences are maintained over a period of more than five years, a recapturing mechanism may apply; in such cases, it is required to regularize the top-up taxes that should have been paid if the firm has not adjusted for the timing difference within the prescribed period.[78]

Nevertheless, the recapture mechanism does not apply to accelerated depreciation on tangible assets, immediate expensing of certain items and several other defined classes of expenditure or assets.[79] Indeed, OECD acknowledging the importance of such incentive to promote investment in infrastructure and tangible capital, exempted it from the above recapturing mechanism. However, enhanced, or super deductions (those incentive which goes beyond the 100% of acquisition cost e.g., a 150% deduction for an investment expenditure) do not qualify because they grant a favorable permanent book tax difference.[80]

Substance-Based Income Exclusion (SBIE)

---

[77] Article 4.4. MR.

[78] Article 4.4.4. MR.

[79] Under Article 4.4.5. MR the Recapture Exception Accrual includes: *"(a) Cost recovery allowances on tangible assets (b) The cost of a licence or similar arrangement from the government for the use of immovable property or exploitation of natural resources that entails significant investment in tangible assets; (c) Research and development expenses; (d) De-commissioning and remediation expenses; (e) Fair value accounting on unrealised net gains; (f) Foreign currency exchange net gains; (g) Insurance reserves and insurance policy deferred acquisition costs; (h) Gains from the sale of tangible property located in the same jurisdiction as the Constituent Entity that are reinvested in tangible property in the same jurisdiction; and (i) Additional amounts accrued as a result of accounting principle changes with respect to categories (a) through (h)"*.

[80] OECD (2022), Tax Incentives and the Global Minimum Corporate Tax, Supra, p. 40.

According to the OECD, the SBIE plays a key role affecting the use of tax incentives.[81] The carve-out, which was designed to give some benefit to real investment,[82] was both a salient point of the "Two-pillars" solution's political agreement and an attempt to remain faithful to the idea of aligning taxation with value creation. Thus, the carve-out - by exempting an amount of income from top-up taxation equal to 5%[83] of the carrying value of tangible assets and payroll costs - seems to indicate the amount the OECD understands to be the actual "value creation" in a jurisdiction.[84]

As underlined by the OECD, the introduction of SBIE reverberates three important effects on tax incentives: *"First, at the jurisdiction level, jurisdictions with higher levels of substance will be on average less extensively affected. Second, firms and particularly those performing activities or operating in sectors that require a greater level of substance in the jurisdiction may be less affected. Third, tax incentives, where the scale of tax relief is more tightly associated with the amount of economic substance, may be less affected".*[85]

Consequently, the GloBE Rules will have a greater effect on the use of tax incentives in jurisdictions with more economic substance and, consequently, tax incentives targeted toward such activities would reasonably be more protected by the SBIE than activities that rely on intangible factors or are otherwise highly profitable. Although SBIE does not distinguish between tax incentive instruments, it can be expected to have a more favorable impact on expenditure-based tax incentives (e.g., investment allowances, training allowances or R&D tax incentives, which typically target current expenditures) than income-based tax incentives (e.g., holidays). SBIE will more likely accrue in cases where expenditure-based incentives are used to promote investment in tangible assets, including infrastructure, and labor in the jurisdiction.

[81] Id., Supra, p. 41. *See* also in literature M. P. Devereux, M. Simmler, J. Vella, and Wardell-Burrus, *What is the Substance-Based Carve-Out under Pillar 2? And How Will it Affect Tax Competition?*, EconPol Policy Brief 39, 17 November 2021; L.E. Schoueri, *Supra*, p. 545.

[82] As stated in the MR Commentary: "*The policy rationale behind a formulaic, substance-based carve-out, based on payroll and tangible assets is to exclude a fixed return for substantive activities within a jurisdiction from the application of the GloBE Rules. The use of Payroll and Tangible Assets as indicators of substantive activities is justified because these factors are generally expected to be less mobile and less likely to lead to tax-induced distortions. Conceptually, excluding a fixed return from substantive activities focuses GloBE on "excess income", such as intangible-related income, which is most susceptible to BEPS risks*" (OECD (2022), Tax Challenges Arising from the Digitalisation of the Economy – Commentary to the Global Anti Base Erosion Model Rules (Pillar Two), OECD, Paris, https://www.oecd.org/tax/beps/tax-challenges-arising-fromthedigitalisation-of-the-economy-global-anti-base-erosion-model-rules-pillar-two-commentary.pdf, Commentary on article 5.3, par. 25, p. 119 ff.).

[83] Higher rates apply during a 10-years transitional period, as described in Sect. 4.

[84] L.E. Schoueri, *Supra*, p. 546.

[85] OECD (2022), Tax Incentives and the Global Minimum Corporate Tax, Supra, p. 42.

6 Conclusions

The last two decades could be considered as period of great changes and reassessment for the international tax framework. The works on harmful tax competition before and the BEPS project together with the "Two-Pillars" Proposal after, are reshaping the international tax system both to curb profit shifting and to mitigate the race to the bottom. This process reverberates significant effects not only on taxpayers but also on the Governments around the world, thereby affecting their tax policies and, to a certain extent, their sovereignty.[86]

The narrowing possibility for jurisdiction to grant certain tax incentives represents indeed a clear consequence of such a process. Despite GloBE Rules do not explicitly prohibit countries from maintaining a system of tax incentives, Pillar Two places a floor to tax competition and is expected to ease the pressures on those jurisdictions less inclined to provide significant tax incentives to attract foreign investments. Where a MNE's jurisdictional ETR falls below the 15% threshold, the jurisdiction granting the incentive would eventually give up taxing rights not in exchange for offering more favorable business environment but to the benefit of the tax revenue of the other jurisdictions. As a result, the incentive's benefits to the investor would simply become payable elsewhere, transferring part of a country's tax base to other countries. Therefore, many governments all over the world will need to rethink their incentives policy, at least those more likely to be affected by GloBE Rules.[87]

In light of the above, certain literature[88] has identified - through a "traffic light system" - three categories of tax incentives to analyze their compatibility with the GloBE rules.

The first group ("red light") includes certain regimes that are expected to be the most affected tax incentives by the GloBE Rules to the extent that they would be fully targeted by the Pillar Two Rules for in-scope situations and excess profits. In general terms, zero-rate corporate tax systems, tax holidays and SEZs form part of this group.

On the other hand, there is another group of incentives ("yellow light") whose interaction with GloBE Rule cannot be easily determined a priori. This group includes, for instance, reduced-rate incentives (e.g., patent boxes regimes), business tax credits (particularly refundable credits), direct funding programs and corporate tax rates below 15 percent. From an Italian perspective certain tax incentives like

[86] M. P. Devereux, J. Vella, H. Wardell-Burrus, *Pillar 2: Rule Order, Incentives, and Tax Competition*, Oxford University Centre for Business Taxation Policy Brief 2022, 14 January 2022.

[87] *See* M. P. Devereux, The OECD Global Anti-Base Erosion ("GloBE") proposal, Oxford University Centre for Business Taxation, January 2020, p. 36 ff.

[88] *See* D. Bunn, *What Do Global Minimum Tax Rules Mean for Corporate Tax Policies?*, Tax Foundation, December 2021, available at https://taxfoundation.org/oecd-global-minimum-tax-rules/; B. Ferreira Liotti, J. Waruguru Ndubai, R. Wamuyu, I. Lazarov, J. Owens, *The treatment of tax incentives under Pillar Two*, UNCTAD World Investment Report (WIR), June 2022. Available at https://papers.ssrn.com/sol3/papers.cfm?abstract_id=4132515; UN Conference on Trade and Development, World Investment Report 2022, June 2022 p. 131 ff. Available at https://unctad.org/webflyer/world-investment-report-2022.

the tax credits for R&D expenses and for investments or (the newly introduced) IP Box regimes are likely to be unaffected by the GloBE Rules because of their structure which can be abstractly referable to that of Qualified Refundable Tax Credits under Article 3.2.4. of Globe MR.[89] However further analysis and clarifications are expected by both the OECD and tax administrations.

The last group comprises those incentives expected to be largely unaffected by the GloBE Rules ("green light") such as accelerated depreciation (full expensing), Last-In-First-Out (LIFO) inventory treatment, unlimited loss carryforwards and participation exemption regimes (aimed at eliminating double taxation). The higher goals of such incentives are indeed recognized by the international community and the OECD, to the extent that GloBE Rules prevent them to be affected by Pillar Two.

However, countries might also explore other forms of competition to attract FDI. As suggested by Englisch,[90] jurisdictions could still compete in respect of taxes and levies out of the GloBE Rules' scope, like expat regimes that offer personal income tax privileges for certain categories of personnel. Countries could also try competing by enacting or endorsing "favorable" financial accounting standards. Furthermore, jurisdictions could also grant direct subsidies in order to attract FDI. Tax competition could even turn into a "tax procedures competition" aimed at attracting FDI by introducing more favorable (e.g., in terms of specific safeguards) and effective tax procedures for residents.[91]

Nevertheless, international tax competition could continue above this minimum rate floor, but its effects would certainly be mitigated, reducing the attractiveness of profit shifting to low or no tax areas.

Furthermore, the very broad scope of GloBE Rules is also likely to affect those tax incentives that have proven to be an effective stimulus to promote certain "substance-based" investments or activities (like, for instance, capital investment and R&D incentives) and at the same time reliable tools to generate key spillovers for the economic system as a whole. To this extent, the SBIE seems not appropriate to keep faith with the paradigm of the effective economic substance that permeated the previous OECD work on harmful tax competition.

In addition, the limitation to GloBE Rules scope may give rise to some doubts and equity concerns with respect to purely domestic firms.

To conclude, such a result is questionable at a glance from a theoretical point of view, if considered the inconsistencies with the previous OECD's works on harmful tax competitions. Nonetheless, from a practical point view, the unclear policy objective on which the global compromise is based on might cast some shadows on the concrete implementation of the project.

---

[89] *See* Assonime, Circular letter n. 19/2022, July 2022.

[90] J. Englisch, Supra, p. 1 ff.; H. Wardell-Burrus, *State Strategic Responses to the GloBE Rules* (December 1, 2022); with reference to the developing countries *see* also, L. Parada, Supra, pp. 12–13.

[91] *See* A. Contrino, Supra.

References

1. Arginelli P (2015) Innovation through R&D tax incentives: some ideas for a fair and transparent tax policy. World Tax J. 7(1):3–71
2. Arginelli P (2017) Chapter 5: the interaction between IP box regimes and compensatory tax measures: a plea for a coherent and balanced approach. In: EU Law and the Building of Global Supranational Tax Law: EU BEPS and State Aid (D. Weber ed., IBFD 2017)
3. Arnold BJ (2019) The evolution of controlled foreign corporations rules and beyond. Bull Int Tax 73:631
4. Avi-Yonah RS (2009) The OECD harmful tax competition report: a tenth anniversary retrospective. Brook J Int'l L 34(3):783–795
5. Avi-Yonah RS, Vallespinos M (2017) Special Tax Zones and the WTO, University of Michigan Public Law Research Paper no. 545
6. Avi-Yonah RS (2000) Globalization, tax competition, and the fiscal crisis of the welfare state. Harv L Rev 113(7):1573–1676
7. Becker J, Englisch J (2021) Implementing an international effective minimum tax in the EU (June 23, 2021)
8. Bicer R (2008) Turkey-an assessment of free trade zones from a transfer pricing perspective-this article considers free trade zones in Turkey from a tax and transfer pricing perspective. The focus is on practice, and some simplified examples are presented to explain significant details of current transfer pricing rules applicable in free trade zones. Int Transf Pricing J. 15(5):236
9. Bjerkestuen HM, Wille HG (2015) Tax holidays in a BEPS-perspective. Intertax 43(1):106
10. Brauner Y (2014) What the BEPS. Fla Tax Rev 16(2):55
11. Brauner Y (2022) Agreement? What Agreement? The 8 October 2021, OECD Statement in Perspective. Intertax 50(1):2–6
12. Bunn D (2021) What Do Global Minimum Tax Rules Mean for Corporate Tax Policies?, Tax Foundation
13. Carpentieri L (2018) La crisi del binomio diritto-territorio e la tassazione delle imprese multinazionali. Rivista di diritto tributario 351
14. Celani A, Dressler L, Hanappi T Corporate Effective Tax Rates in the presence of Investment Tax Incentives. OECD Publishing, Paris
15. Cipollina S (2014) I redditi "nomadi" delle società multinazionali nell'economia globalizzata. Rivista di diritto finanziario e scienza delle finanze (1):21–63
16. Cipollini C (2019) The concept of special tax zones in EU tax law. EC Tax Rev 28(6):307–321
17. Cipollini C (2020) GloBE proposal and possible carve-outs: is there a future for preferential tax regimes? World Tax J 12(2):26
18. Contrino A (2022) La concorrenza fiscale tra gli Stati nel contesto della governance economica europea. Riv Dirit Tribut 24:1–6
19. Dagnino A (2008) Agevolazioni fiscali e potestà normativa, Padova
20. De Jong W (2013) Establishing Free Zones for Regional Development, Library of the European Parliament
21. Devereux MP, Vella J, Wardell-Burrus H (2022) Pillar 2: rule order incentives, and tax competition. Oxford University Centre for Business Taxation, Policy Brief
22. Devereux MP et al (2020) The OECD Global Anti-Base Erosion ("GloBE") proposal, Oxford University Centre for Business Taxation
23. Devereux MP, Simmler, M, Vella, J, Wardell-Burrus H (2021) What is the substance-based carve-out under pillar 2? and how will it affect tax competition?, EconPol Policy Brief no. 39
24. Devereux MP, Vella J (2018) Taxing the digitalised economy: targeted or system-wide reform. Brit Tax Rev 4:301–320
25. Devereux MP, Griffith R (2003) Evaluating tax policy for location decisions. Int Tax Public Financ 10:107–126
26. Dietrich M, Golden C (2022) Consistency versus "gold plating": the EU approach to implementing the OECD pillar two. Bull Intl Tax 4:76(2022)

27. Dourado AP (2019) In search of an international tax system in a post-beps tax competition setting. Intertax 47:2
28. Dourado AP (2020) The global anti-base erosion proposal (GloBE) in pillar II. Intertax 48(2):152
29. Dourado AP (2018) Digital taxation opens the Pandora box: the OECD interim report and the European commission proposals. Intertax 46:565
30. Dourado AP (2018) Taxes and competitiveness: how much competitive is European tax competition. Intertax 46:942
31. Dourado AP (2019) Taxing the digital economy. Intertax 47(2)
32. Englisch J (2022) GloBE Rules and Tax Competition. Intertax 50(12)
33. Englisch J, Becker J (2019) International effective minimum taxation – the GLOBE proposal. World Tax J 4
34. European Commission of the European Communities (1997). Communication From the Commission to the Council towards tax co-ordination in the European Union, COM(97)495final, Brussels, 01.10.1997; European Union Council of Economics and Finance Ministers (ECOFIN)
35. European Commission (2014). Working Paper no. 52, A Study on R&D Tax Incentives, European Commission Taxation Papers, Final Report
36. European Commission (2017). Communication to the European Parliament and the Council, A Fair and Efficient Tax System in the European Union for the Digital Single Market, Brussels, 21 Sept. 2017, COM (2017) 547 Final, at 6
37. European Commission (2021). Communication from the Commission to the European Parliament and the Council, Business Taxation for the 21st Century, COM(2021) 251 final, Brussels
38. European Council (2022). Draft Council Directive on ensuring a global minimum level of taxation for multinational groups in the Union - Compromise text, 12 March 2022, 6975/22
39. Ferreira B (2022) Limits of international cooperation: the concept of "jurisdiction not to tax" from the BEPS project to GloBE. Bull Int Tax 76(2):77
40. Ferreira Liotti B, WaruguruNdubai J, Wamuyu R, Lazarov I, Owens J (2022) The treatment of tax incentives under Pillar Two. UNCTAD World Investment Report
41. Gasparri T (2017) Stati sovrani ed imprese multinazionali alla sfida del Fisco, tra sostanza e trasparenza, Note e Studi Assonime, no. 15
42. Gerzova G, Olejnicka M (2018) Chapter 4: Reduced Tax Rates at a Crossroads: Before and After the OECD BEPS Project in Tax Incentives in the BEPS Era (M. Cotrut et al. eds., IBFD 2018)
43. González Cabral A, Appelt S, Hanappi T (2021) Corporate Effectives Tax Rates for R&D: The case of expenditure-based tax incentives. OECD Taxation Working Papers, No. 54, OECD Publishing, Paris
44. Gropp R, Kostial K, (2000) The Disappearing Tax Base: Is Foreign Direct Investment Eroding Corporate Income Taxes?, IMF Woking Paper WP/00/173, Washington, D.C
45. James S (2013) Effectiveness of tax and non-tax incentives and investments: evidence and policy implications. SSRN Electr J
46. Johansson Å et al (2008) Tax and Economic Growth, Economics Department Working Paper No. 620, ECO/WKP(2008)28 (OECD 2008)
47. Joint Statement (2022) by France, Germany, Italy, Netherlands and Spain
48. Katsushima T (1999) Harmhul tax competition. Intertax 27(11):396
49. Keen M, Konrad K (2012) The Theory of International Tax Competition and Coordination, Max Planck Institute for Tax Law and Public Finance, Working Paper 06
50. Klemm A, Van Parys S (2009) Empirical Evidence on the Effects of Tax Incentives, IMF Working Paper WP/09/136 2009
51. Lampreave P (2011) Fiscal competitiveness versus harmful tax competition in the European union. Bull Intl Tax 65(6):17
52. Laukkanen A (2018) Chapter 5: Special Economic Zones: The Acceptance of Tax Incentives in the BEPS World, in Tax Incentives in the BEPS Era (M. Cotrut et al. eds., IBFD 2018), Books IBFD

53. Laukkanen A (2016) The development aspects of special tax zones. Bull Intl Taxn 70(3):152
54. Laukkanen A, Pistone P, de Goede J (2019) Special Tax Zones in the Era of International Tax Coordination. IBFD
55. Mascagni G, Moore M, McCluskey R (2014) Tax Revenue Mobilisation in Developing Countries: Issues and Challenges. Report to the European Parliament's Committee on Development
56. McLure JCE (1999) Tax holidays and investment incentives a comparative analysis. Bull Intl Tax
57. Micossi S (2017) La fiscalità d'impresa nel nuovo mondo globalizzato, Note e Studi Assonime, no. 1
58. Owens J (1998) Curbing harmful tax competition – recommendations by the committee of fiscal affairs. Intertax 26(8–9):230
59. Ogazón Juárez LG, Calderón Manrique D, Cotrut M, Munyandi K (2018) Tax Incentives in the BEPS Era - Part One: Introduction - Chapter 1: Introduction to Tax Incentives in the BEPS Era, IBFD Tax Research Series
60. da Empoli D (1998) OECD, harmful tax competition–an emerging global issue. J Publ Financ Publ Choice 16(1):73–75
61. OECD (2001). Towards Global Tax Co-operation: Progress in Identifying and Eliminating Harmful Tax Practices (OECD 2001)
62. OECD (2002). The OECD's Project on Harmful Tax Practices: The 2001 Progress Report (OECD 2002)
63. OECD (2004). Harmful Tax Practices: The 2004 Progress Report (OECD 2004)
64. OECD (2006). The OECD's Project on Harmful Tax Practices: 2006 Update on Progress in Member Countries (OECD 2006)
65. OECD (2014). Countering Harmful Tax Practices More Effectively, Taking into Account Transparency and Substance, OECD/G20 Base Erosion and Profit Shifting Project, OECD Publishing
66. OECD (2013). Supporting Investment in Knowledge Capital OECD Publishing Growth and Innovation
67. OECD (2013). Action Plan on Base Erosion and Profit Shifting, OECD Publishing
68. OECD (2015). Addressing the Tax Challenges of the Digital Economy, Action 1 - 2015 Final Report, OECD/G20 Base Erosion and Profit Shifting Project, OECD Publishing
69. OECD (2015). Countering Harmful Tax Practices More Effectively, Taking into Account Transparency and Substance, Action 5 – Final Report (OECD 2015)
70. OECD (2018). Tax Challenges Arising from Digitalisation – Interim Report 2018: Inclusive Framework on BEPS, OECD/G20 Base Erosion and Profit Shifting Project. OECD Publishing, Paris
71. OECD (2019). Addressing the Tax Challenges of the Digitalisation of the Economy – Policy Note, as approved by the Inclusive Framework on BEPS. OECD, Paris
72. OECD (2019). Programme of Work to Develop a Consensus Solution to the Tax Challenges Arising from the Digitalisation of the Economy
73. OECD (2020) .Tax Challenges Arising from Digitalisation – Report on Pillar Two Blueprint: Inclusive Framework on BEPS, OECD/G20 Base Erosion and Profit Shifting Project. OECD Publishing, Paris
74. OECD (2020). Cover Statement by the OECD/G20 Inclusive Framework on BEPS on the Reports on the Blueprints of Pillar One and Pillar Two, OECD/G20 Inclusive Framework on BEPS (OECD 2020)
75. OECD (2021) Tax Challenges Arising from the Digitalisation of the Economy – Global Anti-Base Erosion Model Rules (Pillar Two): Inclusive Framework on BEPS. OECD, Paris
76. OECD (2022). Tax Challenges Arising from the Digitalisation of the Economy – Commentary to the Global AntiBase Erosion Model Rules (Pillar Two). OECD, Paris
77. OECD (2022). Tax Challenges Arising from the Digitalisation of the Economy –Global Anti-Base Erosion Model Rules (Pillar Two) Examples. OECD, Paris

78. OECD (2022) Tax Incentives and the Global Minimum Corporate Tax: Reconsidering Tax Incentives after the GloBE Rules. OECD Publishing, Paris
79. Oguttu A (2022) Preventing international tax competition and the race to the bottom: a critique of the OECD pillar two model rules for taxing the digital economy - a developing country perspective. Bull Intl Tax 76:11. Journal Articles & Opinion Pieces IBFD
80. Palazzi P (2011) Taxation and Innovation, OECD Taxation Working Papers, no. 9. OECD Publishing
81. Parada L (2022) Global Minimum Taxation: A Strategic Approach for Developing Countries
82. Parada L (2020) Taxing somewhere, no matter where: what is the GloBE proposal really about, Multinational Group Tax & Transfer Pricing News
83. Pistone P, Pinto JF, Turina A (2021) Digital services tax: assessing the policy reasons for its introduction in the European union. Intl Tax Stud 4(7):20
84. Perrone A (2019) Tax competition e giustizia sociale nell'Unione Europea, Padova
85. Piantavigna P (2017) Tax competition and tax coordination in aggressive tax planning: a false dichotomy. World Tax J 9(4):477–519
86. Pistone P, Pinto Nogueira JF, Andrade B, Turina A (2020) The OECD public consultation document "global anti-base erosion (GloBE) proposal – pillar two": an assessment. Bull Intl Tax 74(2)
87. Piketty T (2014) Il capitale nel XXI secolo. Milano 776
88. Radaelli CM (1999) Harmful tax competition in the EU: policy narratives and advocacy coalitions. J Common Market Stud 37(4):661–682
89. Saw R (2018) Tax competition: and the winner. In: Cotrut M, Munyandi K, Tax Incentives in the BEPS Era, IBFD Tax Research Series
90. Schoueri LE (2021) Some considerations on the limitation of substance-based carve-out in the income inclusion rule of pillar two. Bull Intl Tax 75:11
91. Schoueri P (2019) Conflicts of International Legal Frameworks in the Area of Harmful Tax Competition, chapter 1 – introduction, WU - Tax Law and Policy Series
92. Spengel C, Stage B, Steinbrenner D (2022) R&D tax incentive regimes – a comparison and evaluation of current country practices. World Tax J. 14(2):331–364
93. Tandon S (2022) The need for global minimum tax: assessing pillar two reform. Intertax 50(5)
94. Traversa E, Flamini A (2015) The impact of BEPS on the fight against harmful tax practices: risks... and opportunities for the EU. British Tax Rev 3:396–407
95. Titus A (2022) Pillar two and African countries: what should their response be? the case for a regional one. Intertax 50(10)
96. United Nations Conference on Trade and Development (2000). Tax Incentives and Foreign Direct Investment: A Global Survey, ASIT Advisory Studies No. 16
97. Wardell-Burrus H (2022) State Strategic Responses to the GloBE Rules https://ssrn.com/abstract=
98. World Bank (2022). The Global Minimum Tax: from agreement to implementation. Overview booklet. Information and Communications for Development. Washington, D.C

Megaproject, Taxation and the (New) Relationship Between the Italian Revenue Agency and Taxpayers

Marco Allena

Abstract The article, once focused on the relationship between taxation and megaprojects, proposes to verify how the Italian legal system allows for a preventive dialogue between tax authorities and taxpayers in order to prevent situations of tax uncertainty that may characterize a given economic transaction. Specifically, particular attention will be paid to the interpello institute - and, precisely, to the so-called "interpello" on the new investment s as a potentially useful tool for those who intend to start megaprojects in Italy - and to the regime of collaborative compliance, as a regime that allows the establishment of a constant dialogue between (large) companies and the tax authorities in order to ensure full and proper tax compliance.

Keywords Megaprojects · Tax · Tax compliance · Investments · Interpello tributario · Tax Risk

1 Megaprojects and Taxation

Megaprojects are large-scale projects developed with the aim to manage public utilities. Among the most common types of megaprojects there are the infrastructure and aerospace ones. The decades-long planning of these projects can require more than §1 billion of investment and has a macroeconomic impact on the overall interested Region. In detail, megaprojects have both positive and negative effects: they create employment and improve productivity, but they also involve diverse stakeholders with conflicting interests, such as contractors, investors, bankers and national Governments. Supporting megaprojects means also preserving the environment: in this regard, the promotion of sustainable activities often contrasts with the pursuit of private interests, such as the stakeholders profit one. Therefore the States have the task of promoting high-level projects and – at the same time – public interests such as the environmental one: in a nutshell, promoting megaprojects in a sustainable way. In this regard, the global experience with Covid-19 disease should lead national

M. Allena (✉)
Università Cattolica del Sacro Cuore di Piacenz, Piacenza, Italy
e-mail: marco.allena@unicatt.it

E. Favari and F. Cantoni (eds.), *Complexity and Sustainability in Megaprojects*, Lecture Notes in Civil Engineering 342,
https://doi.org/10.1007/978-3-031-30879-6_15

Governments to favor an economic restart within the sustainability. rules. If future investments in megaprojects are desirable as they give added value to the belonging Countries, their development is strategic mostly when they are designed to comply with economic, social and environmental sustainability. As stated also by the UN in its 2030 Agenda for Sustainable Development, the promotion of sustainability in its broad sense should be regarded as a priority goal, so that megaprojects definitely will have to be included among those activities suitable to embody this sustainability concept [1].

In this context, one can well understand, therefore, how megaprojects represent huge projects that are significant for the society and community in which they are carried out and are destined, if realised, to have an enormous relevance on the economy as well.

In this sense, megaprojects are both the object of attention of the public administration and the subject of public opinion.

Therefore, precisely in light of this, in the context of the realization of megaprojects, in addition to environmental sustainability - which is, today more than ever, a key factor in guaranteeing infrastructure development that takes into account environmental health and preservation - it should be the transparency of decision-making and, above all, production processes that should be used to strengthen the reliability of decision-makers and to involve the entire community in their planning.

It is precisely in the area of transparency that the fiscal factor also comes into play.

In fact, tax compliance plays an important role in the changing relationship between the tax authorities and businesses.

From the taxpayer's point of view, the latter originates with the discipline of tax risk management and represents a new behavioral model in the relationship with the tax administration.

This procedure, in fact, protects the interests of the Treasury but is equally important for taxpayers.

Firstly, as mentioned, the tax interest of the state or, indeed, the general interest in tax collection as a vital interest of the community is protected because it makes possible the smooth functioning of public services.

This interest is protected, within the Italian legal system, by the Constitution, and specifically by Article 53 of the Constitution, like any individual right. The tax interest, if from the point of view of substantive law it requires the simplicity of the tax, from the point of view of formal law and therefore with regard to the application of taxes, it requires regular and prompt collection.

In this perspective, the preventive dialogue between the Financial Administration and taxpayers can guarantee - also in the context of megaprojects - the fulfilment of these objectives.

Similarly - as will be highlighted below - through the preventive and constant dialogue with the Tax Administration, taxpayers also have the opportunity to have greater certainty and security both in terms of taxation and from a strictly operational perspective [2].

What has just been said will become clearer when reference is made to the institute of the s.c. "interpello" and, specifically, to the so-called '*interpello on the newinvestments*'.

In the light of what has been said, megaprojects - probably more than other economic operations with a lesser social impact - require special attention from the tax authorities so that the planning and execution of the projects do not encounter operational or fiscal problems either during the execution of the works or, above all, after their completion.

2 The Relationship Between Tax and Taxpayer: Tax 'Interpellations' as an Instrument of Dialogue

Tax law - especially within the Italian legal system - is subject to very frequent legislative and interpretative changes.

This profile inevitably leads to great certainty not only for interpreters but also for economic operators wishing to invest in Italy.

Trying to compare the current Taxpayer/Fiscal dynamics with those of the past, it would seem to emerge, today, the existence of a true and proper relationship between different parties that are necessarily called upon to interact and conduct a collaborative confrontation.

The current state of the art, which will be discussed shortly, is the consequence of legislative choices that have affected - and will undoubtedly continue to affect - profoundly the dynamics of the functioning of the tax system to the point of hopefully positively modifying also the role of the parties involved in the tax relationship, that is, essentially, the Tax Administration and the taxpayer.

The key word that comes to the fore with respect to the topic at hand is essentially one: 'tax collaboration'.

The latter, in fact, is configured as an effective tool to pursue a consistent reduction of costs related to tax collection and a certainty, for the taxpayer, of being made the recipient of the 'right' tax levy and not of other illegitimate, arbitrary and disproportionate forms of tax injunction.

In fact, today's tax system - despite the many continuities that continue to characterise it - would seem to be increasingly characterised by a widespread recourse to (preventive) participation and collaboration between the tax authorities and the taxpayer as indispensable and closely connected factors. Talking about confrontation between tax authorities and taxpayers means dealing not with a merely ancillary issue but with attention:

(i) dialectics between the parties;
(ii) the opportunities for dialogue;
(iii) of all those moments that are functional to the issuance of the measure, understood as the final act of the procedure. A procedure that is governed, within the

legal system, by important principles, first and foremost that of good faith and fairness in the relationship between the parties [3].

The need for a revolution in the relationship between tax authorities and taxpayers is a necessity that has been recognized, at various levels, for some time now. Only in this way, in fact, can economic operators and taxpayers operate within the Italian legal system aware of the application of certain tax rules. This is a need that has also been recognized at European level.

In its Communication No. 351 of 27 June 2012 [COM (2012)351], the European Commission, with the aim of countering the phenomenon of aggressive tax planning - with similar effects to those resulting from the fight against tax evasion - proposed, among other measures, to examine ways to improve tax transparency, in a context characterized by fair cooperation between taxpayers and financial administrations.

Also in this perspective, therefore, correct compliance on the part of the taxpayer plays a fundamental role in any kind of, if any, strategy to combat tax fraud and evasion.

However, in order to ensure proper compliance it is necessary that the taxpayer is properly informed about the tax rules in force in the EU and, no doubt, within the Member States in which he operates. Indeed, reducing the costs and complexity associated with tax compliance on the part of taxpayers facilitates compliance on the part of the latter. The administrative costs of tax compliance vary considerably from one State to another [4].

In this context, the Italian legal system would seem to have moved in time by introducing, through the Statute of Taxpayers' Rights and, more specifically, with Article 11 thereof, the general regulation of the institute of the tax appeal, an absolutely central tool - as will be discussed below - for the purposes of a fruitful collaboration between the tax authorities and taxpayers and, therefore, declining the institute in the context of me-gap projects, between subjects who intend to promote projects with a high economic and social impact and the tax authorities.

2.1 *Tax Appeals in the Italian Legal System*

The institution of the interpello finds its general definition in Article 11 of the Italian Statute of Taxpayers' Rights [5].

5] The provision specifies that an interpello is a request that the taxpayer addresses to the tax authorities containing, in essence, a question in order to obtain a preliminary opinion in relation to a concrete and personal case concerning the scope of tax provisions for which there are doubts as to their interpretation, the existence of conditions for the adoption of specific tax regimes, and the application or non-application of various legal provisions relating to taxation.

Generally speaking, four types of interpellation are provided for within the Italian legal system:

(i) ordinary interpellation;

(ii) evidentiary interpellation;
(iii) anti-abuse interpellation;
(iv) disapplication interpellation.

In addition to these classic types of appeals, a fifth type has been added over time, known as the 'new investment appeal' (the content and grounds for inadmissibility of new investment appeals are regulated separately).

As will be discussed below, it is precisely this type that would seem to be the most relevant for those wishing to carry out mega-projects.

With regard to the first type of interpello, i.e., the ordinary interpello, it must be said that this is distinguished, in turn, into an ordinary so-called 'pure' interpello - in which the object of the petition is the application of the tax rule (thus excluding acts that do not have a regulatory character) - and an ordinary so-called 'qualifying' inter-pello, in which the object of the petition is the correct regulatory qualification of the concrete case.

Generally speaking, apart from the differences now referred to, the ordinary interpello may be used when there are conditions of objective uncertainty as to the correct interpretation of provisions, or as to the correct qualification of a case for tax purposes.

Different considerations apply to the so-called 'probatory' interpello The category of the probatory interpello includes all those petitions aimed at obtaining an opinion from the tax authorities on the existence of the conditions for access to a certain tax regime, or on the suitability of the evidence required by the law for access to a certain tax regime.

The possibility of submitting an evidentiary interpello is limited to 'cases expressly provided for', so that, in the present case, the procedure is not activated in relation to any case in which the taxpayer considers an assessment by the tax administration to be useful, but only in the hypotheses in which such an option is expressly provided for by law.

Still different is the so-called 'anti-abuse' type of interpello, which is aimed at obtaining an opinion on the application of the abuse of rights rules to a specific case. More precisely, the petitions submitted cannot be limited to generically requesting the Agency's opinion on the abusiveness of a specific transaction or case, but must define, in detail the qualifying elements of the transaction or transactions; the taxable sector with respect to which the transaction raises the applicative doubt; the precise reference rules, including those liable to be challenged in terms of abuse of law with reference to the transaction represented; the valid extra-tax reasons, not marginal, also of an organisational or managerial nature, which respond to the purpose of structural or functional improvement of the enterprise or of the professional activity of the taxpayer.

Lastly, further different is the so-called 'disapplicative' interpello. The latter, in fact, constitutes the only category of mandatory interpello that remains in the system [6]. With the disapplicative interpello, the taxpayer requests an opinion on the existence of the conditions that legitimise the disapplication of tax rules that, in order to counter evasive behaviour, limit deductions, deductions, tax credits or other subjective positions of the taxpayer.

Without prejudice to the foregoing, from a technical point of view, the content of the interpello petitions, valid for all the types provided for, is expressly discipli-ned by Article 3 of Legislative Decree No. 156/2015. The absence of certain elements determines the inadmissibility of the interpellation petition while other deficiencies may be regularised.

In light of the characteristics of the interpellations now defined, it clearly emerges that even for those who are planning to implement a megaproject within the Italian system the institute can serve as an instrument of cooperation with the tax authorities in order to collect very useful information both to know the exact scope of the tax rules and to understand whether the operations to be carried out can be considered elusive operations and, finally, to know the exact tax regime within which the operating entity would fall under the applicable Italian legislation [7]..

In this context, particular attention must also be paid to the last type of interpellation provided for by Italian law, i.e. the interpellation on new investments. The latter, in fact, as is clear also from its name, represents a tool through which large companies wishing to invest in Italy - for the realisation, for example, of megaprojects - can gather useful information and agree on specific tax treatments by directly interacting with the tax authorities.

2.2 *The New Investment Interpellation as a Tool for Fiscal Planning of Megaprojects*

The s.c. "*interpellation on the new investments*" [8] is a particular type of interpellation usable by companies, both domestic and foreign, which:

(i) intend to make investments in the territory of the State amounting to no less than twenty million euros (ceiling set by Article 01 Decree-Law no. 119/2018 (so-called Tax Decree), converted into Law no. 136 of 17.12.2018. which replaces the previous threshold of thirty million euros) (a total sum that may therefore also cover several years); a provision that applies to interpello petitions filed as from 1 January 2019;
(ii) have significant employment effects, in relation to the activity in which the investment takes place, as well as lasting effects.

Such companies may submit a petition to the Revenue Agency concerning: the tax treatment of their investment plan; the tax treatment of any extraordinary transactions envisaged for the implementation of such plan; the assessment as to whether or not the company is a going concern; the prior assessment as to whether or not there is any abuse of tax law or avoidance; the existence of the conditions for the disapplication of anti-avoidance provisions and access to any regimes or institutions provided for by the tax system.

With regard to taxes that do not fall within the competence of the Revenue Agency, the latter forwards the investor's request to the competent bodies, which independently render a response. In short, it is an institute aimed at incentivising and facilitating important investments in our country. In fact, with such an interpellation, the investor should not have to fear possible assessments or objections by the Administration [9].

The discipline does not seem to place particular limitations on either the subjects or the type of investment to be made.

It can be companies of any legal form: both Italian and foreign.

With regard to the investment, the only requirements seem to be the consequences it must produce and the size of the investment, which can then take any form.

With reference to the nature of the investment, it has been clarified (Ministerial Decree of 29 April 2016) that it may concern both the creation of new economic activities and, alternatively, the expansion of pre-existing economic activities, the diversification of the production of an existing production unit, the restructuring of an existing economic activity in order to allow the company to overcome or prevent a crisis situation, and operations concerning participations in a company.

In order to determine the value of the investment, all the financial resources, including those of third parties, necessary for the undertaking to implement the investment plan must be taken into account [10].

From a formal point of view, as specified by the Italian tax authorities, the application must be submitted in Italian, while the accompanying documentation may be submitted in English, French, Spanish and German [11].

2.2.1 The Content of the Interpellation Request on New Investments

In addition to the above-mentioned profiles, particular attention must be paid by those who intend to plan the implementation of megaprojects within the Italian legal system to the content of the applications.

In this sense, it is useful to briefly go over the essential contents that the application should necessarily contain.

The application must contain.

- the identification elements of the company (name, legal representative, registered office or tax domicile, if different from the registered office, the tax code or VAT number or other identification code of the company, as well as the indication of the contact details, including electronic data transmission, of the domicile for the interlocutory appeal procedure at which it is requested to forward the communications relating to the procedure). Where more than one party is involved in the investment, the application must contain the identification details of all of them;
- a detailed description of the investment plan, on which the Revenue Agency's assessment is requested with reference to the tax treatment of the same and the corporate operations planned for its implementation. The description must necessarily specify: the amount of the investment, not less than EUR 30 million, and the methodology followed for its quantification; the timing and methods of its implementation; the significant employment effects, in terms of increasing or

maintaining the level of employment, to be assessed in relation to the activity in which the investment is made and the lasting effects, also in quantitative terms, of the investment in question on the Italian tax system

On this point, it should be noted that the petitioner must indicate only the effects on the Italian tax system that are directly and immediately deriving from the specific plan referred to in the petition, with no importance being attached to the further and more general effects deriving "in cascade" from the investment on the Italian tax system and economy" (Agenzia delle Entrate Circular 1.6.2016 no. 25, § 4).

– the specific tax provisions whose interpretation is requested or in relation to which the assessment of the possible abusiveness of the transactions related to the investment plan is requested
– the specific anti-avoidance provisions of which the disapplication is requested and the specific regimes or institutions to which access is requested;
– the exposition, in a clear and unambiguous manner, of the tax treatment that the taxpayer considers to be correct in relation to the investment plan, with an explanation of the solutions and conduct that the petitioner intends to adopt in relation to its implementation
– the signature of the petitioner or its legal representative or general or special receiver (in which case the power of attorney must also be attached).

In the light of the above considerations, it is easy to understand both the complexity of the institution and, at the same time, its great usefulness both for those who intend to open up to new businesses characterised by a strong economic and social impact within the Italian territory - characteristics, the latter, which are certainly made their own by those who realise mega-projects - and for the Financial Administration, which will be able to immediately enter into contact with the large eco-nomic players interested in investing in the Italian economic context.

Reasoning along these lines, it becomes clear how the interpello on new investments truly represents the expression of a new way of conceiving the relationship between taxpayers and Tax Administration. A new way because it is aimed not only at preventive dialogue but at identifying the best tax profiles for those wishing to create a multi-level value within the Italian system.

The importance of the interpello on new investments with respect to the subject we are dealing with - and therefore in relation to the evolution of the dialogue between taxpayers and tax authorities in the specific area of megaprojects - emerges even more clearly also in light of the very close connection between the type of interpello now referred to and the so-called regime of ‘collaborative compliance’ to which we briefly referred above and which now deserves greater attention.

3 The So-Called Collaborative Compliance Regime as an Example of a New Dialogue between Tax Authorities and Taxpayers

The decree on legal certainty (No. 128/2015, in force since 2 September 2015), in order to promote the adoption of enhanced forms of communication and co-operation and to foster the prevention and resolution of possible counter-verses, has established the collaborative compliance regime between the Revenue Agency and taxpayers with a system of:

- detection:
- measurement;
- management
- and control of tax risk.

It was emphasized that the regime envisaged by the regulation is entirely characterized by transparency, with duties imposed on both the Revenue Agency and the taxpayer, in order to provide certainty with regard to the company's tax risks.

In particular, through a system of monitoring and exchange of information in advance with the Revenue Agency, the contestable con-duct of the taxpayer who risks operating in violation of tax rules is delimited with certainty, thus allowing him to operate within well-defined margins and minimizing the risk of a possible unintentional violation of tax rules [12].

In terms of the subjective scope of operation of the scheme, access to the scheme is reserved [13]:

- for the years 2022, 2023 and 2024, to resident and non-resident entities (with permanent establishment in Italy) that realise a volume of business or revenues not less than €1 billion. The new entry threshold was thus further lowered by the Ministerial Decree of 31 January 2022 [14];
- to resident and non-resident entities (with permanent establishment in Italy) that realise a turnover or revenues not lower than €1 billion and that have submitted an application to join the Collaborative Compliance Pilot Project;
- companies intending to implement the response of the Agenzia del-le Entrate, provided following a request for an interpellation on new investments (to Article 2 of Legislative Decree No. 147 of 14 September 2015) in-dependently of the volume of business or revenues;
- to persons who are part of the VAT Group of companies already admitted to the scheme, regardless of the volume of business or revenue.

That being said, the cooperative compliance regime is distinguished by the existence of two absolutely significant elements that define its concrete usefulness not only vis-à-vis the Tax Authorities but also, if not above all, for those who, operating through a company falling within the parameters identified above, carry out specific activities in Italy (activities that could well include the direct support or implementation of mega-projects):

(i) firstly, a peculiar feature of the co-laborative compliance regime is the establishment of a constant communication and exchange of information between the adhering taxpayer and the Revenue Agency. These exchanges take place during a series of formalised meetings that are referred to as constant and preventive interlocutions [15].

These constant and preventive interlocutions with the Revenue Agency fulfil the very important function of reaching a common assessment of situations that may generate tax risks prior to the submission of tax returns. Such a provision offers the opportunity to deal with uncertain situations by means of a prior discussion of factual elements that may also include the anticipation of control and thus lends itself to the prevention and early resolution of potential tax disputes.

(ii) secondly, of great importance are the reward aspects arising from adherence to the regime [16]. On this point, in fact, Article 6 of D. Lgs. 128/2015 provides, for companies wishing to join the scheme: 1) an abbreviated procedure of preventive interpello in the context of which the Revenue Agency undertakes to answer the companies' questions en-route to forty-five days from the receipt of the application or any supplementary documentation requested; 2) the application of penalties reduced by half, and in any case to an extent not exceeding the minimum amount, with suspension of collection until the final assessment, for risks communicated in a timely and exhaustive manner, where the Revenue Agency does not agree with the company's position; 3) the exemption from providing guarantees for direct and indirect tax refunds for the entire period of the regime.

In the light of what has been specified so far, therefore, the regime of collaborative compliance would really appear to be an instrument of new collaboration and new dialogue between the tax authorities and the taxpayer, putting the latter in a position to verify, in a more effective and conscious manner, the business risks arising from the implementation of a specific and significant (in quantitative terms) transaction within the Italian territory.

Among the risks that the regime allows to be taken into account is, moreover, precisely the 'fiscal re-risk', i.e., the risk of operating in violation of tax regulations or in contrast with the principles or purposes of the tax system [17].

In this perspective too, it is easy to understand how the regime of collaborative compliance represents a further, not alternative, tool for a new dialogue between tax authorities and taxpayers that can also be used - provided that the requirements set forth in the regulations are met - by those already operating in Italy or intending to open up to the Italian system, who plan to implement projects with a high economic and social impact within the Italian territory, i.e. those programmes that we have defined here as 'megaprojects'.

4 Conclusions

In light of the considerations formulated in the preceding paragraphs, it clearly emerges how the dialogue between the tax authorities and the taxpayer - or the absence of dialogue between them - can be decisive with respect to the possibility of achieving a more widespread legal certainty and tax certainty.

The benefits that derive from a preventive (and endoprocedural) confrontation between the tax authorities and the taxpayer are, therefore, manifold and the various types of institutions referred to move in this direction.

And clearly, all taxpayers, without exclusions, can benefit from the aforementioned benefits. The only exception is the collaborative compliance regime which, as mentioned above, requires the existence of absolutely specific subjective and objective conditions usually possessed by large economic operators with very high turnovers.

Precisely within this category of subjects could fall those who intend to plan in order to launch projects with a high economic and social impact such as megaprojects.

The realisation of large-scale projects, therefore, may well intersect not only with the environmental and sustainability issues, but also with the issue of taxation and, specifically, with the relationship between the tax authorities and taxpayers, especially since this relationship is capable of facilitating and, at the same time, if misdirected, of hindering the realisation of projects within the national territory.

In this perspective, therefore, starting from the consideration that those who realise a mega-project are subjects who - given the scope of the investments to be made and the duration (often not short, neither with reference to the time of realisation nor with regard to the subsequent management of the project realised) - intend to remain within the Italian territory, it would seem that these economic operators, precisely because of these circumstances, can (and must) enjoy a privileged relationship with the financial administration.

This is especially so with regard to the need to establish and cultivate a dialogue aimed at the preventive elimination of any uncertainty of a fiscal nature relating to investments made by those who intend to realise megaprojects in Italy.

References

1. Allena M (2020) Reshaping tax policies for sustainable megaprojects in the post pandemic era. In: Cantoni F, Favari E (eds) Megaprojects, Torino 9–13
2. Melillo C (2020) Compliance fiscale, Milano, p. 9.
3. Marchetto R (2015) Tendenze moderne nel rapporto tra Contribuente e Fisco, in Dialoghi di diritto tributario tra attualità e prospettive
4. Valente P (2016) Il rapporto Fisco-contribuente secondo i modelli di Statuto proposti a livello europeo. In: Il Fisco, 38:3639–3645
5. See L. 212 del (2000)
6. All other types of interpellation, in fact, are merely an option for taxpayers
7. One Fiscale, chapter "Interpello tributario"

8. See art. 2, D.Lgs. 147/2015
9. One Fiscale, chapter "Interpello sui nuovi investimenti"
10. If the investment is realised by groups of companies or clusters of companies, reference should be made to the total value of the unit investment (resulting from the sum of the value of the individual investments of all entities participating in the initiative)
11. See Agenzia delle Entrate, circolare 1° giugno 2016 n. 25
12. See One Fiscale, chapter "Adempimento collaborativo"
13. See art. 7 del D. Lgs 128/2015, punto 2 del Provvedimento dell'Agenzia del 14 aprile 2016 - pdf e DM del 31 gennaio 2022)
14. On this point, it should be noted that the Ministerial Decree of 30 March 2020 had already reduced the threshold for entry into the scheme for the financial years 2020 and 2021 from EUR 10 billion to EUR 5 billion.
15. See art. 4, del Provvedimento direttoriale del 26/05/2017 in attuazione dell'art. 6, comma 1, D.Lgs. n. 128/2015.
16. https://www.agenziaentrate.gov.it/portale/schede/agevolazioni/regime-di-adempimento-collaborativo - Agenzia delle Entrate (Sito Web Ufficiale)
17. See Ferroni B (2016) Il tax control framework nel regime di adempimento collaborativo, In: Il Fisco, 38:3621–3631

Managing Inflation Risk in Megaprojects: Contract Law and International Best Practices

Francesco Zecchin

Abstract The writing deals with the problem of the current inflation's level in megaprojects, in order to avoid that it leads to their failure. Firstly, the paper focuses its attention on the megaprojects' contracts concluded when the purchasing power has been low and stable for a long time. In relation to these cases, the present devaluation of money is an extraordinary and unpredictable event, at least because it is due mainly to the war between Russia and Ukraine, and partly to the long Covid-19 economics effects too. As a result, both under many European contract laws and international models it can be addressed in terms that involving, directly or indirectly, the client in a renegotiation, whose goal is to rebalance the megaprojects contract's performances considering the new reality. Secondly, the author tries to identify best practices in relation to contracts that has been concluded when the inflation was already at current level, or it was predictable that it would have been such a strong rise. In these situations a good solution would be the so-called stabilization clauses, as foreseen by a recent Italian statute in the field of public procurement, and for megaprojects contracts by NEC and Fidic models.

Keywords Project Success Factors · Megaprojects Contracts · Inflation · Contract Law · Best Practices · Stabilization Clauses · Private Autonomy

1 Inflation, Law and Megaprojects

After an event that has not happened for nearly a century, namely a global pandemic like the Covid-19 one, megaprojects must deal with another change of circumstance that was a distant memory, at least in the western world, that is a high inflation. While in the last two decades it was anchored of low levels, during the last months inflation has peaked at 10%, indeed [2].

Of course, this is not a problem considered by the law in cases of immediate performance contracts. If the price of a goods drops a lot after a purchase is performed,

F. Zecchin (✉)
Università Cattolica del Sacro Cuore (Sede di Piacenza-Cremona), Via Emiliana Parmense, 84 – 29122 Piacenza PC, Italy
e-mail: francesco.zecchin@unicatt.it

E. Favari and F. Cantoni (eds.), *Complexity and Sustainability in Megaprojects*, Lecture Notes in Civil Engineering 342,
https://doi.org/10.1007/978-3-031-30879-6_16

the buyer does not have ground for a legal remedy, and he has only himself to blame, eventually. On the opposite, if the contract is one of continuous, periodic, or deferred performance, so called long-terms contracts, rise in costs is something that might put a strain on a juridical relationship which is still ongoing, and therefore the law cannot fail to take an interest in it [26]. In other words, in case of a contract which is not one of instantaneous execution but long-lasting, the law takes into consideration the time between its conclusion and its fulfilment, precisely from the changes of circumstances that distort the economic balance agreed by the parties.

One of these, and maybe the most important, is a strong devaluation of money [30]. Following the ancient nominalism's principle [3], debtors have to pay the amount provided by the contract, although its actual value has meanwhile changed. The law choses that this kind of risk is borne by the creditor, and not by the debtor [20].

However, there are some exceptions, often linked to the inflation, precisely. It was born as problem during the middle-age due to a choice of the authority about the physical weight of the coin, but it became the topic as we know today after the rethinking of the link between currency and gold made in the nineteen century [13, 14]. Inflation means not just a simple increase of some products' price, but a general increase in prices related to the purchasing value of money [14]. Of course, rules different than the nominalism's principle are related to inflation's high levels, for instance the so-called "double digit", as evidenced both by the hotly debate and by the several regulatory actions aim to managing this problem after the first world war [15] and in the seventies and eighties of the last century [30].

As it has been shown, strong devaluation of money is particularly harmful to constructions agreements [30], even more if they are megaprojects' contract, whose performances are referred over a long period of time. Increase in the price of energies, materials, labor, or equipment means uncertainty of the parties' obligation, and this makes true the risk of slowing down or even stopping the realization of megaprojects. For instance, because the costs agreed during negotiation will not cover the real value of the contractor's performance, or the client will not be able to pay the new final cost, if the choice was that it will be determined at the end of the works. One could perhaps say that inflation's current level is one of the main reasons that confirms the truthfulness of so-called iron law of megaprojects: "Over budget, over time, under benefits, over and over again" [17].

2 Current Inflation as an Extraordinary and Unpredictable Event

A first approach to address this issue is considering the high inflation level registered in these months as an extraordinary and unpredictable event. After all, it relates mainly to the war between Russia and Ukraine, but partly for the long Covid-19 economics effects, too.

As a result, the question is not a merely problem of prices' increase, which are normally solved with more outlay for the parties. The answer is much more complicated, because the contract must deal with an inconvenience that could not be imagined by the parties at the time of the contract's conclusion. Right this circumstance opens up chance for not leaving of megaprojects, both in the light of law contract and international best practices.

In the former point of view, tradition of civil and common law has long since developed rules designed to face contingencies like an increase of the prices due to an extraordinary and unpredictable event, such as inflation [23]. However, there are substantial differences between these two law systems.

Into the European one, we have no identical solutions, of course. But in general it can be said, that almost all of them recognize the costs' change of the work, because of the occurrence of extraordinary and unpredictable devaluation of money between the stipulation of the contract and its performance, as a basis for modifying the contract [5], although the situation is evolving.

The German law, for instance, pays traditionally more attention than others to this issue [33], and it admits that the contractual relationship may be terminated or modified when - as expressed in a famous case - "to maintain the original contract would produce intolerable results incompatible with law and justice" (BGH 25 May 1977, NJW 1977, 2262, 2263; BGH 13 November 1975, NJW 1976, 565, 566). This rule was originally based on the good faith principle, but today it expressly regulated by the new § 313 of BGB: in a case of *Störung der Geschäftsgrundlage* both parties have a right to claim an agreement which adapts the contract to the new reality (1). If the adaptation results not reasonable, the contract can be terminated by the withdrawal of the party aggrieved by the change of circumstances (3), but it is not clear if judge can *ex officio* dissolve or change the contract [16]. However, it quite sure that the § 313 is applicable in case of extraordinary inflation caused by an unexpected event [19]. After all, the legislative history of the 2002 reform of BGB expressly refers to distortion of the contract's economic balance due to strong devaluation of money [18].

On the opposite, into the French law the change of circumstances used to be irrelevant, unless in the case of administrative contracts: following the *imprévision* rule a *bouleversement de l'économie du contract* might lead to termination or to a modification by the court of this kind of contract [6, 23]. Nevertheless, this is not an exceptional rule anymore [8], because the new Article 1195 of the *code civil* provides that the party, whose obligation becomes excessively onerous because of an unforeseeable change of circumstances, can ask to the other one a renegotiation of the contract (par. 1). If there is a refusal or the renegotiation fails, parties can agree on termination of the contract or asking to a court its adaptation. In case there is no agreement within a reasonable period, the same demands can made by only one party (par. 2). As French legal scholars has shown, one of the typical cases that the legislator wanted to be ruled was the one related to a performance became more expensive due to high inflation [11].

The Italian model is someway in the middle [26]. In case of extraordinary and unpredictable event which affects the economic contractual balance, the disadvantaged party has the right to request the dissolution of the contract (Article 1467, para 1, *codice civile*). On the other hand, its termination may be avoided if the other party offers an equitable modify of the contract that removes the quantitative contingencies' effects (Article 1467, para 3, *codice civile*).

It has been said that the application of this rule in cases of strong devaluation of money would have a negative influence on the distribution of wealth between social classes, actually: from entrepreneurs, who are creditors of money, to consumers, who are creditors of goods or services [3, 29]. These are insightful remarks - because it is true that applying the Art. 1467 inflation risk is not deleted, but just relocated on consumer, which is not the party who can best manage it and can sustain its consequences - but they have sociological or legislative politics nature [14, 22], and are of less value in case of contract between enterprises. It is not by chance that case law has been applying this rule since 1948 (Cass. 23rd January 1948, n. 88, in *Foro it.*, 1948, I, c. 298). And after all, under the previous *codice civile*, which did not have a norm like the Art. 1467, the problem of inflation's high level was solved recalling the Latin principle *rebus sic stantibus* to make a dissolution of a contract whose economic balance had been distorted by an unpredictable event [20].

As a result, in the Italian legal system duty to renegotiate is not the general remedy for an extraordinary and unpredictable level of inflation. But this is the rule indirectly provided for specific contracts, for instance procurements ones. In cases of an increase beyond a certain limit (10%) of the agreed cost, and not prices [24], of materials or workforce due to unforeseen circumstances, it is possible to request a revision of the price (Article 1664, para 1, *codice civile*). The norm means that risk related to the costs inherent in procurement contract is on the shoulders of the contractor up to an increase of 10%, beyond which it does not be moved at all to the client, but there is a right to renegotiation [12]. Of course, this mechanism is provided only for the procurements whose costs have been specified during the negotiation, while it is out of the picture if the parties agreed that the costs will be determined at the end of the woks.

To be more precise, originally the Art. 1664 was not applied in the case of strong devaluation of money, but this opinion has been overruled since the last century 80 s [23]. And today it is the rule which can help the contractor not just in the case of an extraordinary and unpredictable increase of some costs, but in case of inflation with the same features, too [22].

Similar regulation, at least in terms of their *ratio*, are provided by the Article 106, par. 1, lett. c), of Legislative Decree No. 50 of 18th April 2016 for public procurement [25]. According to this norm, public procurement contracts can be renegotiated, without a new assignment procedure, if the change's need does not distort the nature of the contract and it is due to an unexpected and unpredictable event for the parties. And following the Article 7, par. 2-*ter*, of Legislative Decree No. 36 of 30th April 2022, the mechanism has to be construed as meaning that it is also a matter of inconveniences which significantly affect materials' cost of the work.

Moreover, but again only for public procurement, the Italian Parliament introduced some rules of price review in relation to the high material cost's increase, similarly to what happened both after the First World Was and in the last century 70 s [14]. An example is the Art. 1-*septies* of Law Decree No. 73 of 25th May 2021, which provides for some compensation prices mechanisms "in order to cope with the exceptional price increases of some construction materials in 2021". The disposition was thought for the expenditure increased during the Covid-19 pandemic, but the Ministerial Decree No. 241 of 28th July of 2022 has extended its validity in the 2022.

Moving to the English common law, things are quite different. It would not be useful to recall the change of circumstances, because in that legal system a major expenditure for the debtor not amounting to impossibility [4]. From this point of view, the so called "sanctity of contract" is untouchable, and the major costs due to an extraordinary and unpredictable inflation's level traditionally do not represent something relevant in terms of debtor's obligation. On the other hand, it is conceivable that expenditure for one of the parties would be increased to the point that performance is "radically different". In other words, something closed to the so-called frustration of the contract [33]. For instance, in *Metropolitan Water Board v Dick Kerr* [1918, AC 119] the House of Lords held that the contractor was not obliged to complete the work, discontinuing during the First World War due to the wartime government restrictions, because in the new and unknown economic circumstances it would be "radically different". Nowadays, there would be a chance to save the contract, although the parties do not make a new agreement: an intervention of the judge in the light of the "implied term theory" or imposing upon the parties the just and reasonable solution [33]. *Metropolitan Water Board v Dick Kerr* has something in common to the situations caused by current inflation, but in that case the big change in the contract's economic balance was due to a very long period of temporary impossibility of performance. This is the reason why it can be led back to the frustration rule, otherwise it should be ruled as an inconvenience that does not have any consequence to the parties' obligations [7]. In short, the principle laid down by the House of Lords in the 1918 seems no applicable to our issue because one of its condition is not fulfilled.

Anyway, it is possible, and maybe probable, that outstanding contracts have a clause which can deal with the risk of extraordinary and unpredictable inflation, applicable instead of general or specific contract law rules. The reference is not to a clause expressly dedicated to an unusual costs' increase, but to a hardship one. Clauses such as these are frequently used, precisely with the aim to not leave the megaprojects in case of contingencies which make the performance more onerous for one of the parties, such that it leads to a fundamental alteration of the original contract's equilibrium.

For instance, two of the most important international models, like Fidic and NEC ones, suggest adopting hardship clauses which can be useful to deal with this kind of setbacks. Both provide for the employer's broad discretion, albeit based on somewhat different criteria, in the requests for changes, and an obligatory renegotiation procedure to be pursued in the event of variations not related to the will of the parties [9, 31]. Furthermore, they identify a person assigned with the task of mediating between the parties: the Engineer for the FIDICs and the Project Manager for the NECs. These clauses are based on the criterion of economic efficiency and their purpose is precisely to get to the solution of the problem as quickly as possible, and without disputes, protecting both parties' interests to the maximum extent possible [1].

To summarize, if today's level of inflation problem is seen as an extraordinary and unpredictable event - as indeed it is in relation to the contracts concluded when the devaluation of money was not so strong, but it has been low and stable for a long time, and not just the war between Russia and Ukraine was unpredictable, but its aftermaths too, like the ones of Covid-19 pandemics - it can be addressed in terms that does not lead necessarily to the failure of megaprojects. Both under contract law and international models, the trend is not to leave all the risk on the shoulders of the contractor, but to involve the client in a renegotiation whose goal is to rebalance the megaprojects contract's performances considering the new reality.

3 Best Practices for New Contracts: Stabilization Clauses?

The scenario is completely different for contract that has been concluded when the inflation was already at current level, or it was predictable that it would have been such a strong rise. These months have shown that inflation near 10% is not a contingence of past ages, but something that can happens in our days, too. The events that we have been living since the winter of 2022 makes current costs' increase like an ordinary and predictable event: this is the inflation's levels of the present time. As a result, both contract rules and contract's clauses - that we have just seen - will not be applicable, because they are related to an extraordinary and unpredictable event [22]. Then, major costs will remain on the shoulders of the party who has to fulfill, as it happens in relation to a normal and foreseeable inflation [20].

In other words, while in the last years the question of costs' increase, in the negotiation too, was frequently a secondary topic, because inflation's levels were very low and stable, it will be a crucial for the success of megaprojects from now on. Providing clauses dedicated to inflation, which can better deal with this problem than the hardship ones, will mean the safety of megaproject. After all, one of the aims of the so-called "freedom of contract" is that parties predict and manage future occurrences not just better than the law, but also as better as they can.

This means - in general and first - that stipulating contract by which the final cost of the work is not agreed in advance, but contractor has simply a right to the amount actually necessary for carrying out or managing the works, is quite dangerous. The cost of purchasing plant, materials, and other resources such as people, equipment and fuel would be included in the schedule of cost components. Then, their increase would lead to an increase in the calculation of the final cost of the work,s and client would carry this risk, without controlling it. On the opposite, one of the main goals of a contract is to manage risks which cannot be removed. It is, therefore, essential taking measures to prevent inflation from distorting the megaprojects contract balance, that is going beyond the nominalism principle and trying to spread the inflation's risk more widely across client and contractor [13]. After all, legal instruments, like contract, aim to a better exercise of business activity and its success [10, 32], in megaprojects too [21].

In this perspective, a good starting point could be the index clauses, also known as stabilization clauses. They aim to protect the contracting parties against inflation linking the value of the performance to a broadly-based price index, which could have monetary nature, if it is related to the value, for instance, of gold, or economic nature, if it is related to a parameter which expresses the general increase of the costs in a specific field, like living cost or cost of building homes [13]. As a result, "the real economic value of the parties' bargain from changes in the value of money" is maintained [30].

Thanks to a clause such as this, the cost of megaprojects will be specified at the end of the works, but it will not be an astonishment for the contractor, because the index is already known. Moreover, what is due by the contractor is not the sum of the single component's final cost, but a total calculated using an index that expresses the overall increase of the works' cost [13]. It is not by a chance that the Italian law lays down for the public procurement stipulated in 2023 to provide a specific clause of a price's revisions (Art. 29 Law Decree No. 4 of 27th January 2022).

From a practical point of view, in case of international megaprojects it is essential checking if this kind of clauses are in general valid following the law applicable to the contracts, as it is in Italian law, but for instance in some cases it was not in France [13]. If the answer is yes, as it has been shown, they should be kept as simple as possible; it should be related to the most appropriate index to needs of the parties; the index should be described with considerable care and specificity; the manner of adjustment should be drafted with great care, and the index does not have to be usurious [30] or, more generally, the clause content cannot be against the Italian law or, if the megaprojects involved foreign parties, the law applicable to the contract.

In the field of megaprojects, the stabilizations clauses proposed by New Engineering Contract (NEC) and FIDIC models deserve particular attention.

In the first one, the risk of increasing of costs is provided by the Secondary Option X1 (Price adjustment for inflation), which "involves setting a base date before the tender date, then calculating the price adjustment before each assessment date based on the changing value of an agreed prices index or indices" [28].

If using this option in relation to priced contracts meaning (called A and B) - under which, absent any changes to or breaches of contract during the works, the contract comprising the activities or measure must be delivered at the agreed price or rate - a contractor's interim payments are adjusted for inflation by calculating the "Price Adjustment Factor" (PAF), and the client pays an extra amount for inflation in each assessment. For example, if the interim amount was 500.000 Euro and the calculated PAF in the month was +8,5%, then the calculated amount for inflation would be 42.500 Euro, which would be applied to the interim payment. This should recompense a contractor for inflation, but only at amounts calculated through the indices chosen [27].

If using this option for target cost contracts (called C and D) - under which the contractor will initially be reimbursed the defined cost for delivering each element of the works, which relates to the total of the prices (the target) - the "Price Adjustment Factor" (PAF) is calculated in a similar way but is then deflated by itself before being factored against the interim amount. For example, if the interim amount was 500.000 Euro and PAF was +8,5%, the calculated adjustment for the total of the prices would be £39,170.51. In this instance, the target is adjusted to allow for the inflation contained within the defined cost at the time of the interim payment [27].

To summarize, following the Secondary Option X1 client pays an extra amount for inflation in each assessment in the case of priced contracts, and adds an amount for inflation to the total of the prices in target contracts [28].

In the second one - that is Fidic models - the mechanism is provided into the clause 13.8 of Red Book:

$$Pn = a + b\, Ln/Lo + c\, En/Eo + d\, Mn/Mo + \ldots$$

where:

"Pn" is the adjustment multiplier to be applied to the estimated contract value in the relevant currency of the work carried out in period "n", this period being a month unless otherwise stated in the Appendix to Tender;

"a" is a fixed coefficient, stated in the relevant table of adjustment data, representing the non-adjustable portion in contractual payments;

"b", "c", "d", ... are coefficients representing the estimated proportion of each cost element related to the execution of the Works, as stated in the relevant table of adjustment data; such tabulated cost elements may be indicative of resources such as labour, equipment and materials;

"Ln", "En", "Mn", ... are the current cost indices or reference prices for period "n", expressed in the relevant currency of payment, each of which is applicable to the relevant tabulated cost element on the date 49 days prior to the last day of the period (to which the particular Payment Certificate relates); and

"Lo", "Eo", "Mo", ... are the base cost indices or reference prices, expressed in the relevant currency of payment, each of which is applicable to the relevant tabulated cost element on the Base Date.

In other words, the amounts payable to the contractor shall be adjusted for rises or falls in the cost of labour, goods, and other inputs to the works, by the addition or deduction of the amounts determined by a formula. It is based on an adjustment multiplier, calculated considering a fixed coefficient, *plus* a coefficient representing the non-adjustable portion in contractual payments, *plus* other coefficients representing the estimated proportion of each cost element related to the execution of the works related to the current cost indices or reference prices for period *n* and cost indices or reference prices, expressed in the relevant currency of payment. This adjustment multiplier must be applied to the amount otherwise payable to the contractor, as valued in accordance with the appropriate schedule and certified in payment certificates, shall be determined from formulae for each of the currencies in which the contract price is payable.

Of course, following the NEC and Fidic model clauses, inflation's risk is on the shoulders of contractors, but it is a risk quite managed. This means keeping alive the chance of megaproject's success, because the danger of strong devaluation of money would have a rule, so that megaproject won't be a leap in the dark.

For the same reason, the case of inflation higher than the current one - which is not a ruled-out event, but it would be again an extraordinary and unpredictable event covered by contract law - should have a specific rule into the contract, at least as inconvenience that implements the hardship clause.

References

1. Amore G (2007) Appalto e claim. Padova, Cedam
2. Arnold P. https://www.ft.com/content/0a7d5bba-4355-4e4c-9f2a-d42b2298fed3. Accessed 6 Nov 2022
3. Ascarelli T (1968) Obbligazioni pecuniarie. Bologna-Roma
4. Atiyah PS (1995) An introduction to the law of contract. Clarendon Press, Oxford
5. von Bar C, Clive E, Schulte-Nölke H (2009) Principles, definitions and model rules of European private law: draft common frame of reference (DCFR), 2. Obligations and corresponding rights (contd.), Sellier. European Law Publisher, München
6. Barcellona M, Maugeri MR (2007) Il mutamento di circostanze e l'obbligo di rinegoziazione. In: Castronovo C, Mazzamuto S (eds) Manuale di diritto privato europeo, II, pp 521–542, Giuffré, Milano
7. Beale H, Twigg-Flessner C (2021) Covid-19 and the English contract law. In: Hondius E, Silva S, Nicolussi M, Coderch A, Wendehorst PS, Zoll-F C (eds) Coronavirus and the Law in Europe, pp 461–490, Intersentia, Cambridge
8. Benatti F (2018) L'imprévision nel *code civil* riformato. Giur. it., I: 1302–1309
9. Besaiso H, Fenn P, Emsley M, Wright D (2018) A comparison of the suitability of FIDIC and NEC conditions of contract in Palestine. Eng Constr Archit Manag 25(2):241–254
10. Betti E (1953) Teoria generale delle obbligazioni I. Prolegomeni: funzione economico-sociale dei rapporti di obbligazione. Giuffrè, Milano
11. Borghetti JS (2021) Non-performance and the change of circumstances under French Law. In: Hondius E, Silva S, Nicolussi M, Coderch A, Wendehorst PS, Zoll-F C (eds) Coronavirus and the Law in Europe, pp 509–525. Intersentia, Cambridge
12. Boselli A (1948) Rischio, alea ed alea normale del contratto. Riv. trim. dir. proc. civ., II:769–795
13. Carbonetti F (1980) Clausole di indicizzazione. In: Irti N (eds) Dizionario del diritto privato. Diritto civile, pp 125–156. Giuffré, Milano
14. Carli G (1981) L'inflazione e i suoi riflessi sull'ordinamento giuridico. Giustizia civile, 6: pp 298–307
15. Dawson JP (1934) Effects of inflation on private contracts: Germany, 1914–1924. Mich Law Rev 33(2):171–238
16. di Majo A (2011) Modernisierung delle obbligazioni nel diritto tedesco. In: Zweigert K, Kötz H (eds) Introduzione al diritto comparato. II. Istituti, pp 269–279, Giuffré, Milano
17. Flyvbjerg B (2011) Over budget, over time, over and over again: managing major projects. In: Morris PWG, Pinto JK, Söderlund J (eds) The oxford handbook of project management. Oxford University Press, Oxford, pp 321–344
18. Grigolet C, Hondius E (2011) Unexpected circumstances in European contract law. Cambridge University Press, Cambridge
19. Grüneberg C (2022) §§ 313, 314. In: Grüneberg. Bürgerliches Gesetzbuch: BGB, pp 532–542, Beck, München

20. Irti N (1981) Inflazione e rapporti tra privati. Giustizia civile, 6: pp 298–307
21. Jobling PE, Smith NJ, del Rey F (2019) Discussion: experience of the role of contracts in megaproject execution. Proc. Inst. Civ. Eng.-Manag. Procurement Law 172:36–37
22. Lipari M (1987) La risoluzione del contratto per eccessiva onerosità: la struttura del giudizio di prevedibilità e la rilevanza dell'inflazione. Giustizia civile, 11: pp. 2795–2808
23. Marinelli F (1983) L'equilibrio finanziario nel contratto di appalto. Riv. giur. ed., 2: pp 159–165
24. Moscarini LV (1989) L'appalto. In: Rescigno P (ed) Trattato di diritto privato, 11, Obbligazioni e contratti, III, pp 703–743, Utet, Torino
25. Musolino G (2016) L'esecuzione dell'appalto nel nuovo codice dei contratti pubblici. Rivista trimestrale degli appalti, 3:457–502
26. Nicolussi A (2021) The pandemic and change of circumstances in Italy. In: Hondius E, Silva S, Nicolussi M, Coderch A, Wendehorst PS, Zoll-F C (eds) Coronavirus and the Law in Europe, pp 793–812. Intersentia, Cambridge
27. Parham L. https://www.neccontract.com/news/managing-inflation-risk-in-nec4-and-the-effect-on-compensation-events. Accessed 6 Nov 2022
28. Patterson P. https://www.neccontract.com/news/how-to-use-nec-secondary-option-x1-on-price-adjustment-for-inflation. Accessed 6 Nov 2022
29. Quadri E (1975) Congiuntura economica e svalutazione monetaria: osservazioni in tema di risoluzione per eccessiva onerosità. Dir. giur., 6: 809–825
30. Rosenn KS (1978) Protecting contracts from inflation. Bus Lawyer, 33(2):729–747
31. Rubino-Sammartano M (2008) Appalti di opere e contratti di servizi (in diritto privato). 2nd ed. Cedam, Padova
32. Sacco R (2005) Contratto, autonomia, mercato. In: Sacco R, De Nova G (eds) IL CONTRATTO, 1, 3rd edn, pp 16–48, Giuffrè, Milano
33. Zweigert K, Kötz HT (2011) Introduzione al Diritto Comparato. II. Istituti, Giuffré, Milano

Megaproject, ESG and Taxation

Andrea Purpura

Abstract This contribution, investigates the relationship between megaprojects, environmental sustainability, sustainable development, ESG factors and taxation. More precisely, the article proposes to highlight how taxation nowadays represents a factor that could be used to assess the sustainability of a company and its investments, also in light of recent EU regulations.

Keywords ESG · Megaproject · Tax Risk · Tax Control Framework · Sustainability

1 Introduction

As underlined within the literature, megaprojects are large-scale projects developed with the aim to manage public utilities. Among the most common types of megaprojects there are the infrastructure and aerospace ones. The decades-long planning of these projects can require more than §1 billion of investment and has a macroeconomic impact on the overall interested Region [1].

More precisely, megaprojects are complex ventures that typically cost a billion dollars or more, take many years to develop and build, involve multiple public and private stakeholders, are transformational, and impact millions of people. When we talk about "megaprojects" we are inevitably referring to "privileged particles of the development process" designed to ambitiously change the structure of society.

Megaprojects, therefore, are not just magnified versions of smaller projects. Megaprojects are a completely different breed of project in terms of their level of aspiration, lead times, complexity, and stakeholder involvement. Consequently, they are also a very different type of project to manage [2].

In this sense, specialized doctrine underlined in recent times that megaprojects have been found to play an important role in the economic development of a nation directly and indirectly, through different multiplier effects. Some of the most

A. Purpura (✉)
Postoctoral Research Fellow in Tax Law, Università Cattolica del Sacro Cuore, Piacenza, Italy
e-mail: andrea.purpura@unicatt.it

E. Favari and F. Cantoni (eds.), *Complexity and Sustainability in Megaprojects*, Lecture Notes in Civil Engineering 342,
https://doi.org/10.1007/978-3-031-30879-6_17

commonly cited socio-economic benefits of mega infrastructure investments include improved global connectivity, access to natural resources, competitive markets, and better job opportunities [3].

Due to their great complexity, megaprojects require specific competences and require much more pervasive attention than "simple" projects that do not requests the deployment of large amounts of capital and resources as is the case with megaprojects.

In other words, the proper planning, execution and management of mega-projects requires the implementation of an all-embracing governance including, among others, at least two factors:

(i) sustainability management;
(ii) and the management of risks (including, as will be underlined after, tax risks), that may characterize the realization of a mega-project.

In this perspective, it is necessary to:

(i) first of all, try to understand how the environment and sustainability can, today, become part of the management and decision-making mechanisms of those who are going to implement, or are already operating, mega-projects that, due to the characteristics highlighted above, can produce extremely beneficial or, potentially, disastrous effects on the social community.
(ii) Secondly, to understand how taxation fits into the above-mentioned relationship.

2 Megaproject, Environmental Sustainability and Sustainable Development

2.1 Environmental Protection

The reconstruction of the concept of the environment, ever since the subject became the focus of attention in major international fora up to the present day, has been delineated by an evolutionary path marked by increasing 'complexity'.

In fact, there has been a migration from the analysis of individual environmental elements and the phenomena of degradation related to them, to a way of conceiving the environment as an autonomous complex entity, endowed with relevance in its own right, and therefore in need of a unitary and functional notion for the formulation of adequate protection policies, both at a technical and regulatory level [4].

It is possible to find precise evidence of this profound evolution within the fundamental changes introduced by the EU legislature within the European Union Treaties (as well as, in the Italian experience, within the domestic Constitution) [5].

On this point we could refer, for example, of what the European legislature provided for in article 191 of the TFEU, a provision that filled the absence, within the 1957 Treaties of Rome, of specific provisions concerning environmental matters [6].

Moreover, mention should also be made to Article 11 of the same Treaty, an assumption by which, today, the centrality and transversality of environmental protection objectives has been recognized, and by which it has been sanctioned that the requirements connected with environmental protection must be integrated into the definition and action of the Union's policies and actions, in particular with a view to promoting sustainable development [7].

Furthermore, even looking at the issue from an Italian perspective, it can be seen how, in the wake of the efforts made at European level, the domestic legal system has also undergone a similar, albeit slower, evolution [8].

In the light of what has been said so far, it would immediately appear that the concept of environmental protection is absolutely complex and:

(i) requires, in terms of both analysis and protective action, the contribution of knowledge from multiple disciplines, both technical and legal. Among these, tax law, a subject area within which important regulatory instruments for environmental protection have been configured, has gradually taken its rightful place. So much so, in fact, that today is possible to refer properly to "*environmental taxation*" as a tool to grant environmental protection and as a means through which to sanction offending agents [9];

(ii) directly involve economic operators who, in carrying out their activities - especially with reference to works such as megaprojects, i.e. very large-scale action plans with enormous effects for the entire community—must consider the 'environmental protection' factor and the 'environmental sustainability' factor as key elements of their actions.

2.2 *Environmental Sustainability, Sustainable Development and Taxation*

Equally complex is the concept of "*environmental sustainability*" and "*sustainable development*".

Over the years, the term sustainability has become increasingly central to human culture.

From an environmental origin, through economic and social spheres to institutional issues. The epistemological path of this notion has been constantly evolving, and still is today. Its interdisciplinary character has thus led to the development of numerous different definitions in the merits of which—for the sake of thematic requirements—we will not go into here but for the identification of which we refer to the literature. Indeed, the latter has shown how, over time, the concept of environmental sustainability has come to encompass new aspects in addition to natural biodiversity and the protection of natural ecosystems. A context of conceptual evolution that is well illustrated with the testimony, for example, of the Kyoto Protocol, within which attention was also paid to the energy aspects and consumption of cities, which have now become central to the implementation of sustainable policies [10].

In such a briefly outlined context, taxation is also called upon to play a very important role.

From the standpoint of promoting legislative, economic and entrepreneurial policies that are environmentally sustainable and consequently capable of guaranteeing the achievement of sustainable social and economic development, taxation can and must necessarily play a central role, incentivising the transition of public and private activities towards the new internationally shared sustainable development goals.

Taxation, on the other hand, has always played a crucial role in promoting or deterring collective behaviour.

Tax systems are structured to reflect the 'dual role' of taxation (in addition to its main role, which is, of course, that of securing revenue): on the one hand, tax concessions, usually introduced to support activities considered worthy of attention (e.g. energy-efficient building renovations); on the other hand, tax policies, aimed at preventing activities that damage fundamental public interests (such as environmental protection). In the case of environmental protection, the aim is usually achieved by introducing levies on polluting events. Through environmental taxation, it would be possible to correct false price signals in the market by adding the costs of pollution and other ecological factors, thus helping to establish 'the right price' and implement the Community's 'polluter pays' principle. In order to achieve the necessary reallocation of economic resources to achieve sustainable development, all social and environmental costs must be incorporated into economic activities by internalising external environmental costs. It is clear, therefore, that taxation must be one of the main instruments to ensure the transition to a sustainable economy, given its transversal function of directing the behaviour of citizens. In particular, it will be up to governments (not just the Italian one) to implement tax policies within the legal system aimed at promoting sustainability in the broadest sense: in other words, it will be necessary—also through taxation—to incentivise economic, socio-cultural and, of course, environmental sustainability [11].

Clearly, this also concerns the promotion of policies that, by paying attention to environmental sustainability and implementing sustainable attitudes, target economic operators—such as those, for example, specialised in the realisation of megaprojects—that, due to the type of activity carried out, may produce potentially harmful effects on the community and the environment.

In this perspective, environmental taxation and so-called '*green taxes*' can play an essential role and the approach to the issue—as highlighted in the literatureshould take place at a European level.

On this point was underlined that "*it is key to reform tax systems to remove any tax measures that would have a negative effect on the environment, for example tax measures that encourage polluting activities and polluting behaviours (...) As part of the EU's Recovery and Resilience strategy, the EU should encourage Member States to avoid such policies. The elimination of environmentally harmful tax subsidies would not only improve the consistency of tax regimes, but it would also generate additional revenues that could be used for the EU's recovery. To sum up, green taxes might not be the best instrument to generate revenues to cover the costs of the health crisis, but they should certainly be part of the political agenda to build and rebuild our economy in a more sustainable way*" [12].

In such a defined context, it is easy to understand how:

(i) on the one hand, attention to the environment, to the sustainability of production - and thus to the effects that economic activities have on the environment and the surrounding community - and taxation as a tool for environmental preservation and the adoption of green-oriented actions has evolved over time and has now become an absolutely central theme. Witness to this is also the United Nations' Agenda 2030 where taxation is emphasised as a tool for environmental protection and the promotion of, for example, circular eco-nomies [13];

(ii) on the other hand, the need for economic operators to adopt corporate and entrepreneurial governance tools that take into account the environmental factor and, as will be discussed in the next section, ESG factors, aware of the fact that in today's economic and social context, being environmentally friendly and the effects that one's entrepreneurial and fiscal conduct can have on the surrounding community are no longer a limitation but a true business value.

3 ESG Factors and Megaproject

The term "ESG" stands for "*Environmental*", "*Social*" and "*Governance*": these are the factors that are used to assess companies in terms of sustainability.

ESG criteria are very specific in defining the actual environmental, social and governance impact, providing a more precise score, the so-called "*ESG Score*". More precisely, the three factors (E) Environmental, (S) Social and (G) Governance stand for [14]:

- **(E)**: a company policy aimed at energy efficiency, reducing carbon dioxide emissions, with a focus on wastage of natural resources and pollution;
- **(S)**: a socially-conscious company policy on the quality of the working environment, which safeguards the diversity and inclusion of employees, with an equal distribution of tasks. This parameter also takes into account safety in the workplace and the social responsibility of the company in a global perspective.;
- **(G)**: for a corporate policy that guarantees transparency in corporate governance, with the presence of independent or non-executive directors, that applies diversity criteria in the composition of Boards, with sustainability plans and objectives linked to Board remuneration, as well as policies and, more generally, the behaviour of top management and the company in terms of ethics and compliance.

While it is true that the ESG phenomenon in the terms now described has very distant roots in time, it should be pointed out that for some years now there has been an acceleration of interest on the part of companies, investors and financial institutions. These organizations, in fact, are today called upon to face a profound transformation based on the above-mentioned consideration that respect for ESG factors is no longer a necessity, but represents an opportunity capable of creating value and capable of having a significant impact on the present and future of organizations.

This is clearly even more the case in the context of megaprojects where, as mentioned, profiles relating to the environment, the community and corporate governance are of absolute importance, also in light of the repercussions on the community and the size of the investments they make.

The growing interest and attention paid to ESG factors would seem to be borne out by the interventions expressed on this point at a European level.

In this regard, it was recalled [15] that on 2 August 2021 specific measures implementing the so-called European Green Deal were published in the Official Journal of the European Union:

- **Commission Regulation (EU) 2021/1253 of 21 April 2021** amending Delegated Regulation (EU) 2017/565 with regard to the integration of sustainability factors, sustainability risks and sustainability preferences into certain organisational requirements and operating conditions for investment firms;
- **Commission Regulation (EU) 2021/1255 of 21 April 2021** amending Delegated Regulation (EU) No 231/2013 as regards sustainability risks and sustainability factors to be taken into account by alternative investment fund managers;
- **Commission Regulation (EU) 2021/1256 of 21 April 2021** amending Delegated Regulation (EU) 2015/35 as regards the integration of sustainability risks into the governance of insurance and reinsurance undertakings;
- **Commission Regulation (EU) 2021/1257 of 21 April 2021** amending Delegated Regulations (EU) 2017/2358 and (EU) 2017/2359 as regards the integration of sustainability factors, sustainability risks and sustainability preferences in the product control and governance requirements for insurance undertakings and distributors of insurance products and in the conduct of business rules and investment advice for insurance investment products;
- **Commission Directive (EU) 2021/1269 of 21 April 2021** amending the Delegated Directive (EU) 2017/593 with regard to the integration of sustainability factors into product governance requirements;
- **Commission Directive (EU) 2021/1270 of 21 April 2021** amending Directive 2010/43/EU as regards sustainability risks and sustainability factors to be taken into account for undertakings for collective investment in transferable securities (UCITS).

While not entering, for reasons of space and subject matter, into the merits of all the above-mentioned provisions, it is sufficient here to point out that they all have a direct impact on ESG logic.

If that is true, in addition to these measures, particular attention should also be paid to Regulation 2019/2088 on sustainability disclosures in the financial services sector [16], which, as will be discussed in the previous section, is particularly important in that it creates a very close link between ESG-oriented actions taken by economic operators and taxation.

3.1 Regulation 2019/2088 of November 27, 2019 (Sustainable Finance Disclosure Regulation)

The Regulation 2019/2088 of November 27, 2019 would seem to be relevant in order to identify a relationship between ESG and taxation.

This is essentially because the regulation, when defining the notion of 'sustainable investment', not only stipulates that it can only be considered as such where "*an investment in an economic activity that contributes to an environmental objective, as measured, for example, by key resource efficiency indicators on the use of energy, renewable energy, raw materials, water and land, on the production of waste, and greenhouse gas emissions, or on its impact on biodiversity and the circular economy, or an investment in an economic activity that contributes to a social objective, in particular an investment that contributes to tackling inequality or that fosters social cohesion, social integration and labour relations, or an investment in human capital or economically or socially disadvantaged communities*" stipulates that the sustainability of an investment is also defined on the basis of the circumstance for which "*the companies follow good governance practices, in particular with respect to (…) tax compliance*" [17].

In the perspective shared by the European regulation, therefore, it clearly emerges that taxation and, precisely, tax compliance is fully within the parameters that economic operators must follow in order to be able to claim:

(i) is that it is supporting sustainable investments;
(ii) is to be a sustainable enterprise.

In this perspective, therefore, tax compliance becomes no longer a constraint but a tool through which to create value both within the company and for the community.

What is being argued here finds confirmation in the very functioning of taxation, which, in its strictest sense, is aimed at achieving social and solidaristic ends. This is the case, as much as anything, in the perspective of Italian analysis where Article 53 of the Italian Constitution [18], by providing for a generalized duty to contribute, introduces the latter in general terms, thus declining it to be borne by all those who come into contact with the Italian legal system.

Therefore, if it is true that:

(i) the duty to contribute to public expenditure,, has no basis in the possible satisfaction of individual interests either by the State or, certainly, by the Treasury or any other public entity;
(ii) the duty imposed by Article 53 of the Constitution is based on the principle whereby the duty to contribute to the public not only takes the form of a concrete declination and application of the fulfilment of the mandatory duties of political, economic and social solidarity but, in manifesting its profound solidaristic matrix, fully expresses its meta-individual and—as highlighted by the most authoritative doctrine—redistributive function [19].

Then all economic operators, indiscriminately, are called upon to adopt conducts that, being tax-compliant, give rise to:

(i) on the one hand, through cooperation and fiscal transparency, to the correct collection of taxes by the State and thus to the achievement of greater fiscal justice, since the taxes collected can be used by States for welfare policies (such as education, health care, environmental protection, urban regeneration, etc.) and, in more general terms, for the implementation of legislative policies oriented towards the achievement of a more widespread and sustainable social welfare;
(ii) on the other hand, to the adoption of internal value creation processes that are also sustainable from a tax point of view and thus, as such, oriented towards the adoption of appropriate instruments to control that corporate tax risks—and thus the risks of engaging in tax avoidance or tax evasion— are curbed.

With regard to the latter perspective—which, however, should certainly not be considered more valuable than the former by economic agents—it is therefore necessary for economic agents to adopt appropriate instruments to verify their coherence:

(i) with respect to ESG parameters;
(ii) with respect to the sustainability assessment criteria (in the terms referred to, for example, in Regulation 2019/2088 of November 27, 2019;
(iii) with respect to corporate tax compliance, a factor that, as has been pointed out, is fully among the elements that contribute to classifying a company as sustainable or not.

In this sense, one has to ask oneself what instrument could be considered to be capable of meeting these parameters from an operational perspective.

A possible answer to this question would seem to be to be found in the so-called Tax Control Framework [20], which, as will be highlighted in the concluding section of this paper, makes it possible to monitor the management of tax risk (as well as other risks, including, for example, the 'reputational' risk) and thus to keep under control at least one of the fundamental parameters through which to verify the sustainability of the business activity both when considered ex se and in relation to the surrounding community.

4 The Tax Control Framework as a Business Sustainability Assessment Tool

According to the OECD [21], the *Tax Control Framework* is that tool that constitutes the element through which an internal control system for tax risk management can be put in place that allows for the preventive self-assessment of tax risks, providing a continuous and up-to-date view of one's own position that, if supported by transparency in dealings with the tax authorities, can help to avoid, or at least mitigate, the uncertainties associated with the management of tax risk, first and foremost those related to transactions that could give rise to the case of avoidance/abuse of rights [22].

From a technical and operational point of view, the TCF allows, therefore, both to control, monitor and prevent tax risk and, consequently, to implement a set of decision-making choices taken at a high level on the overall tax position of the company through a systemic and structured approach. On the other hand, the implementation of such conduct requires an analysis of the environment in which it operates in order to identify emerging risk and it is precisely for this purpose that the adoption of the TCF [23] responds.

So, in this scenario, what is the relationship between the adoption of a fiscal control instrument and ESG in the specific context of megaprojects?

On this point it has been pointed out that if it is true:

- that the adoption of a properly structured Tax Control Framework gives the company effective control over tax risk, ensuring compliance with relevant regulatory requirements
- that the TCF goes beyond mere compliance monitoring, where, in the declination of tax governance that is inherent in this tool, it enables the company to intercept the impacts of upcoming – and future – regulations, allowing it to ensure business sustainability,

then the TCF can validly be taken as a paradigm with which to measure the sustainable all-round approach to tax management and the presence of the TCF can be appropriately reflected in the representations that the company is called upon to give in its sustainability communication both externally – and therefore towards the social community where the company operates – and within the company itself as an element of value creation that characterizes it [24].

In this perspective, therefore, the TCF can become an important tool for controlling tax compliance and thus, by responding to the verification of the consistency of the company's operations with tax regulations, it fulfils the function of attesting in a direct way, once shared, to the company's state of sustainability and thus, ultimately, to its compliance with ESG parameters. What has been said would seem to find, moreover, perfect applicability in relation to large companies. It has been pointed out, in fact, how the Tax Control Framework constitutes – probably due to the complexity of the risks to which it intends to give a preventive resolution – an organizational model and a more easily accessible and suitable control for companies, such as those operating in the sphere of megaprojects, which are truly significant in terms of size [25].

5 Conclusions

With the above considerations in mind, there are essentially two conclusions that it is deemed appropriate to share, briefly, here.

From a technical-operational point of view, the hope is that companies already operating in Italy or intending to open up to our system in order to implement projects with a high economic, social and environmental impact will consider the environmental factor and the 'sustainability' factor as key elements of any type of activity.

A prerequisite for doing so would seem to be awareness both of the changed global economic scenario – now more than ever particularly attentive to sustainability and the need to advocate actions aimed at achieving sustainable development - and, above all, of the competition of adopting conduct consistent with ESG factors to create value for the company and the surrounding community.

On the other hand, from a systemic and legislative point of view, there is a need for an ever-increasing awareness among national legislators so that they accelerate a shared path of sustainable development aimed at adopting tax policies that promote economic growth and balance the instruments of taxation with those of incentives, and thus succeed in generating sustainable development through an effective mix of taxation and tax benefits [26].

It is therefore a question of implementing measures that, also from a fiscal point of view, enhance the most virtuous companies by taxing more significantly those who - although they can adopt behaviour more oriented towards environmental protection and sustainable investment – carry out actions that run counter to sustainable development.

References

1. Allena M (2020) Reshaping tax policies for sustainable megaprojects in the post pandemic era. In: Cantoni F, Favari E (eds) Megaprojects, pp 9–13. Torin
2. Flyvbjerg B (2014) What you should know about megaprojects, and why: an overview.Proj. Manag. J. **45**(2):6–19 (2014)
3. Nyarirangwe M, Babatunde OK (2019) Megaproject complexity attributes and competences: lessons from IT and construction projects. Int J Inf Syst Proj Manag 7(4):77–99. The Authors remember that "the term "megaprojects" has been construed as an old phenomenon which dates back to the prehistoric times when ventures such as the Pyramids of Giza in Egypt, the Roman Colosseum in 80AD and the Great Wall of China were built. The term "mega" originated from the Greek word "megas" which means "great, large, vast, big, high, tall, mighty or important". The term "mega" was found to have been initially used in relation to megacities in 1968. Subsequently, from around 1982, the term has been used loosely as a stand-alone adjective to refer to any large project. Scientifically, the term refers to a measurement of worth that is expressed in millions. Using this scientific logic, it can be argued that megaprojects are endeavors worth millions of, notably, dollars, pounds, and euros".
4. Giannini MS (1971) Difesa dell'ambiente e del patrimonio naturale e culturale Rivista Trimestrale di Diritto Pubblico p 1122
5. See Dell'Anno P (2009) La tutela dell'ambiente come materia e come valore costituzionale di solidarietà e di elevata protezione Ambiente Sviluppo 7:585
6. Costato L, Manservisi S (2012) Profili di diritto ambientale nell'Unione Europea, in Quaderni del Centro di documentazione e Studi sulle Comunità europee. Università degli Studi di Ferrara, Milano
7. See Urbani (1999) La difficile realizzazione di uno "sviluppo sostenibile a livello comunitario. Ambiente&Sviluppo 12:1153, where the Author identified the restructuring of EU tax systems as an incentive for companies with a view to sustainable development. A reform of taxation,

in the author's view, would lead to a reduction in indirect labour costs and the incorporation of environmental and resource costs in the market prices of goods and services. Moreover, see also Salanitro U, I principi generali nel codice dell'ambiente, in Giornale di diritto amministrativo, 2009, 1, p103, in his opinion, the objective of sustainable development is difficult to implement whenever exhaustible and non-renewable resources (e.g. hydrocarbons) are at stake, where a policy should be promoted to replace the depleting resource with another resource with a similar function or to promote a management policy to prevent its definitive disappearance. In this sense, Meli M., Il principio comunitario "chi inquina paga", Giuffré Editore, Milano, 1996. Cosi anche Manservisi S., Il principio dello sviluppo sostenibile: da Rio+20 al diritto dell'Unione europea e il suo fondamentale ruolo nel diritto agrario, in Convegno organizzato in onore del Prof. Casadei E., Bologna-Rovigo, 25–26 ottobre 2012., Ed. Giuffré, 2014, Milano

8. In confirmation of this, in fact, Constitutional Law No 3 of 2001 provided for filling the long-lasting gap in the constitutional protection explicitly reserved for the environment, a deficiency already noted, in the pre-reform phase, by the Constitutional Court, which, in making up for these shortcomings, tended to resort to an original interpretation of the provisions contained in Articles 3, 9, 32 and 41 of the Italian constitution. On this point, see, for all, Migiarra, M., Lo sviluppo e la tutela dei beni ambientali, tra proprietà ed uso per la salvaguardia dei diritti delle presenti e future generazioni, in RivistAmbiente 2(3–4)
9. Costato L, Manservisi S *Op. cit.* [6]
10. Malara L (2020–21) Il fenomeno degli ecoquartieri: storia, concetti e proposta di un nuovo modello valutativo. Politecnico di Torino 9–35
11. In these terms, see Allena M (2021) Pnrr, economia circolare e sostenibilità: il ruolo del fisco. We Wealth 29th of November
12. See Pirlot A (2021) FISC Subcommittee of the European Parliament Public Hearing Green Taxation Panel 2, 12 July 2021
13. The 2030 Agenda for Sustainable Development is an action programme for people, planet and prosperity signed in September 2015 by the governments of the 193 UN member states. It encompasses 17 Sustainable Development Goals (SDGs) into a grand programme of action with a total of 169 'targets' or goals. The official launch of the Sustainable Development Goals coincided with the beginning of 2016, guiding the world on the road ahead over the next 15 years: countries have committed to achieving them by 2030. See: https://unric.org/it/agenda-2030/
14. This description. https://finscience.com/blog/esg/
15. Salomone E (2021) ESG: inquadramento normativo e scenario italiano. NT Plus, del 20 dicembre 2021
16. Regulation (EU) 2019/2088 of the European Parliament and of the Council of 27 November 2019
17. See art. 2, par. 17. Regulation 2019/2088 of November 27 2019
18. The ability to pay principle is introduced by Article 53 of the Italian Constitution, which is divided into two paragraphs and states, respectively, that: "Everyone is te-needed to contribute to public expenditure according to his or her ability to contribute" (para. 1). "The tax system is informed by criteria of progressiveness" (par. 2)
19. Which finds its fullest expression in the second paragraph of Article 53 of the Constitution, which states, as mentioned, that the tax system is informed by criteria of progressiveness. In this sense we refer to Moschetti F (1965) Il principio della capacità contributiva, Cedam, Padova, 1973, pagg. 59 e ss.; Manzoni I., Il principio della capacità contributiva nell'ordinamento costituzionale italiano, Giappichelli, Torino p 185
20. Also known with the acronymous "TCF"
21. See the (2016) Co-operative tax compliance: building better tax control frameworks OECD
22. In these terms, see Ferroni B (2016) Il tax control framework nel regime di adempimento collaborativo. Il *Fisco* 38:3621–3631
23. Della Rovere A, Vincenti F (2017) Il tax risk management quale strumento di governance per le imprese Amministrazione Finanza 8–9:53–60

24. Lio M (2022) La fiscalità nelle metriche ESG: rischi, opportunità e prospettive, del 6 settembre 2022 www.dirittobancario.it
25. In these terms, see Marino G (2022) La metamorfosi della sanzione amministrativa tributaria: dalla teoria della afflittività alla pratica della premialità (nell'insegnamento di Giuseppe Dragonetti), in Diritto e Pratica Tributaria 4:1189
26. In this sense, see Turi A (2020) Come il fisco può agevolare lo sviluppo sostenibile. In: Garofalo D, Pardolesi P, Rinaldi A (eds) I Simposio dei dottorandi sul tema dello sviluppo sostenibile, Bari, pp 141–149

Communication Screen Shot in Projects of Multicultural Diversity

Walid S. S. Nassar

Abstract Organizations are turning their projects towards multicultural environment, which adds value to these projects. On the other hand, this diversity may push the project towards the abyss as well. In this essay, I will try to collect the pros and cons of working in a multicultural environment and provide a literature review of the hottest managerial topics related to multiculturalism and diversity through the teamwork, with an intensive focus on the communication tool between management and teams, following its impact on projects and organizations, whether negatively or positively. Then I will provide a roadmap for each manager working in an organization that contains multicultural team projects, to take advantage of this diversity and twist it to an opportunity that can be developed through the evolution of the project towards more success. The competence of project managers in managing communication in a multicultural work environment depends on a set of knowledge skills and tools, the most prominent of which are related to communication in such a situation will be addressed in a progressive and coherent manner.

Keywords communication · multicultural diversity · multicultural team projects · project management

Reviewed by: Prof. Dr. Constanta-Nicoleta Bodea, bodea@pm.org.ro.

W. S. S. Nassar (✉)
ECM Alma Mater Europaea, Salzburg, Austria
e-mail: walid.nassar@almamater.si

E. Favari and F. Cantoni (eds.), *Complexity and Sustainability in Megaprojects*, Lecture Notes in Civil Engineering 342,
https://doi.org/10.1007/978-3-031-30879-6_18

1 Communication Screen Shot in Projects of Multicultural Diversity (Form of an Essay)

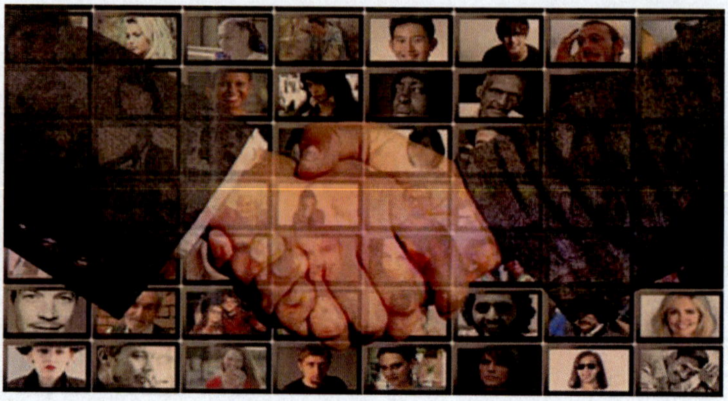

Source: GoDaddy | License No.: 446358758

2 Introduction: Background: Project Management in Multicultural Environment

Have you ever thought about the nationalities of the gadgets in your office? Look well, they are multinational, your computer in front of you or your mobile phone, several global teams have produced and delivered it to your desk or hand in the form and efficiency you usually aspire to.

The environments in which multicultural management occurs have grown more complicated. Now, rather than having a single foreign manager act as the international shield for an entire department or even company, employees must deal with global virtual communication, data exchange in shared server environments, or collaboration in multinational and divided up teams [31].

At the beginning of writing this essay, I asked myself several questions: Why are cultural differences important to a project manager? What administrative classifications intersect with multiculturalism? How might project leadership be able to address the cultural differences of teams working on multicultural projects? How they communicate with less conflicts?

Due to the fact that business relationships are ultimately instances of cultural interaction, the future of global business depends on people's capacity to deal with diversity in an honest and transparent manner [8].

It is necessary to delve into the managerial competence associated with multicultural projects in which include the "skills and the knowledge needed, together

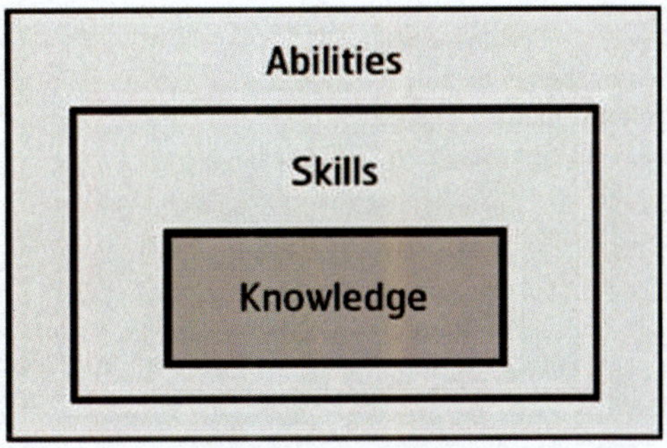

Fig. 1 Individual competence component. *Source* [16]

with abilities" ([16]:15) to understand the value that this human diversity adds to the project, and to avoid the negatives that may be caused by a decline in the performance of culturally different teams and thus the decline in the opportunities of success for the project as a whole, see Fig. 1.

In addition to managing time, money, and quality, a project manager is also in charge of managing integration, scope, human resources, communication, risk, and procurement [30].

Among the most prominent management activities associated with multicultural projects are communication, evaluation, leadership, decision-making and trust between the team and project stakeholders [25].

communication is defined by many authors as the exchange of information between recipient and sender by which a message is directed from a point to another point and the communicators are associated all together through communication channels and it is the process by which information is exchanged and understood by two or more people, usually with the intent to influence or motivate behavior [12, 29].

In this essay I will focus on Multicultural communication which is defined as the exchange of information between individuals with various cultural backgrounds, studying intercultural communication is important to create increased self-awareness [22].

Communication is one of the most important activities that shorten the deep scene between the individual personality with its "cultural" values, passing through the individual behavior covered by professionalism [15], as shown in the Fig. 2, and when individuals are intertwined in many collective work, the scene becomes more complex.

In order to understand communicative activity in a multicultural environment, it is necessary first to identify the advantages and disadvantages of the any Multicultural workplace, this understanding turns into keys of communication that connect the team of the project [34].

Fig. 2 Individual "cultural" values, behavior and professionalism *Source*: self – design based on [15]

3 The Pros and Cons of a Multicultural Project

The pros and cons of a multicultural project environment can be summed up as they are in the following table from different angles upon to (Table 1) [11]:

4 Communication within Multicultural Teams

56% of your project budget might be at risk due to poor communication (PMI, 2013), therefore communication skills within a multicultural workplace or project necessarily need a conscious management of the context range and dimensions of multiculturalism, and how they can transform this diversity into an added value to the project, this can be achieved through conscious communication messages, able to combine the emerging differences in multicultural projects [7].

I have an interesting story about miscommunication. After one session that was discussing "radiofrequencies remotely linking" in California 2014, I visited the representatives of the corporate that was liable for the session, to ask about some details, during my conversation with the US man in his forties who was the team manager, I asked him about the difference in electrical environments from one country to a different, and whether his product is compatible with countries which have unstable electrical current, he asked me why this instability happens in some countries? I told him: Your fingers don't seem to be like mine (I used my left hand to precise my body language), he surprised me with his response when he asked me: do you mean that you are married? Immediately i noticed that he checked out the ring with my left hand and failed to understand what I meant, actually I meant that individuals are different from one to another, countries are likewise. He took my sentence literally and recognized that I'm a married man. This phrase is commonly utilized by

Table 1 Pros and cons of a multicultural project environment

Disadvantages of multiculturalism	Advantages of multiculturalism
Communication problems In a workplace with a diverse workforce, communication barriers can be the most challenging to overcome. This is caused, for instance, by some team members' limited proficiency in the language spoken most frequently or by the manner in which they communicate in their home nation **Different work cultures** Indeed, while certain cultures are adapted to hierarchical hierarchies, others are fine with flat organizational systems. A behavior that is perceived as respectful in one culture could not be in another For instance, team members from hierarchical cultures anticipate varying treatment depending on their position within the company. Workers from egalitarian cultures don't, on the other hand. If some members don't live up to those expectations, they risk being humiliated or losing respect and credibility **Decision-making conflicts** Working on a multicultural team also means dealing with differences in decision-making processes and how much forethought is required of teammates. Conflicts arise when some of them act quickly versus slowly or analytically versus instinctively. Someone who enjoys making decisions quickly might become frustrated with those who require more time. For instance, when compared to managers in other nations, American managers are known to prefer making decisions quickly and with little thought. Asian managers, on the other hand, invest more time in analysis **Negative cultural stereotypes and prejudices** Unconscious cultural beliefs and stereotypes eventually result from interactions between individuals from different countries. They can be particularly challenging to overcome if they make some team members less motivated to collaborate. Negative cultural stereotypes can have a significant negative impact on employee and productivity **Increasing of stress and diverging of opinion** There may be too many opinions at work, which can lead to conflicts and added stress. As a result of poorer social integration within the team and a potential slowdown in project speed, complaints do indeed frequently increase when working in a multicultural team	**Productivity** Different cultural perspectives and backgrounds can inspire creativity and productivity. Every team member brings a unique set of experiences and perspectives, which can enhance the range of services and goods that a company can provide to the wider public. An organization with a global workforce can be more considerate of audiences in other countries by creating, for example, targeted marketing campaigns and materials that meet their needs **Improve of creativity** If someone is only exposed to people from, say, Asia, south America, and Europe, there is a good possibility that the flow of ideas will be unrestricted. Along with sharing their own experiences, colleagues will also instruct others on what is effective in their setting and vice versa **Improve of problem-solving skills** International employees contribute their personal histories and experiences to the team. Due to this, when a solution is required, they are able to think of concepts that a group with a single culture mindset would never have considered. Additionally, teammates are exposed to various points of view and are given the opportunity to learn how to think creatively, which improves their soft skills **Personal growth and flexibility** Being a part of a global team is similar to going to an exchange program every day at work. Members of a multicultural team are exposed to different cultures while working together, and this can easily affect how they think and how they behave. Setting aside expectations and prejudices is necessary for adjusting to new circumstances and challenges. A person can initiate deep personal growth in this way, becoming more adaptable, tolerant, and dedicated to cooperation. They can also develop a more open mind **Thinking quickly** Being aware of the rules is the best way to increase group agility. Particularly in multicultural teams, where members will be more effective, spontaneous, and quick thinkers the sooner they begin to feel comfortable

Source: [11]

the people of the Middle East to precise attitudes of differences, they assert "your fingers aren't like mine". The US man and I totally misunderstood each other.

5 Low Context and High Context of Communication

Erin [25] defined a measure of communication within the project community and made it at two opposite levels, the first he called: low context and the second he called: high context, see Fig. 3.

Typically, cultures cannot be categorized strictly into high or low context. The majority of civilizations fall somewhere in the middle of the range and can, to varied degrees, exhibit both high context and low context qualities (SHINNERS 2017).

The question of whether a culture has high context or low context can be complicated, yet it can influence many other facets of that culture. For instance, similarity is a key attribute in high-context cultures. This is because people in high context cultures tend to share similar educational levels, as well as a common race, religion, and history [25].

By presuming that the audience will think similarly to the speaker or writer and understand the underlying message latent in their speech or writing, communication can be contextualized through these shared experiences [3].

The opposite is accurate in low-context cultures. They typically focus on the person rather than the group and are diverse [25]. Communication in a low-context culture must be simple enough for the greatest number of individuals to grasp it because there are so many disparities between people there [3].

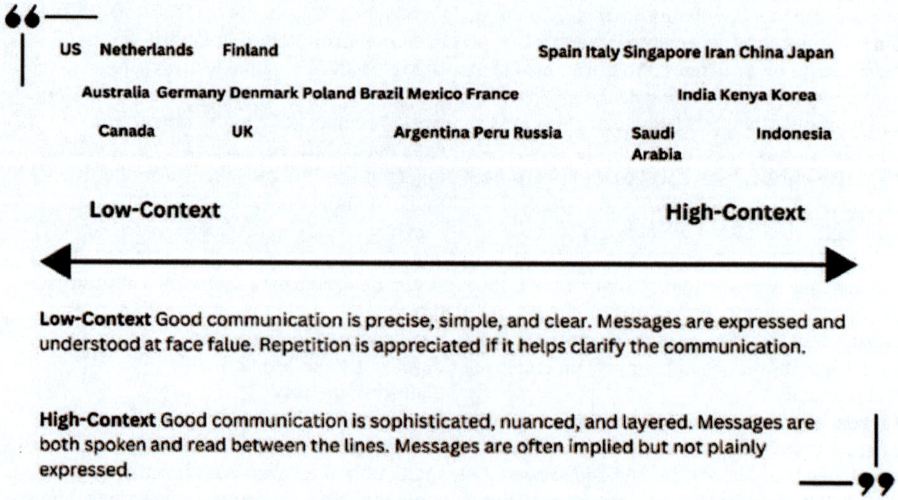

Fig. 3 Low context and high context of communication *Source:* [25]

Scaling communication from low to high, it takes the communication context into account. Low context communicators convey their messages using simple, direct language [5]. Background information, past experiences, body language, speech tonality, etc. are largely irrelevant [20]. Say what you think and think what you say with clear, precise terminology, as opposed to high context communication, where the meaning is inferred from the context and what is said is not necessarily as important as what is meant. The two communicators must share a basic knowledge in order to observe the subtleties [25]. A good comparison would be between a long-married couple who complete each other's sentences and a pair of strangers who can only communicate through language. (SHINNERS 2017). Reverting to the mentioned story with the US man, I was very "high context", and he was very "low context" while we had our conversation.

6 Deep in Project Management Communication Through the Dimensions of Culture

In a recent study was based on a large survey of more than 1100 project managers in 7 countries, [35] discovered that performance-oriented organizations place a greater emphasis on project communication management. Communication is essential since information exchange is a key driver of project performance. They emphasized that communication in a project not only encourages various sorts of information gathering and exchanging, but it also lowers problems caused by misunderstanding and information imbalance.

Collaboration with people from many cultures is essential for effective global business strategies. As a result, the challenge for business is to create ways for people who think differently to collaborate in a positive way [4]. As a result, it is critical for the organizations to be aware of these differences and to have measures in place to deal with any multicultural communication challenges that may occur.

The main questions in this essay are what are these knowledge and skills that project managers must have to communicate effectively within a multicultural teamwork in order to obtain the advantages of diversity in such an environment? And how he/she can improve his / her knowledge and skills?

Both effective communication methods and communication message can assist leaders in anticipating and responding to possible crises that threaten, disrupt, or imperil a business and the stakeholders it serves [9]. In order to act better in a project of cultural diversity, the project management administrations and teams have to have an enough knowledge of which this diversity can shape the dimensions of the culture embedded in the Organisation [15].

These dimensions of cultural are summed up briefly in a recent essay for [18] based on - Hofstede's 6 Cultural Dimensions theory as shown in Fig. 4.

[18] explains that: Power Distance Index (PDI): A team's acceptance of an unequal, hierarchical distribution of power is indicated by a high PDI score. Low PDI

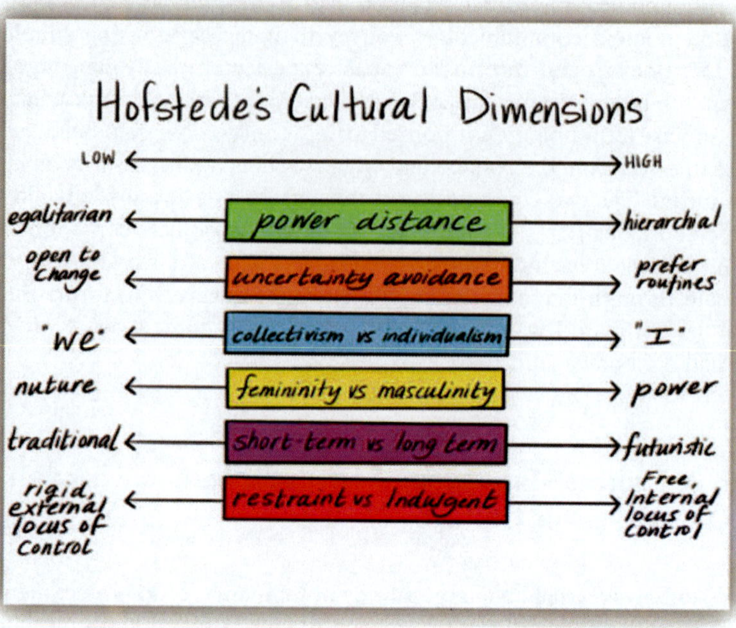

Fig. 4 Hofstede's 6 Cultural Dimensions theory *Source*: (Maryland 2019)

scores indicate that power is shared and that team members do not accept situations in which power is not distributed fairly.

1. Uncertainty Avoidance Index (UAI): People who have a high Uncertainty Avoidance score try to make life as predictable and manageable as they can. People in nations with low UAI scores tend to be more relaxed, inclusive, or open.
2. Individualism Vs Collectivism: In places with a high Individualism score, people bear less responsibility for the decisions and actions of others. People are expected to be loyal to the group they belong to, in collectivist teams.
3. Masculinity Vs Femininity: Men are expected to act assertively in communities where gender roles are less overlapping. However, there is a lot of crossover between male and female duties in feminine societies, and modesty is viewed as a virtue.
4. Long Vs Short Term Orientation: Long-term oriented nations typically have a pragmatic, modest, and more self-sufficient outlook. People in short-term oriented nations are typically religious and nationalistic and tend to place more value on principles, consistency, and truth.
5. Indulgence Vs Restraint (IVR): High scoring "IVR" nations support relatively unrestricted satisfaction of an individual's inner desires and feelings. People's behavior is more strictly regulated and social norms are more rigid in a nation with a low "IVR" score.

I believe that any project manager who moderates a multicultural team has to understand these six cultural dimensions, in order to create a personal and organizational strategy to communicate with the team, which makes the working atmosphere more suitable for the nature of the team, the project and the organization itself, to takes advantage of the high value of diversity that exists in such circumstances in order to achieve a sufficient level of less conflicts and more efficiency and success.

7 Understanding the Project Manager Role in Communication

The many task operators in the various activity must be kept up to date on all activities, progress reports, issues, solutions, and other pertinent information. All of them come together thanks to communication, which also enables various units to coordinate their efforts, enable the necessary growth, a communication strategy is needed [34].

A project manager is impacted by both his or her own country culture and the multicultural context in which he or she has grown up. His or her individual attitudes and prejudices affect how diversely he or she manages and have a significant impact on how well the team manages cultural diversity [7]. This is frequently best expressed in how decisions are made at various phases of project management, in how tasks are coordinated, in leadership and authority, as well as in time management, project communication, and project risk [1].

The management style of a project manager in a multicultural setting is distinguished by a higher degree of adaptability, prompt problem-solving, open communication, the pursuit of consensus and compromise, as well as an effort to prevent or lessen confrontation [7]. One of the variables that, despite bringing value to the projects, may persuade as one of the finest obstacles in the practice of project management in a multicultural setting is cultural diversity [27].

Due to the evolution of the global business environment, the introduction of new technologies, and the diversity of cultures, project management has required the fusion of a wide range of talents and individuals. The project manager is in charge of those elements, and it is his or her job to utilize all environmental influences, including cultural variances [7]. When the development of a strong, inclusive culture and values is promoted, when the project participants are not only participating, but also empowered, which leads to more effective teams and greater performance, there is a strong multicultural team [1].

With reference to the abovementioned, a significant step was taken for organizations in terms of project implementation in a global, multicultural setting. And today more than ever, there is a need for a wider and more varied collection of tools, competencies, abilities, and strategies in our unpredictable, contentious, and constantly changing environment [1, 9, 17, 25, 27].

To achieve this, "communication" must be the greatest common denominator of all these knowledge that a successful manager needs in a multicultural project, in the

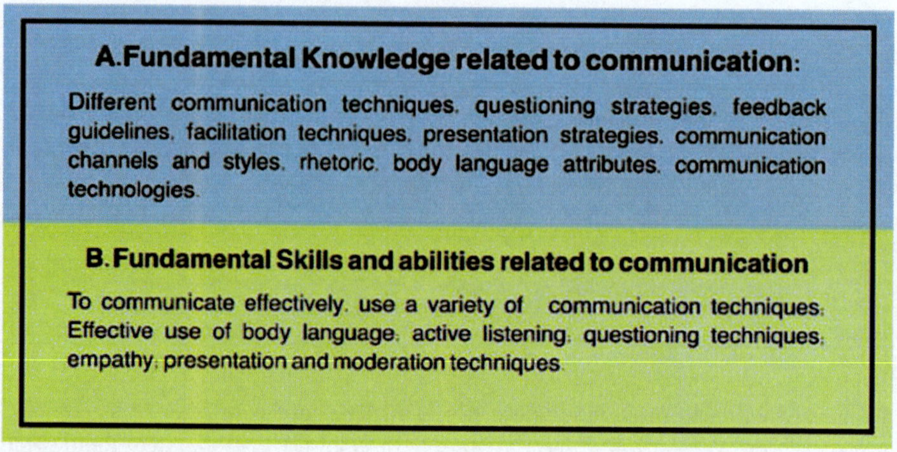

Fig. 5 Fundamental components of effective communication *Source*: [16]

near future, having effective multicultural communication skills will be more and more crucial [20].

There has to be a method that communication is managed in a multicultural organization, what is the mindset in which helps project managers do that in Practical and scalable way?

The fundamental components of effective communication are described by personal communication. Both the message and the medium must be understandable and suitable for the intended audience [16]:76). These components are divided into two levels as listed below in Fig. 5:

8 Communication Management in Multicultural Environment

Project communication management, at its most basic level, makes ensuring that important information is effectively and purposefully shared among many parties involved in (or influenced by) a project (PMI 2013). Different people use information differently (remember my short story with the US manager), and some people may only need to be aware of it rather than acting in a certain way as a result. The success of a project depends on establishing a plan for information sharing throughout the project's lifecycle [13].

One of the most important lessons learned is that project communication management needs to happen at every early stage of a project's lifecycle, especially when a transition takes place, this can be clearly noticed we understand that the project communication management is a project by itself upon to [13], she emphasized that competent project managers must realize the lesson that communication management

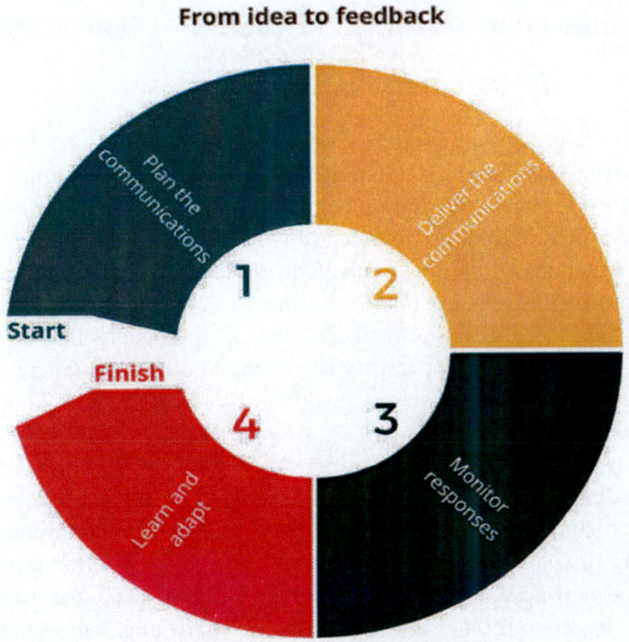

Fig. 6 Project communication process (Plan, deliver, monitor and learn). *Source:* [13]

happens in a well-defined and repeatable cycle as indicated in the Fig. 6., managers often communicate in one way or another in a cycle of (Plan communication, deliver communication, monitor responses and then learn and adapt).

I really like this way of thinking, but there is a central question on my mind: What culture must project managers have in order to run a multicultural project?

9 Communication Culture

Effective communication is crucial to project management because it links together the various stakeholders that are participating in a project and who have varied cultural and organizational backgrounds, degrees of knowledge, perspectives, and interests in how the project is carried out [34].

Up until the goal is accomplished, communication is a process that includes numerous additional steps and actions. It's crucial to remember that communication's main goal is to transfer information from a source to a receiver [21]. The source needs to deliver information to a certain recipient, who may or may not be expecting it. The contact between practitioners at various levels needs to be changed in order to improve the communication culture [34].

10 Communicate in Person or Virtually? or Something Else?

It is crucial to emphasize that internal communication shouldn't be viewed as a task that is begun at a specific time, completed, and then abandoned. Internal communication is a continuous process that has no beginning or end; this illustrates the complex relationship between internal communication and organizational performance [2].

Despite the technological development in the means of communication, face-to-face communication remains one of the most important administrative communication tools because of its emotional and mental dimensions that carry messages and deliver them to the opposite person with the least potential for misunderstanding, considering the culture to which the corresponding person belongs. This claim meets the numbers of a recent report in which was published by [10] that face to-face communication channels are still considered the most impactful, with more than 90% of respondents rating most of these as 'very' or 'quite' effective, see Table 2.

On the other hand, managers can use many virtual tools to communicate, such as Text messaging, Instant messaging, Video chat, Voice over internet protocol (VoIP), social media, Email and/or Web conferencing. Upon to [10], the effectiveness of these channels is generally less than the face – to – face channels but still effective, see Table 3.

We can conclude from the above, that communication using digital means (tools) are effective which save time and effort, in the meanwhile it cannot match face-to-face communication, so the use of hybrid communication means, especially in a multicultural work environment, may be a way to increase productivity and reduce conflicts in the organization.

Table 2 Face to face communication effectiveness

Item	Use	Effectiveness
Team meetings run by people managers	82%	89%
Face-to-face or virtual conferences/Roadshows/Town halls (all employees)	76%	94%
One-to-ones run by people managers	66%	92%
Face-to-face or virtual conferences/Roadshows/Town halls (senior leaders and/or people managers only)	60%	92%
Informal get-togethers and social events (In person or virtual)	57%	91%
Web calls/Conference calls (all employees)	52%	90%
Web calls/Conference calls	49%	90%
Knowledge-sharing / Lunch and learn / Brown bag sessions	46%	85%
Employee or staff forum/Work councils	35%	78%
Ambassadors/Communications champions	31%	77%

Source: [10]

Table 3 Digital communication tools effectiveness

Item	Use	Effectiveness
Email announcements	94%	78%
Intranet	83%	60%
Employee portal(s) e.g., for benefits. wellbeing. learning and development	77%	75%
Enterprise chat tools	72%	91%
E-newsletters	64%	64%
Enterprise social networks (Workplace from Meta, Yammer. etc.)	39%	59%
Mobile app(s)	31%	70%
Online peer-to-peer recognition platforms	30%	68%
Messaging apps (text. WhatsApp. etc.)	24%	87%
Artificial Intelligence (Al) e.g., chatbots, etc.	10%	52%
Extranet	8%	62%

Source: [10]

11 Communication Skills (Communicate as a Manager)

The core of everything you do as a manager is communication. Managers communicate for up to 80% of their working hours [24]. You interact with people all day long by talking, listening, presenting, and exchanging knowledge. You will understand your team's needs more and be a more effective manager if you are good at sharing ideas and communicating as a manager [26]. Table 4 shows 7 communication skills that every manager needs to have specially when acting in a multicultural environment.

12 Discussion

It is clear that the idea of a management in multicultural environment has been subjected to many scientific discussions, including those who supported dealing with the multicultural reality, and tried to lay the foundations and theories for it as [14] did, after that they began framing the transactions through understanding the form and content that result from the multiple cultures in an organization as in Hofstede's 6 Cultural Dimensions theory and its development until 2020.

The future of global business depends on people's ability to deal with diversity in an honest and transparent way because business relationships are ultimately instances of cultural interaction [8]. This balance entails being conscious of one's cultural identity in relation to diversity as a whole, since "It is only when you start to identify what is typical in your culture, but different from others, that you can begin to open a dialogue of sharing, learning, and ultimately understanding" ([25]:203).

Table 4 shows 7 communication skills every manager needs

Communication skill	Description
Be a positive communicator	Discourage complaining and negativity among the employees Make personal connections with employees Be positive in the nonverbal ways you communicate Be aware of the messages you send with body language, tone of voice and eye contact Create an atmosphere of open communication
Be a careful communicator	If you don't know the answer to a question, be honest and say so Speak simply, clearly and avoid Jargon Explain your decisions as much as possible Never communicate when you are angry or feeling highly emotional Don't make promises you can't keep Apologize Keep people Informed
Actively listen	Practice active listening and make it a habit Paraphrase what the other person said using your own words to be sure you understood Be curious and ask people what they think Listen for the meaning behind the words and make sure you watch the nonverbal's as well
Meet regularly with each ofyour Direct reports	Take the time to explain and review goals Listen and ask questions at your meetings with employees Coach your employees to help them reach their full potential
Powerful questions - be curious	What is your vision for this? Would you say more about that? What have you already tried? What has worked well for you in the past? Tell me what happened next What should be the results? What steps have you taken to get to the results? What small steps can you take to get you closer to your vision? What needs to happen next? What needs to happen differently? How much energy are you willing to put into that?
Give and ask for frequent feedback	Provide specific appreciative feedback on a regular basis Provide opportunities for employees to share ideas and concerns Ask for feedback from employees Follow through both in answering questions and responding to feedback

(continued)

Table 4 (continued)

Communication skill	Description
Effective Ways to Handle Conflict	Take conflict seriously Where there is a conflict, face it directly and promptly Help employees understand that resolving conflict requires give and take from everyone

Source: [26]

On the other hand, some scholars have tried to interpret their ideas related to multiculturalism based on the results of economic projects, and the extent of profit or loss achieved by an organization, whether it is multicultural or monocultural, and their fears cannot be denied, since a multicultural environment is difficult to manage in a purely local way. This was expressed by [32], by highlighting the negatives that may reflected in the projects in which a multicultural teams work together.

In the recent study of [32], several correlation coefficients emerged that showed a negative relationship between culturally diverse work teams and the desired results of the work and warned all organizations and managers against taking the step of entering into projects of diverse cultures before seriously considering its dimensions and setting a prior strategy to deal with the caveats. They concluded that this needs time before any desired change.

Despite my understanding of the warnings issued by some scholars regarding the cultural diversity of the work team in an organization or a project, I tend to the idea of considering diversity as an opportunity for development as [5, 7, 14],SHINNERS, 2017; [34] argued, cultural diversity may be a qualitative addition to the project or organization, if there is a strategy in the institution to deal with this diversity, through prior understanding and analysis of the basics of cultural diversity that I have briefly introduced throughout this essay.

Do you remember I invited you to look at the computer in front of you in the office, or your mobile phone in your hand? My invitation was not free, and it took me a one month to write this essay, but the good news is that I got a statistic in the Fig. 7 below that may make you rethink the topic of diversity, and may change your prejudices in judging others, this is one of the fundamentals that made the giants of Silicon Valley believe in multiculturalism and they take advantage of it to the extent that you can now close this essay with one **click** on a button, someone in the world may have made button for you, you both speak a different languages and live in a completely different place of culture, but finally you communicate each other, albeit indirectly.

This notable success in managing cultural difference and diversity in the work teams of giant hi-tech companies, makes me more inclined to the opinions of scholars who see cultural difference as an opportunity to increase creativity in understanding the other, which can be a new opportunity for a successful project.

Enabling project managers to manage multicultural teams has become a duty for heads of companies and organizations seeking to internationalize their projects and

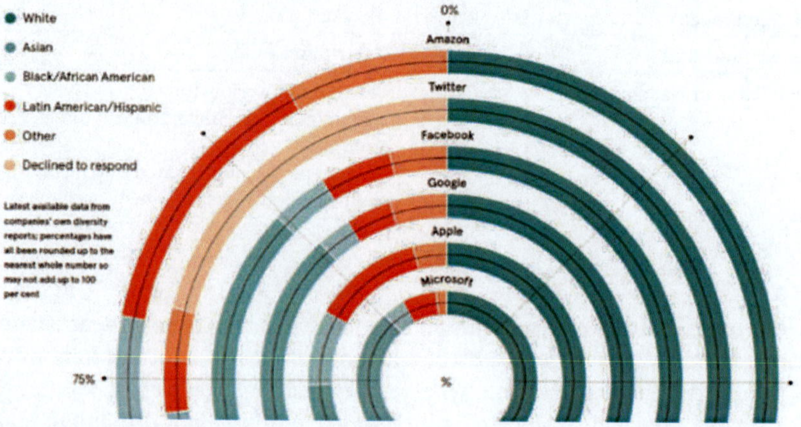

Fig. 7 Race/ethnical, non-technical leadership positions combined in USA onl (Silicon Valley) *Source*: [6]

expand their profits, and these ideas must be clear before the planning process of any multicultural project. The consequences could be catastrophic if this preamble does not take place.

I believe after this explanation of the processes of communication in cross-cultural work environments, that the culture of difference, and the difference of culture creates new environments of work, not necessarily positive, but at least the negatives must be understood to avoid them as much as possible, here lies the strength of multicultural human teams, they understand their differences and turn it into a new culture that engages their professionalism, to produce the best that can be produced within the framework of human capacity. Professional communication knowledge and skills in which are covered by the ability will be one of the most distinguished milestones of successfully managing a project of a multicultural human environment.

References

1. Alexandrova M (2018) A principal component analysis of project portfolio management practices. Ekonomicko-manazerske spektrum 12(2):96–105
2. Ali BJ, et al (2021) Business communication strategies: analysis of internal communication processes. J Hum Educ Dev 3(3):16–38
3. Alizadeh Afrouzi O (2021) Humanitarian behavior across high-/low-context cultures: a comparative analysis between Switzerland and Colombia. J Int Human Act 6(1):1–10
4. Andersen ES (2016) Do project managers have different perspectives on project management? Int J Project Manage 34(1):58–65
5. Bansal V (2021) High context culture vs low context culture: communication design for avoiding uncertainty- category: communication/leadership https://www.techtello.com/high-context-culture-vs-low-context-culture/

6. Barton L (2019) Is Silicon Valley using culture fit to disguise discrimination? Raconter- Content for business decision-makers - Diversity and inclusion. https://www.raconteur.net/hr/diversity-inclusion/silicon-valley-discrimination/
7. Bogoeva B (2020) Project manager competencies in multicultural environment. In: International conference on research in business, management and finance - Oxford- Uk (iCRBMF), pp 27–29 March
8. Dumbravă G (2018) Cultural boundaries in business communication. Ann. Univ. Petrosani Econ. 18(2):37–44
9. Fortunato JA, Gigliotti RA, Ruben BD (2017) Racial incidents at the University of Missouri: the value of leadership communication and stakeholder relationships. Int J Bus Commun 54(2):199–209
10. Gallagher (2022) State of the sector 2022: global internal communication and employee engagement insights and trends. AJG. https://www.dropbox.com/s/sp4fjqxrugay8ah/state-of-the-sector-2022.pdf?dl=0
11. Gambardella GB (2021, July 29). Advantages and challenges of working in a multicultural team. https://www.teambuildingincentive.com/working-multicultural-team/
12. Ganeshkumar C, Prabhu M, Abdullah N (2019) Business analytics and supply chain performance: partial least squares-structural equation modeling (PLS-SEM) approach. Int J Manag Bus Res 9(1):91–96
13. Harrin E (2021) project communication management: what is it all about? https://rebelsguidetopm.com/project-communication-management-what-is-it-all-about/
14. Hofstede G, Fink G (2007) Culture: organisations, personalities and nations. Gerhard Fink interviews Geert Hofstede. Eur J Int Manag 1(1–2):14–22
15. Hofstede G, Hofstede GJ, Minkov M (2005) Cultures and organizations: software of the mind (vol 2). Mcgraw-hill, New York
16. IPMA (2015) Individual competence baseline for project management, version 4.0. . International Project Management Association
17. Kim YY (2017) Cross-cultural adaptation. Oxford Research Encyclopedia Of Communication
18. Kobiruzzaman MM (2021, December 30th). Hofstede's cultural dimensions- Hofstede's 6 cultural dimensions examples. Word Press
19. https://newsmoor.com/cultural-dimensions-hofstedes-cultural-dimensions-theory-with-six-dimension/
20. Lifintsev D, Wellbrock W (2019) Cross-cultural communication in the digital age. Estudos em Comunicação 1(28):93–104
21. Llopis-Lorente A, Díez P, Sánchez A, Marcos MD, Sancenón F, Martínez-Ruiz P (2017) Interactive models of communication at the nanoscale using nanoparticles that talk to one another. Nat Commun 8(1):1–7
22. Martin JN, Nakayama TK (2010) Intercultural communication in contexts, 5fth edn. McGraw Hill
23. Maryland UO (2019, November 1) Hofstede's cultural dimensions and assessment of cultural and linguistically diverse clients. https://hespinterpretation.wordpress.com/2019/11/01/hofstedes-cultural-dimensions-and-assessment-of-cultural-and-linguistically-diverse-clients/
24. Jasson F, Mazor AH (2017) The employee experience: culture, engagement, and beyond. deloitte global. Deloitte University Press, 144p. https://www2.deloitte.com/us/en/insights/focus/human-capital-trends/2017/improving-the-employee-experience-culture-engagement.html
25. Meyer E (2016) The culture map (INTL ED): Decoding how people think, lead, and get things done across cultures. PublicAffairs
26. OSU (2022) Communicating as a manager. The Ohio State University. https://gatewaytolearning.osu.edu/leadership-development/building-relationships/communicating-as-manager/
27. PMI (2021) A guide to the project management body of knowledge (PMBOK guide) and the standard for project management (vol 7). Project Management Institute, 7th edn
28. PMI, P. (2013). The Essential Role of Communication.

29. Prabhu M, Nambirajan T, Abdullah N (2020) Operating competitive priorities of manufacturing firms: an analytical study. J Indust Eng. Manag 13(1):38–55
30. Radujković M, Sjekavica M (2017) Project management success factors. Procedia Eng 196:607–615
31. Romani L, Mahadevan J, Primecz H (2018) Critical cross-cultural management: outline and emerging contributions. Int Stud Manag Organ 48(4):403–418
32. Schommer M, Richter A, Karna A (2019) Does the diversification–firm performance relationship change over time? A meta-analytical review. J Manage Stud 56(1):270–298
33. SHINNERS, C. M. (2017) Communication, culture and effective teams. Three Swords Mag 31:79–85
34. Zita T (2020). Communication as a tool for effective project execution at selected construction sites in Cape Town, South Africa. (Doctoral dissertation, Cape Peninsula University of Technology). https://etd.cput.ac.za/handle/20.500.11838/3107
35. Zwikael O, Pathak R, Ling F, et al (2021) Variation in project management practices across borders. Prod Plan Control 33:1–13. https://doi.org/10.1080/09537287.2020.1858362

The Edge of Megaproject Case Studies

A New Path for Green Hydrogen

Marta Gobbi, Paola Bongiorni, Andrea Bricchi, and Giovanni Cantoni

Abstract Green hydrogen is one of the keys for meeting the European Union's target of net zero emissions by 2050. It is classified as green if produced through electrolysis utilizing the energy produced by renewable sources such as solar or wind energy. The idea is to build up a compact system to be placed in containers able to produce and store green hydrogen. Stored hydrogen can then be utilized where needed. This product can be of interest for small-medium customers with renewable energy production, such as farm owners and companies. It's not a huge investment and the risk is low, but if widespread it can contribute both to planet's health and customer's wallet.

Keywords Green Hydrogen · Renewable sources · Energy · Megaproject · Electrolyzer

1 Introduction

Climate change and ecological degradation pose worrisome threats to economic, social and environmental sustainability; therefore, ecological transition has become a central focus of modern economic policies. This phenomenon is mainly related to the rapid expansion of the global economy, particularly the increase in emissions and overexploitation of natural resources. The consistent growth of economies, however, is associated with a continuous rise in energy consumption, which has led to an

M. Gobbi (✉) · P. Bongiorni · A. Bricchi (✉) · G. Cantoni
Piacenza, Italy
e-mail: m.gobbi@brianandpartners.it

A. Bricchi
e-mail: a.bricchi@brianandpartners.it

P. Bongiorni
e-mail: p.bongiorni@brianandpartners.it

G. Cantoni
e-mail: g.cantoni@brianandpartners.it

E. Favari and F. Cantoni (eds.), *Complexity and Sustainability in Megaprojects*, Lecture Notes in Civil Engineering 342,
https://doi.org/10.1007/978-3-031-30879-6_19

increase in environmental impacts raising the issue of ecological decline on a local and global scale.

Ensuring clean hydrogen and universal access to integrated energy systems is defined as the new key vector for achieving climate neutrality goals. There are different types of hydrogen; however, only some of them are truly sustainable. Grey hydrogen is the one obtained by reforming of fossil fuels without CCS, which stands for carbon capture and storage. Blue hydrogen is the one still obtained by reforming of fossil fuels but with the application of CCS, having so low impact on the CO2 emissions. Green hydrogen instead is the one produced from renewable energy sources only, through water electrolysis, and grants zero CO2 emissions. Thus, green hydrogen represents the green solution par excellence and also the most viable route at the moment. For a country like Italy, investing in green hydrogen means becoming a modern and functional, climate-neutral and internationally competitive economy even in terms of resources. Therefore, green hydrogen is a potential zero-emission energy tool. Its production generates clean electricity, and its storage is a viable alternative to increase the flexibility of the energy system, enabling the integration of large amounts of non-programmable renewable energy.

Green hydrogen represents a sustainable energy carrier that could provide the perfect solution for energy production. In fact, if produced cleanly, hydrogen can be a viable sustainable alternative for energy production at the expense of energy sources that have a greater impact on the environment, such as fossil fuels.

Governments and public–private institutions are paying increasing attention to environmental preservation. Recently, even in the European sphere, several plans have been arranged to allocate funds to encourage the production of hydrogen and production of infrastructures operating by means of this valuable source of sustainable energy.

Thanks to the 3.6 billion euros allocated by the PNRR, Italy is one of the countries that has decided to focus on this sector by initiating a series of investments aimed at the development of Italian supply chains.

In this context, both at the macroeconomic and technical level, we decided to invest in a particular niche related to hydrogen production that would allow anyone with a renewable energy source, even a very small one, thus farmers, small industrialists and the like, to produce green hydrogen.

2 Project Description

2.1 The Different Components

This project is aimed at employing used containers that have typically made only one trip, usually by ship, equipping them with everything needed to produce hydrogen. Specifically, a modular electrolyzer of the size needed from time to time, control software we produced and tailored, an industrial PLC system with related software.

Since one of the renewable sources' main characteristic is the unsteadiness, some storage batteries can be placed to temporarily gather the excess of energy that cannot be processed while the electrolyzer is working full. In the case renewable sources can be both external photovoltaic, as well as photovoltaic placed on the roof of the container itself, together with any other form of generation such as wind, micro-wind (on which Brian and partners is working through an investee company), biomass etc. Lastly, a storage or distribution system for the hydrogen produced that can be either stored for later sale in cylinders or similar, or used on site (for example to run a hydrogen boiler or otherwise an engine or fuel cell that uses hydrogen on site).

In a process known as electrolysis, an electrolyzer uses electricity to split water into hydrogen and oxygen. The electrolyzer system produces hydrogen gas by electrolysis. The excess oxygen is either discharged into the environment, or it can be trapped or kept to provide other industrial operations or, in some situations, medicinal gases.

Since hydrogen is an energy carrier, it can be used to power any hydrogen fuel cell electric application, including trains, buses, trucks, and data centers. The hydrogen gas can be either compressed or liquefied for storage.

The Electrolyzer. An electrolyzer has a cathode (negative charge), an anode (positive charge), and a membrane in its most basic configuration. Pumps, vents, storage tanks, a power source, a separator, and other parts are also included in the overall system. Within the cell stacks, there is an electrochemical reaction called water electrolysis. Water (H20) is split into its component molecules, hydrogen (H2) and oxygen (O2), when electricity is delivered to the anode (H2) and cathode (O2).

Size and functionality of electrolyzers vary. The size of these electrolyzers can be adjusted to accommodate different ranges of input and output, from tiny industrial plants installed in shipping containers to massive centralized production facilities that can provide hydrogen via trucks or be connected to pipelines.

Proton exchange membrane (PEM), alkaline, and solid oxide electrolyzers are the three primary categories of electrolyzers. Depending on the type of electrolyte being used, these various electrolyzers operate slightly differently. Alkaline and PEM electrolyzers can both produce 99.999% pure, dry, and carbon-free hydrogen. They can also produce pressurized hydrogen without a compressor on-site and on demand.

The three primary types of electrolyzers vary in the following ways:

- Proton Exchange Membrane (PEM) electrolyzers use a solid polymer electrolyte. Water splits into hydrogen and oxygen when current is delivered to the cell stack, and the hydrogen protons then flow through the membrane to create H2 gas on the cathode side [5, 6].
- In the Alkaline Electrolyzer the hydrogen is produced in a "cell" that is made up of an anode, a cathode, and a membrane. It uses a liquid electrolyte solution, such as potassium hydroxide (KOH) or sodium hydroxide (NAOH), and water. Typically, the cells are connected in series to form a "cell stack" that, as the number of cells increases, produces more hydrogen and oxygen. Hydroxide ions (OH-) flow through the electrolyte from the cathode to the anode of each cell when current

is given to the cell stack, as shown here. On the cathode side of the electrolyzer, hydrogen gas bubbles are produced, and oxygen gas is produced on the anode [2].

- Solid Oxide Electrolyzer Cell uses as the electrolyte solid ceramic material. At the cathode, water and electrons from the external circuit mix to produce hydrogen gas and negatively charged ions. Then, oxygen travels through the solid ceramic membrane before reacting at the anode to produce oxygen gas and produce electrons for the external circuit. Alkaline and PEM electrolyzers run at temperatures up to 80 degrees Celsius while SOECs function at over 500 degrees Celsius. SOECs have the potential to be significantly more efficient than PEM and Alkaline.
- A low temperature technology is the Anion Exchange Membrane Electrolysis (AEMEL). Even if less known and recently developed, it has shown great progress and the interest on this technology is growing. It works at low temperature (30–60 °C) and can combine the advantage of alkaline environment (absence or heavy limitation of the presence of noble materials) together with the solid electrolyte (polymer membranes able to selectively transfer OH^- ions). An advantage with respect to Alkaline Electrolyzer is the reduction of corrosive fluid's presence, whilst with respect to PEM lower costs are retrieved for membranes and other materials.

Actually, the state of the art seems to be moving on a new direction [3]. Indeed, even if currently most efforts to reduce the cost of green hydrogen focus on electrolyzer capex, the electricity makes up most of the cost of green hydrogen. The more energy-efficient the electrolyzer, the less electricity is wasted, and the lower the cost of the green hydrogen. That's the idea that brought an Australian firm, Hysata [4], to completely reinvent electrolyzers, coming to a new model whose name goes by Capillary-Fed Electrolysis (CFE) Cell. The CFE works by a thin, low resistance membrane placed between the electrodes which sucks up like a sponge the electrolyte and the water from the reservoir thanks to capillarity. The production of the two gases comes without bubbles, which are present in the other electrolysis technologies, and this leads to an efficiency of the cell of 98%, and of the system of 95% (Fig. 1). The presence of bubbles makes mandatory to place separators to divide hydrogen from the electrolyte, in this configuration they are not needed.

To have a measure of the improvement that this new electrolyzer can provide, the existing ones are able to get an efficiency of 75%. This means that, for example, to produce one million tons of green hydrogen there would be a 3 GW of renewable energy saving.

Also, since the major issue with efficiency in electrolyzers is the energy wasted in heating, the higher the efficiency, the lower the need for cooling systems, thus another cost is avoided.

Hydrogen fuel cells may be powered by electrolyzed water, which is ideal. Fuel cells operate similarly to batteries in that they provide energy and heat as long as fuel is available. They do not deplete or require recharging. Hydrogen is used by the fuel cells to produce power with no emissions at the point of usage. This indicates that there are no harmful emissions or use of fossil fuels.

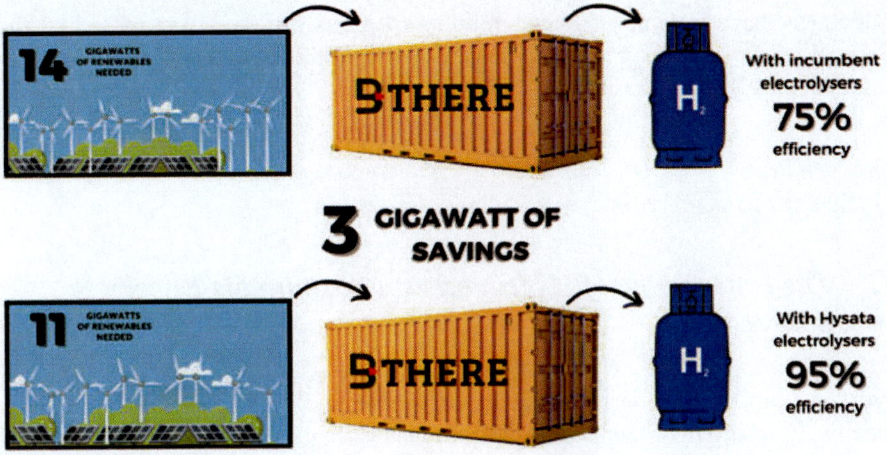

Fig. 1 Estimated savings increasing the efficiency of the state-of-the-art electrolyzers

Even better, the hydrogen produced is regarded as renewable and CO2-free from well to wheel when the electrolyzer system is driven by a renewable energy source.

The Boiler. A lot of companies producing boilers are working on the transition from natural gas to hydrogen. The goal is of course to get to a 100% hydrogen boiler, but at the moment the only way to have it working is to have an own production of hydrogen, which is our case. Indeed, the state of the art is the so called "H2 Ready" boiler, commercialized by different companies, which can work with a maximum 20% hydrogen blend. The mixture entering the boiler is analysed by the boiler itself and settings are properly electronically modified in order to keep the highest efficiency at any hydrogen percentage.

The Software. The tailored and produced software mentioned before has the aim to increase the client's awareness about the process of hydrogen production starting from the renewable source's monitoring. Indeed, by having a complete overview of the process it is possible to detect but more importantly prevent failures and machine downtimes. Taking as an example the photovoltaic technology, the software is able to track and store Key Performance Indicators of the correlation between the irradiation and temperature and the effective electricity production, together with a comparative analysis of the efficiency of each inverter. All this easily on your smartphone.

2.2 The Project as a Megaproject

Megaprojects are generally considered as massive investment projects costing $1 billion or more. This project falls into the megaproject category because it has a very high starting value. Specifically, the idea is to develop a series of systems that collectively have a value of many millions of euros each year composed of the summation of many small modular projects from €75,000–100,000.00 euros up to €250.000,00–300.000,00 euros.

3 Scenario Analysis

3.1 *The Importance of Hydrogen as a Sustainable Energy Source*

Hydrogen can be considered a viable sustainable energy carrier in its own right. Specifically, green hydrogen represents the clean variant of hydrogen, is not naturally occurring, and is produced through renewable sources by a process of electrolysis. It can be transported, stored, and used just like a gas, but without generating polluting effects. Institutional, retail, and government investors are increasingly demanding climate-related investment strategies. This phenomenon is due to both the benefit of investing to generate a positive impact for society as a whole and the opportunity to increase portfolio diversification. Recent studies have shown numerous benefits in the use of hydrogen in the Italian energy system. In the current context of 95% decarbonization, within 30 years, almost a quarter (23%) of all energy demand in Italy could be covered by hydrogen, mainly in the sectors of transportation, building heating, and some industrial applications in which there is already wide use of grey hydrogen. This percentage potentially exceeds the current total share of electricity produced from both fossil fuels and renewable sources.

As far as the Italian territory is concerned, the extensive gas infrastructure already in place in the country would make it possible to connect the substantial amount of renewables that lie in the South of the country, with the North of the country hosting large demand. In the islands, dependent and fully renewable energy systems can be built. Moreover, thanks to the richness of renewable sources, Italy could be ahead of many European countries in terms of the competitiveness of the cost of hydrogen. The cost could be competitive as early as 2030, about five to 10 years ahead of Germany for example.

In detail, green hydrogen may have potential applications in multiple areas:

- *Power to power*. Through an electrolysis process, energy is converted into hydrogen that will then be stored in an underground tank. This is a long-term solution, from weeks to months, during which the hydrogen can be turned into electricity at any time through the use of a fuel cell or gas turbine.
- *Power to gas*. Electrolysis plays a key role in the transformation of electricity into hydrogen. Once hydrogen is generated, it will be introduced to the gas grid. This is referred to as direct injection when a compound of natural gas combined with hydrogen is produced by inserting the hydrogen itself directly into the natural gas

grid; and methanation when the hydrogen is used to create synthetic methane to be mixed into the gas grid.

- *Power to industry.* Hydrogen generated by the electrolysis process can play a relevant role as a feedstock in industries that need high-quality hydrogen with low carbon emissions. To name a few, power plants, refineries, chemical industry, or steel and metal production plants.
- *Power to fuel.* In refineries, there is still a substantial demand for fuel and a not insignificant use of hydrogen to remove sulfur from fossil fuels. Therefore, using green hydrogen in this process is a viable alternative to considerably reduce the amount of carbon produced by refineries.
- *Power to mobility.* Hydrogen generated from excess power generation from renewable sources is fed into the fuelling systems of road/rail vehicles. In detail, the electric motor is powered by the electricity generated from hydrogen in the fuel cells contained in the vehicles [1].

The energy sector has a chance for widespread change thanks to hydrogen. Globally, energy systems are undergoing a major shift to put more of an emphasis on reduced emissions and adverse environmental effects.

3.2 *The Importance of Hydrogen in the Energy Storage System*

Renewable technologies like wind and solar have emerged as essential components of a solution to reduce the adverse effects of climate change and decarbonize the electricity sector. However, incorporating these sporadic energy sources into the electrical system can be difficult.

In fact, the above-mentioned unsteadiness of the renewable source, often climate dependent, is something that need to be overcome in order to use at the maximum the green energy and avoid any possible waste. The energy storage is a crucial problem in meeting the net zero requirements, but also has been a historic problem. Indeed, energy storage involves converting energy from forms that are difficult to store to more conveniently or economically storable forms, thus looking for the most efficient way to exploit energy. Some technologies provide short-term energy storage, while others can endure for much longer. For example, the dams are a clear example of potential energy storage, which is utilized when the electrical grid has peaks of demand by letting the water flow through the turbines. If there was no dam, the water could be flowing through the turbine even during the night, when the demand is generally low, and all that energy would have been wasted.

In order to overcome these grid-related difficulties, hydrogen can serve as an energy storage medium, making it simpler to use renewable energy off the electric power grid. Indeed, with the electrolyzer the energy is spent to create hydrogen, and then with the fuel cell the hydrogen is burnt to produce energy, all this with a higher round-trip efficiency with respect to batteries. Round-trip efficiency is the percentage of electricity put into storage that is later retrieved. The higher the round-trip efficiency, the less energy is lost in the storage process. For efficient long-term storage and transportation of renewable electricity, hydrogen is a solid option. In this way renewable electricity produced by the sun and wind can be stored and used at a later time or location. Due to its capacity for energy storage and transportation, hydrogen is an essential component in a global switch to renewable energy.

3.3 On Site Utilization

Of course, hydrogen can be also utilized immediately after its production. One idea can be for example to burn the hydrogen in a boiler to heat up a surrounding building. Also, installing a green hydrogen production system on vehicles can be a way to have a fuel production on board, in the vehicle's engine is an electric motor. Let's take as example a ship. If solar panel are placed on the deck and connected to an electrolyzer and a hydrogen store system, when the ship needs to be fuelled the current goes straight to the ship's engine and fuels it. When instead the engine is not requesting any power, as an example when the ship is moored, the energy gathered form solar panels produces hydrogen. In this way it is possible to avoid any waste while storing energy in hydrogen form. Such model can be applied on any other vehicle, and it's simply an infinite energy gathering.

4 Demand Analysis

4.1 Who are the Users

As mentioned earlier, users are all those who have energy generated from renewable sources. We will focus on small, medium-small customers so industry, farmers and private individuals with excess power generation available from photovoltaic or any other type of generation for example pyrolysis, molecular cracking, wind, mini-wind, biomass etc.

4.2 What are the Costs for the Users

Costs for users are variable depending on the size of the system and thus particularly depending on the Megawatts produced on a case-by-case basis. However, costs are estimated to vary from a minimum of €75,000.00 to a maximum of €250,000.00-€300,000.00.

5 Financial Analysis

5.1 Investment Costs

Initial investment costs are called investment costs. It contains the capital expenses of all physical assets, both fixed and non-fixed, broken down into three categories:

a) Work start-up and supervision, which comprises project management and technical support activities.
b) Overall management costs, which include the management of the action, tender, procedure, technical monitoring activities, certifications, and safety procedures.
c) Expenditures associated with construction, which include labor costs (skilled and unskilled) and building supplies.

Investment costs of the project are very limited because it is assumed that work will be done on projects only as a result of timely orders. Consequently, investments are to be found primarily in the use of specialized personnel and in the work itself, which, however, given the purchase of materials subsequent to an acquired order, is already covered by a consequent payment. Therefore, the investment costs are considerably lower than the expected revenues.

5.2 The Break-Even Analysis

The break-even analysis is hard to do at this stage because the break-even should be reached already in the first year for the reasons mentioned above. There are no big investments, we work on a project basis, so we are already making profit from time to time.

5.3 Source of Finance

The source of finance is partially internal, therefore we are self-financed. However, part of the project's funding sources could be actors operating in the private equity. It indicates that we and the fictitious person in charge of the project might each contribute the necessary financial resources to complete the project. A private corporation, a public organization, or a private–public collaboration could carry out this investment initiative (PPP). A PPP is a collaboration between a public organization and a private business to perform a service that is typically handled by the public sector. Risk is transferred from the public to the private entity in a PPP.

6 Economic Analysis

6.1 Impact and Benefits of the Project

The purpose of the economic analysis in the CBA (Costs-Benefits Analysis) is to assess the project's contribution to social welfare and to look into how the initiative will affect society. The cost and revenues of the financial analysis are the starting point for the socio-economic analysis, which also involves correcting market values for tax and price distortions. Direct effects are measured by substituting user willingness to pay for revenues and adding market (and where appropriate, non-market) impacts for all stakeholders in the project's core market. To reduce the potential of duplicate counting, the impact on secondary markets is disregarded; however, externalities are assessed and included in the study.

After removing the distortions brought on by market imperfections and applying a suitable intertemporal discount rate, it is possible to quantify the project's economic outcomes and assess its Economic Net Present Value (NPV-E). This indicator, which is the difference between the project's expenses and economic benefits that have been appropriately discounted, demonstrates if the project ultimately results in an incremental benefit for the reference community. Calculating the Economic Internal Rate of Return on Investment will yield the project's average economic return on capital (IRR-E). This formula results in an NPV-E of zero, which is the discount rate. If the NPV-E is positive and the IRR-E exceeds the discount rate, the project is deemed economically acceptable by the community. If the NPV-E is positive, the discounted benefits alone will ensure that all expenditures will be covered for the whole period of time taken into account.

The long-term relationship between infrastructure investments and progress or economic growth has been the subject of numerous quantitative studies by academics. This research demonstrated that infrastructure and wellbeing have a beneficial relationship. For this reason, we strongly believe that implementing a megaproject focused on green energy produced using hydrogen can have numerous benefits for stakeholders and the growth of the local territorial system can benefit from the availability of sustainable infrastructure.

There are many aspects of benefit that can be derived from the implementation of the megaproject, and in particular are those that enable individual operators-private customers whether farmers, industrialists, or ordinary citizens to optimize energy production by channelling everything in excess into hydrogen. This prepares us for a future in which hydrogen will be increasingly important. Indeed, remember that hydrogen is the element most present in the universe, so in the future we will undoubtedly move toward hydrogen. This project makes possible to start moving toward that direction using renewable sources and excess energy.

The impacts on the local area are extremely useful. Meanwhile, we would like to make Piacenza an experimental Italian hub of reference: starting in Piacenza with real and concrete projects, the goal is to become a driving force for the rest of Italy.

In general, this is a project with a high redeemability from which it is expected a fair margin.

7 Risk Analysis

Sensitivity Analysis, Qualitative Risk Analysis, Risk Prevention and Mitigation are the suggested processes for assessing the project risks.

7.1 *Sensitive Analysis*

Sensitivity analysis makes it possible to determine the project's crucial variables. These are the variables that have the biggest effects on the project's financial and/or economic success, whether they be favorable or adverse. One variable at a time is changed in the analysis in order to see how it affects the Net Present Value. The recommendation is to use the condition that a change of $\pm 1\%$ from the value used in the base case results in a change of more than 1% in the value of the Net Present Value as a guideline.

7.2 *Qualitative Risk Analysis*

A risk matrix is used to convey the qualitative risk analysis. It accounts for the primary potential uncertainties in relation to every facet of the project. The primary dangers include:

a) Administrative dangers (building permit acquisitions, utilities, approvals...)
b) Acquiring land (costs, delays of land purchasing...)
c) The project's design (inadequate site surveys, inadequate design cost estimates...)
d) Construction-related risks (construction cost estimates, construction quality...)

To lessen or eliminate the negative effects, it is required to suggest preventative and/or mitigating measures for each one.

Project's estimations shown not relevant economic and financial risk because individual projects are developed on the basis of orders already placed. Therefore, there is no particular material production, but it is mainly engineering and EPC work that through its own software assembles systems of others and runs them. It is therefore purely engineering work with no particular risk.

8 Conclusions

In conclusion, considering the macro-economic framework described above along with all the changes that are happening in the world, gas and energy increasingly in trouble, it is considered necessary to accelerate the transition to renewable technologies, especially hydrogen. We have a niche project to develop hydrogen on small and medium-small customers and therefore very widespread in the area. So, the goal is to create a capillary structure where those customers install these highly profitable experimental systems in a relatively short time, without big investments and therefore without big risks.

References

1. Atilhan S, Park S, El-Halwagi MM, Atilhan M, Moore M, Nielsen RB (2021) Green hydrogen as an alternative fuel for the shipping industry. Curr Opin Chem Eng 31:100668. https://doi.org/10.1016/j.coche.2020.100668
2. David M, Ocampo-Martínez C, Sánchez-Peña R (2019) Advances in alkaline water electrolyzers: a review. J Energy Storage 23:392–403. https://doi.org/10.1016/j.est.2019.03.001
3. Esposito DV (2017) Membraneless electrolyzers for low-cost hydrogen production in a renewable energy future. Joule 1(4):651–658. https://doi.org/10.1016/j.joule.2017.07.003
4. Hodges A, Hoang AL, Tsekouras G, Wagner K, Lee C-Y, Swiegers GF, Wallace GG (2022) A high-performance capillary-fed electrolysis cell promises more cost-competitive renewable hydrogen. Nat Commun 13(1):1304. https://doi.org/10.1038/s41467-022-28953-x
5. Lindquist GA, Xu Q, Oener SZ, Boettcher SW (2020) Membrane electrolyzers for impure-water splitting. Joule 4(12):2549–2561. https://doi.org/10.1016/j.joule.2020.09.020
6. Millet P et al (2010) PEM water electrolyzers: from electrocatalysis to stack development. Int J Hydrogen Energy 35(10):5043–5052. https://doi.org/10.1016/j.ijhydene.2009.09.015

Infrastructure and Value Creation for Local Systems: The Case of Brescia-Padua A4

Paolo Rizzi, Lorenzo Turci, and Edoardo Favari

Abstract First, the correlation between per-capita GDP and infrastructural endowment is investigated, at the national level, at the regional level and at the provincial level. A simple regression is then set up at the provincial level in Italy: the role of the highway index is confirmed as significant in explaining economic growth. A case study is then presented: the development of the Brescia-Padua A4 highway in the last 70 years, in parallel with the socio-economic evolution of the provinces where the highway was built ("Alta Padana"). A counterfactual exercise is carried out: the same analysis is undertaken on "Bassa Padana", a neighboring area where no highway was built. Several themes are analyzed: demography, the value of production, the dynamics of enterprises, tourist attractiveness, quality of life and the environmental system. It is concluded that the A4 highway has "accompanied" development of the "Alta Padana" area, by fostering its physical and economic interconnection.

Keywords Megaprojects · Road infrastructure · Territorial development · Value creation · Local systems · Case study

1 Infrastructure Projects and Value Generation

The impact that infrastructure projects have on the economic development of local systems can be categorized, according to their impact, into three groups: the direct and measurable impact related to employment and revenues during the project implementation; the indirect impact, related to the chain of links upstream and downstream

---

P. Rizzi
Department of Economic and Social Sciences, Università Cattolica di Piacenza, Piacenza, Italy

L. Turci (✉)
Local Economy Laboratory, Università Cattolica di Piacenza, Piacenza, Italy
e-mail: lorenzo.turci@unicatt.it

E. Favari
Department of Management, Economics and Industrial Engineering, Politecnico di Milano, Milan, Italy

E. Favari and F. Cantoni (eds.), *Complexity and Sustainability in Megaprojects*, Lecture Notes in Civil Engineering 342,
https://doi.org/10.1007/978-3-031-30879-6_20

the specific project implementation; the measurable positive and negative externalities related to the economic growth of the area such as value added, enterprises, export, touristic flow (positive) and traffic congestion and pollution.

The value generation of infrastructure megaprojects go beyond the mere financial efficiency [7, 8]. In the past decades, the theoretical frameworks of project management were just based on respecting schedule and budget, on value chains based on customers and suppliers, and on a benefit–cost analysis in which the willingness to pay of users was above the costs incurred [21].

Today the vision on large infrastructure projects has evolved in many ways [7]. The goal of megaprojects is more and more recognized as broader than the specific technical traffic problem (in the case of a transportation project), but includes the tackling of climate change, loss of biodiversity, inequality of salaries, lack of homes and loss of social mobility, and are exacerbated by pandemics and wars. The goal of enhancing "economic development" of such capital projects is often referred to in public debates because of potential social impact of private and public–private large projects. The more capital investments become tools for generating social value, the more the goal of such projects moves away from the mere financial profit and targets the social value creation to "shape the future" [22, 23].

The new scenario requires a new approach in engaging stakeholders, leaving the old logic based on the questions "Who is in and who is out?" and "Who gets what?" [15].

Nonetheless, in parallel with this change of perspective in the definition of the goals of a megaproject's construction and its organizational governance, the traditional approach of the feasibility evaluation and management of large infrastructure projects seems to endure. In fact, the benefit–cost analysis is still the main—often the only—decision-making tool for public resource allocation, despite the obvious difficulty in monetizing most of the externalities caused by a megaproject [4]. In addition, the iron triangle "on time, on budget, in scope" is still referred to as the only indicator for good performance of a megaproject's management and implementation, even if it is clear that such endeavors are subject to feedback cycles and non-linear relationships (Shapiro & Lorenz 2000), and the opportunism of the unavoidable political decisions underlain to large public investments (referred to as optimistic bias and strategic misrepresentation) are commonplace anywhere in the world [6].

The next step is asking ourselves how topics such as climate change, biodiversity and social impacts can be included into the traditional investment evaluation, and what alternative can be used for that purpose [5].

These new needs in the decision-making process for assessing megaprojects' feasibility and implementation rise challenging questions on how the value is co-created-independently if the promotors look for this co-creation -, on "who is in" and "who is out", on "who gets what" and "who pays what".

In this new scenario, the study of how a major European infrastructure, such as the Italian highway A4 Brescia-Padua [2, 14], generated and distributed value in its 70-years life span, since its original construction in the 1950s, up to its enlargement from 2 to 3 lanes, and adding new exits, can definitely help in understanding created and distributed value in the territory where it is built. Moreover, this study can add knowledge on how this value has been distributed among the broad group of stakeholders, including power stakeholder, and both users and non-users (market and non-market) ones, and on which scale this phenomenon has occurred. Such a comprehension is of utmost interest for the understanding of mechanisms of those phenomena, and can lead future planning and management processes of new infrastructure projects, helping promoters to implement a project while achieving the maximum of value creation and distribution to the communities—present and future—involved.

2 Territorial Development and Road Infrastructures: Empirical Evidence

The first significant correlation at an empirical level is between per-capita GDP of European countries and the infrastructural endowment measured in terms of motorway network density. Eurostat Motorway index measures the kilometers of motorways in each country with respect to the total area (thousands of square kilometers). Among the countries with the highest motorway endowment, the Netherlands, Belgium and Luxembourg (Benelux) emerge, followed by Germany, Denmark and Spain (see Fig. 1). Countries such as Portugal and Slovenia are also positioned before Italy, while Eastern European countries such as Romania, Bulgaria, Poland, Slovakia, Estonia, Lithuania are still relatively penalized. The lower infrastructural endowment of some Nordic countries such as Sweden and Finland appears peculiar, but this positioning is partly explained by the lower population density of vast natural areas of these states.

The link with per-capita gross domestic product appears to be growing and significant, albeit with some outliers, attributable to Scandinavian areas (high level of GDP per capita and low motorway endowment). Already in this first empirical analysis, the positioning of the two Italian regions of Lombardy and Veneto is higher than the national average and in line with the richest countries in Europe. Moving on to a regional analysis scale, the correlation between per-capita GDP and infrastructural endowment becomes even more evident (see Fig. 2). In this case, the index is elaborated in the Regional Competitiveness Report by the European Commission and concerns the road network as a whole. The most competitive European regions such as Hamburg, Brussels, Paris, Bavaria, Utrecht are still observed in the first places both for income per inhabitant and for the Road Infrastructure index. Also in this ranking, Lombardy and Veneto are aligned among the most gifted and productive regions (upper right quadrant of the scatter plot). However, it should be emphasized that the relationship does not occur with regional GDP growth rates, which - as is well known - were significantly higher in some areas of Eastern European countries, thanks to the massive foreign industrial investments of the last two decades, beyond the infrastructural constraints still present today in this part of Europe.

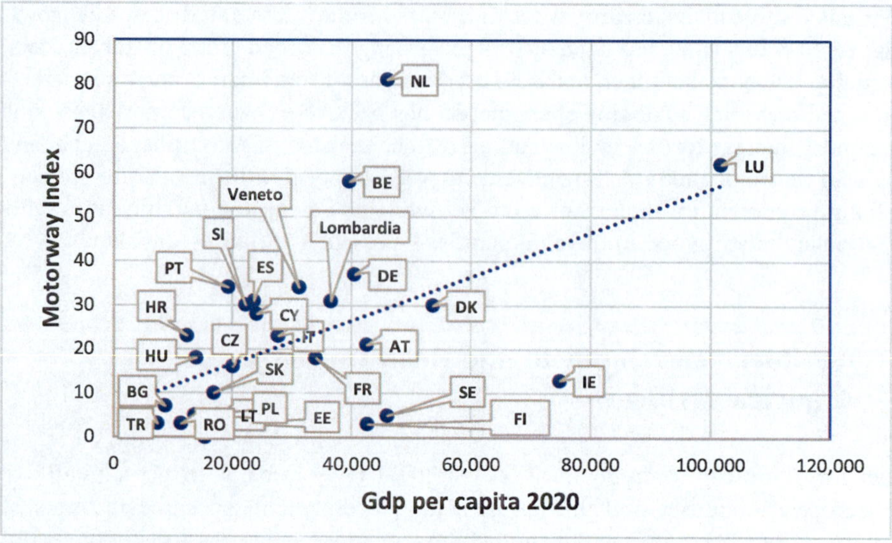

Fig. 1 The relationship between GDP per capita and motorway infrastructures in Europe (per-capita GDP current values in euro 2020 and Motorway - KM index for TKM2). *Source* Eurostat

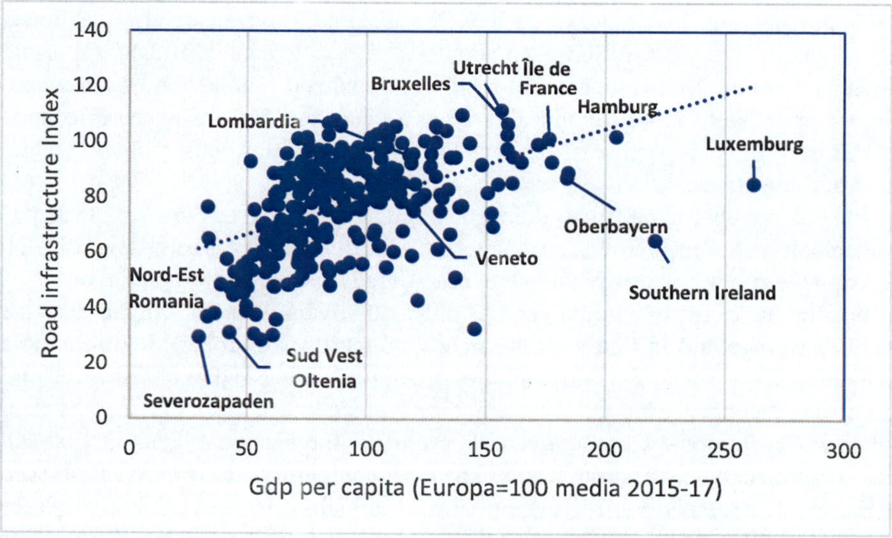

Fig. 2 The relationship between GDP per capita and infrastructural endowment in the EU regions (average GDP per capita values out of 100 Europe and Road Infrastructure index). *Source* European Regional Competitiveness Report 2019

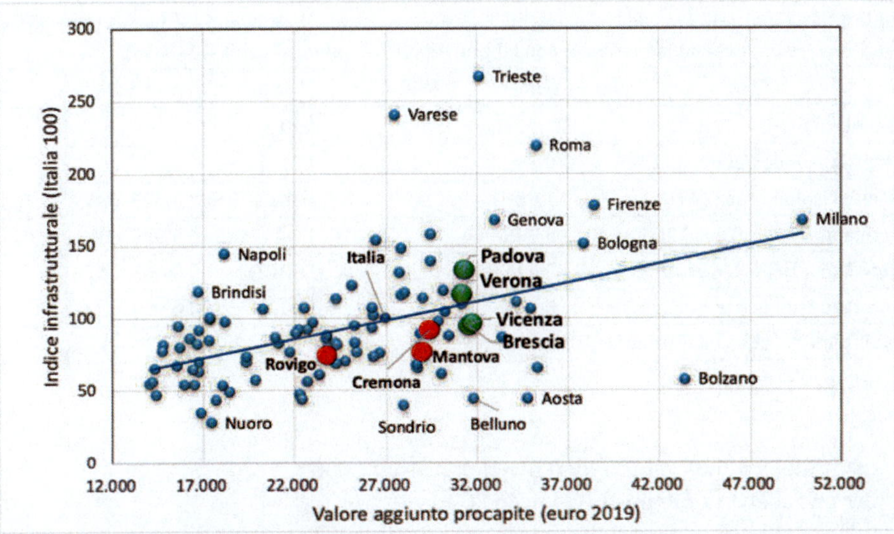

Fig. 3 The relationship between per-capita value added and infrastructural endowment in the Italian provinces. (value added per capita in euro 2019 and infrastructure endowment index 2011 Tagliacarne). *Source* ISTAT, Tagliacarne

The analysis of the relationship between infrastructural equipment and the level of economic development is also confirmed on a provincial scale in Italy (see Fig. 3). The index was developed by the Tagliacarne Institute, a historic Italian territorial research center, today the Unioncamere Foundation, which measured the provision of economic and social infrastructures in the provinces, not limiting the data to physical road networks only. Even in this case, the correlation between provincial per-capita value added and infrastructural endowment appears positive and significant, with some provinces showing excellent positions both in terms of per-capita income and infrastructure index, such as Milan, Florence, Bologna, Genoa.

Then there are areas of the country with an oversizing of the infrastructural endowment with respect to per-capita value added, such as Trieste, Varese and Rome, or on the contrary provinces such as Bolzano, Aosta and Belluno, which record high levels of income even exhibiting infrastructural indices below the national average.

If we consider the four provinces that are crossed by the Brescia-Padua motorway (therefore also Verona and Vicenza), we can observe both the good positioning with respect to the two indicators analyzed, and with respect to the Lower Padana area (Cremona, Mantua, Rovigo), which is inserted as an area of counterfactual comparison due to the absence of the motorway infrastructure in the period considered, or in the last 70 years.

An econometric verification was therefore elaborated, through a multidimensional linear regression (see Table 1), which places the growth rate of the provincial value added in the last twenty years as a dependent variable and the following variables as independent variables (regressors):

Table 1 The multidimensional relationship between economic growth and explanatory variables (OLS regression; dependent variable: annual growth rate of value added 2000–2019)

	Coeff	Std. Err	T Student	p-value	
(Intercept)	21,72	2,07	10,516	<2E−16	***
Highways Index 2006 (ISTAT)	0,39	0,18	2111	0,037	*
Social Capital Index [17]	1,98	0,52	3825	2,32E−04	***
Per-capita Value Added 2000	−5,20E−04	1,24E−04	−4198	5,98E−05	***
Per-Capita Export Value 2000	3,42E−10	1,13E−10	3012	0,003	**
Touristic Nights per capita 2008	0,12	0,04	2976	0,004	**
Multiple R^2: 0,2735 Adjusted R^2: 0,236 Number obs: 103					

Signif.: 0 "***" 0,001 "**" 0,01 "*" 0,05

- the initial level of value added per capita (Istat 2000)
- the initial level of exports (Coeweb 2000)
- the initial level of tourist presences per inhabitant (Istat 2000)
- a composite indicator of share capital (Rizzi and Popara 2006)
- a motorway infrastructure index: Km per surface (Istat 2006)
- the annual rate of population growth (Istat 2000–2021)

The initial value of value added is generally used in the literature on economic development to verify convergence processes [1, 16, 19]. If in the regression the coefficient is negative in sign, it is confirmed that the lower richest ones the initial level of per-capita output, the faster should be its growth, i.e. the more backward areas approach ("converge") to the richest ones. At the international level, the processes of economic convergence have been empirically confirmed for decades and they explain the global rebalancing trend of emerging economies, led by China, India and East Asia, which are indeed approaching the income levels of rich countries. Within the Western world, some bottlenecks persist that slow down the growth of some less organized countries from a technical-productive and institutional point of view, such as Italy and the Mediterranean countries. In Europe, however, convergence prevails, with accelerated development paths of eastern economies (former Soviet Union), thanks to the contribution of foreign direct investments and aid related to the EU Structural Funds and to industrial and fiscal strategies conducive to growth [11].

But in Italy convergence, the more accelerated growth mechanism of the less developed territories (i.e. the South) no longer occurs, at least it has not occurred in recent decades, when, on the contrary, we have witnessed to processes of relative divergence. Even in our econometric exercise, which concerns growth over the last twenty years, the coefficient relating to the initial value of the per-capita value added is not significant, therefore no convergence is observed between the Italian provinces.

On the contrary, the other factors that help explain the production growth of the Italian provinces are all positive and statistically very significant (Student's t greater than 2).

Of course, these include exports, which have been the country's lifeline in recent decades and a lever for development in Italian industrial districts, but also tourist attractiveness, which before the pandemic represented a consistent source of influx of not-solely-economic resources in many areas suited from a naturalistic and/or historical-artistic point of view.

Again, the composite indicator of social capital, a measure of the capacity for cooperation and positive relationship between companies and people (from volunteering to business networks to electoral participation [17]) takes a positive and significant sign.

Also the specific indicator of our study, the provision of highway infrastructures, which represents the role of physical capital, a prerequisite for economic and entrepreneurial activities, takes a significant and positive sign.

3 Development of the Brescia-Padua Highway

The relationship between the development of the highway infrastructure and the economic growth of the provinces crossed is not direct cause-and-effect, but of "accompanying" the processes of social and economic evolution. The A4 highway can be considered as a structural prerequisite that has favored the paths of population growth, development of production, growth of companies and employees, as analyzed in the following paragraph.

Comparing the number of vehicles transited with the value added of the "Alta Padana" (provinces of Brescia, Verona, Vicenza and Padua) in the period 1961–2021, the correlation of these two quantities appears clear and strong (the correlation between total vehicles and value added is 0.98). This figure should be considered with caution, only to testify how the evolution of the highway network and the production growth of the areas is symmetrical over time (see Fig. 4).

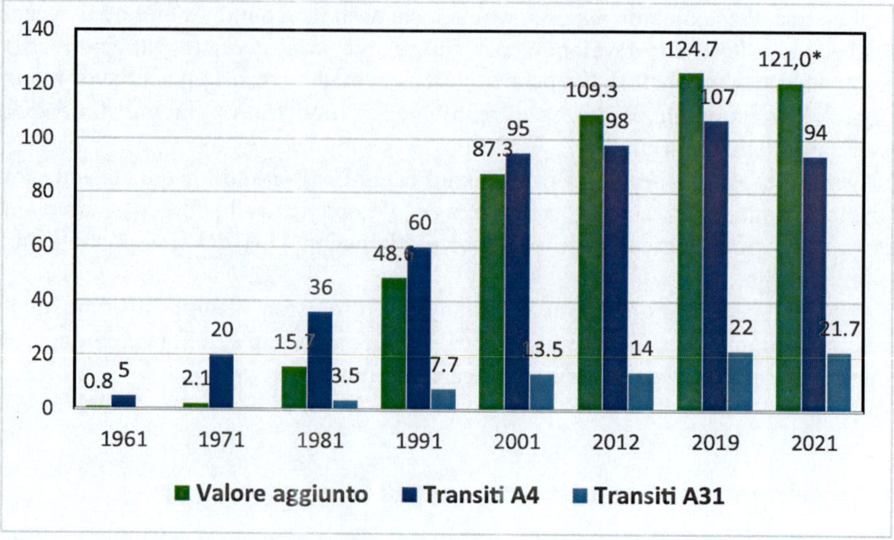

Fig. 4 The dynamics of transiting vehicles and value added (Alta Padana value added, billions of €; vehicles transited, millions of vehicles) *estimated data

4 The Evolution of the Socio-Economic System of the Provinces of Brescia, Vicenza and Padua as Compared to Bassa Padana

4.1 Demographic Dynamics

The demographics of the analyzed provinces have followed national and regional dynamics in recent decades, but at consistently higher rates. Four different periods can be distinguished: the economic boom period, when population growth went hand in hand with the soaring development of the 1960s (consumption explosion, rising living standards of the middle class, widespread industrialization in the North); the slowdown period of the 1980s, following the major social and industrial transformations of the previous decade (energy crisis, stagflation, terrorism, labor struggles, industrial reconversions); the strong recovery in the 1990s, driven by the exports of "pocket" multinationals in the Northeast and the influx of immigrants from abroad (Eastern Europe and African countries), which more than compensated for the negative birth-mortality; and finally, the stabilization starting in 2010, when the negative natural balance is no longer fully offset by the positive migration balance. The different trajectory of the Bassa Padana area is evident in Fig. 8: this area is characterized by demographic stasis since 1951 and it is now threatened by slow processes of population decline (Table 2).

This is the first major split in development processes between the area traversed by the "Serenissima" highway and an area with seemingly similar characteristics

Table 2 Population dynamics in Alta Padana and Bassa Padana (residents, 1961–2019)

	1961	2022	Var. % 1951–2022
Alta Padana	2.859.990	3.965.189	+38,6%
Bassa Padana	1.016.226	984.824	−3,1%
Italy	50.623.569	58.983.122	+16,5%

Source ISTAT [12]

in terms of history, culture and economics such as the Bassa Padana: of course, it cannot be said that an increase or a decrease in population directly results from the presence of a new road or highway network, but the first statistical association in our counterfactual exercise is displayed here (see Fig. 5).

Considering the seventy years from 1951 to 2022, this difference appears impressive: Brescia grew by 46.3%, Verona by 43.7%, Vicenza by 40.6%, and Padua by 30.4% (the average for Alta Padana is 40.5%), while Cremona declined by 7.7%, Mantua by 4.4%, and Rovigo by 35.5% (the average for Bassa Padana is -15.1%). At the national level, the population grew (+24.7% since 1951), at least until the last decade, when for the first time there was a slight population reduction in the country.

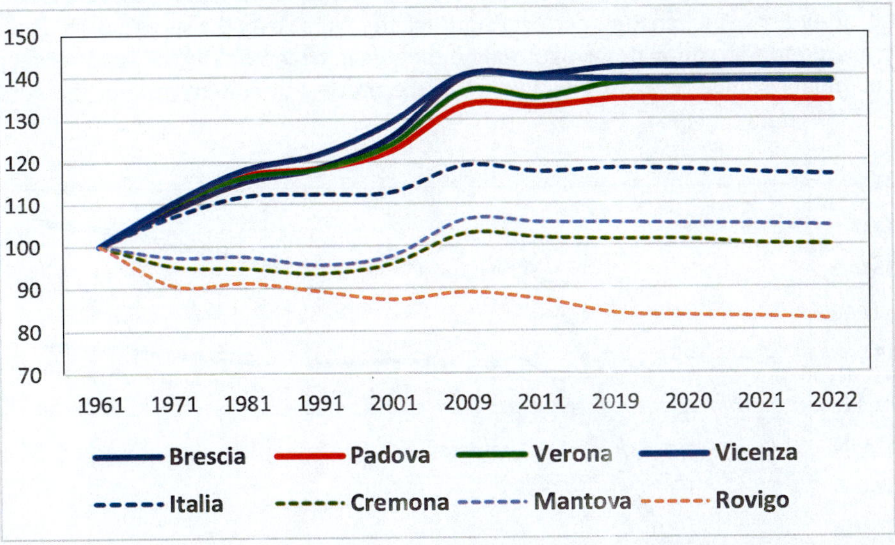

Fig. 5 Population dynamics in the provinces of Alta Padana and Bassa Padana (index numbers 1961 = 100). *Source* ISTAT, Population Censuses [12]

4.2 The Value of Production

The value added dynamics follows the demographic trend, with the boom of the 1960s; the crises of the 1970s, caused by the war in the Middle East and the twofold rise in oil prices (1973 and 1979); the periodic crises of the economic cycle (early 1990s and the Twin Towers crisis); the Great Recession of 2008. Since then, Italy has experienced a decade of zero growth, comparable only to Greece and a few African countries. Figure 6 shows the value added recorded every 10 years up to 2008 and every year from 2008 to 2019: therefore, the fluctuations of the cycle do not appear. Nonetheless, the chart clearly shows how the value added evolution of the four provinces crossed by the A4 highway has been stronger than both the three provinces of the Bassa Padana and the Italian average. After the 2008 crisis, a second recessionary wave hit in 2012–13.

In conclusion, analyzing the dynamics of population and value added over the long period, since the construction of the A4 highway, it is evident that the four provinces crossed by the A4 highway stand out both for higher population growth (right-hand position in the graph of Fig. 7) and higher value-added growth (top position), as compared to the national average and especially to the three counterfactual comparison provinces. The territories of the Bassa Padana, although geographically and culturally homogeneous, perform significantly worse over the same period, even compared to the national average: today, they have become "slower" and weaker economic systems. Figure 7 also illustrates the process of convergence of the Alta

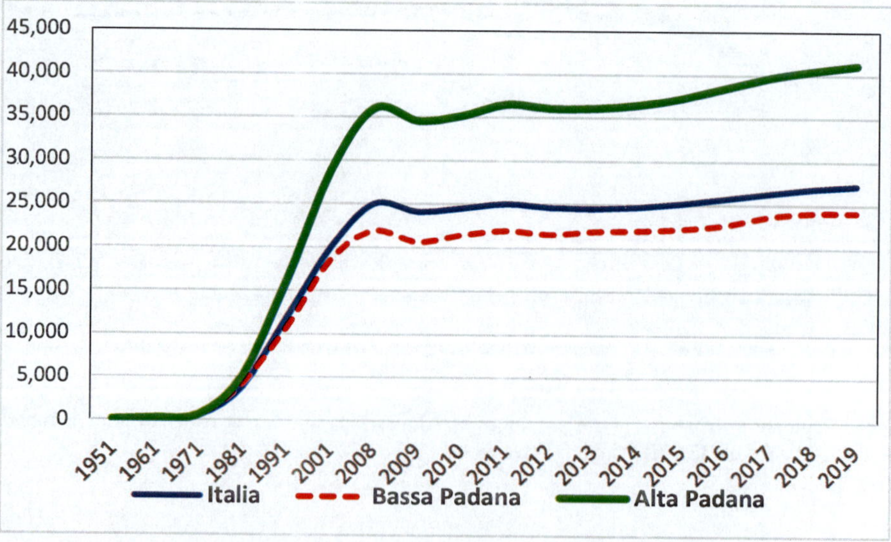

Fig. 6 Total value added in the provinces of Alta Padama and Bassa Padana (total value added 1951–2019; index numbers 1951 = 100). *Source* Istituto Tagliacarne, ISTAT

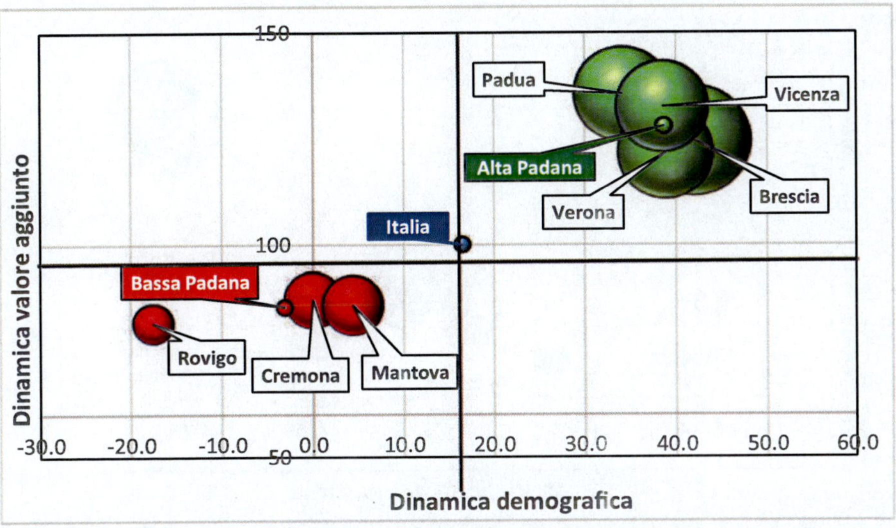

Fig. 7 Demographic and value-added dynamics in the provinces of Alta and Bassa Padana (yearly value-added growth rate 1961–2021; bubble size: value-added 2019). *Source* Istituto Tagliacarne, ISTAT

Padana provinces, which are characterized by considerable value-added and population magnitudes (121 billion euros and 4 million inhabitants), with growing trends in both dimensions.

4.3 The Dynamics of Enterprises

The third factor of economic development, together with population and value-added growth, is the dynamics of enterprises. Even in terms of local units, the economic growth of the four provinces crossed by the A4 highway appears impressive when analyzed over the long period, from 1961 to the present day. The number of local units grows in the Alta Padana provinces from 124 to 351 thousand, while in the Bassa Padana provinces it grows from 49 to 74 thousand (Table 3). However, the dynamics of local units, after the boom of the 1960s–1970s, seems to have come to a halt since 2001. After decades of continuous growth, first in 2009 as a result of the global financial crisis and then since 2012 with the second recessionary wave, terminations have begun to overtake births of new enterprises.

The slowdown of the businesses birth rate seems structural and therefore not only related to the economic conjuncture, but linked to deeper cultural and value processes. Recent GEM reports (Global Entrepreneurship Monitor 2021) highlight the Italian lag in the birth of new businesses due to a lack of incentive policies, weakness of the incubator business environment but also cultural factors of low drive for autonomy and entrepreneurship by the younger generations.

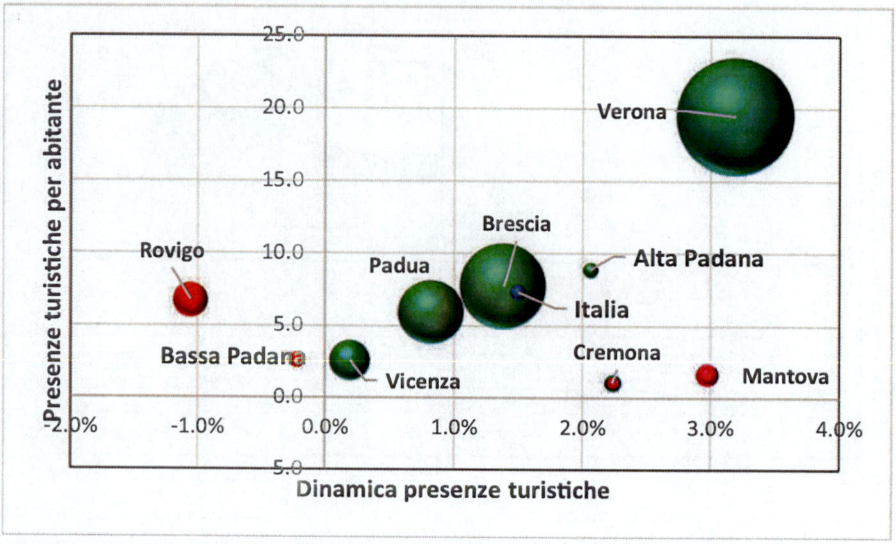

Fig. 8 Density and dynamics of nights spent in the provinces of Alta and Bassa Padana (x-axis: yearly growth rate of nights spent 2003–2019; y-axis: nights spent per capita 2019; bubble size: nights spent 2019). *Source* ISTAT

4.4 Tourist Attractiveness

Analyzing the dynamics of tourist attractiveness of the Alta Padana and Bassa Padana provinces, first it is useful to frame the proportions of this phenomenon in these two areas (Table 4). On the one hand, the 4 provinces of the Alta Padana area fluctuate between 24 and 35 million annual nights spent at tourist accommodation establishmments (ignoring the *annus horribilis* of the pandemic, 2020, when they touched the minimum of 15 million); on the other hand, the Bassa Padana area settles on a lower order of magnitude, fluctuating around 2.5 million annual nights spent. Even as a percentage of total nights spent in Italy, the Alta Padana area starts from 7.2% in 2003, reaches a pre-pandemic high of 8.4% in 2016 and 2017, drops to 7.3% in 2020 and rebounds to 9.0% in 2021, exhibiting above-average resilience. The Bassa Padana area, on the other hand, never exceeds 0.8% of total nights spent in Italy: this maximum was reached both in 2003 and in 2021. Between these two years this variable stalled for 7 years at 0.6% (2013–2019). Thus, we are considering two non-comparable realities in terms of tourist attractiveness. It is important to mention that the highway infrastructure cannot be considered the cause of this success, but it is certainly an important prerequisite, along with many others.

If we relativize the dynamics of nights spent by setting the value in 2003 equal to 100, we notice how Alta and Bassa Padana, which started from values of a different order of magnitude, followed two different paths. On the one hand, Alta Padana has experienced growth in different phases: a first growth phase between 2003 and 2008 (109.9), a second growth phase between 2008 and 2011 (119.7), followed by a

Table 3 Local units in Alta Padana and Bassa Padana (local units, 1961–2019)

	1961	1971	1981	1991	2001	2011	2019	Var. % 1961–2019
Brescia	40.421	41.628	78.280	88.091	97.253	112.810	108.755	169%
Padua	30.542	35.503	57.417	58.874	74.534	81.987	80.227	163%
Verona	25.348	33.572	55.472	58.717	71.869	77.023	74.047	192%
Vicenza	27.516	36.536	58.295	64.243	82.874	91.389	88.038	220%
Alta P	123.827	147.239	249.464	269.925	326.530	363.209	351.067	184%
Cremona	16.874	18.022	25.536	25.088	24.912	27.495	25.131	49%
Mantoa	19.886	21.589	31.423	32.203	32.622	32.712	31.488	58%
Rovigo	11.972	12.623	18.373	19.312	20.799	21.711	17.723	48%
Bassa P	48.732	52.234	75.332	76.603	78.333	81.918	74.342	53%
Italy	2.079.540	2.425.988	3.513.523	3.633.194	4.403.431	4.806.014	4.718.444	127%

Source ISTAT [13]

Table 4 Nights spent at tourist accomodation establisments in Alta Padana and Bassa Padana (nights spent, 2003–2021)

	2003	2019	Var. % '03–'19	2020	2021	Var. % '03–'21
Alta Padana	24.680.565	35.476.228	+43,7%	15.301.149	25.907.746	+5,0%
Bassa Padana	2.588.044	2.599.329	+0,4%	1.469.183	2.230.999	−13,8%
Italy	344.413.317	436.739.271	+26,8%	208.447.085	289.178.142	−16,0%

Source ISTAT

period of stability until 2014 (120.9) and a small boom from 2014 to 2019 (143.7). The worst year of the pandemic was a record low (62.0), as was the case for Bassa Padana (56.8) and Italy (60.5), but the rebound of Alta Padana was stronger, topping 105.0 in 2021 (as compared to 86.2 for Alta Padana and 84.0 for Italy). On the other hand, the Bassa Padana area followed a path of substantial stability until the pandemic, with a parenthesis of decline between 2012 and 2018. The national trend fits exactly between the two curves.

Plotting the dynamics of nights spent on the horizontal axis and nights spent per inhabitant on the vertical axis, it becomes clear that it is misleading to group the provinces into the two groups of Alta and Bassa Padana when we investigate tourist dynamics (Fig. 8). The only dimension for which these seven provinces can meaningfully be attributed to two groups—as already discussed—is the absolute number of nights spent, represented by the size of the "bubbles". This variable is much higher for Alta Padana provinces. As for the other dimensions, the seven provinces appear scattered around. Verona, for instance, stands apart both in terms of strong growth (a compound annual growth rate between 2003 and 2019 of 3.2%) and high nights spent per inhabitant. One figure that appears more clearly here is the significant growth rate of Mantua (3.0%) and Cremona (2.2%), whose absolute size, however, remains very small. The group of Brescia, Padua and Vicenza has a slightly positive growth rate (1.4%, 0.8% and 0.2%, respectively), but the nights spent per inhabitant are much lower than those of Verona. Rovigo appears to be an outlier: even if it is categorized as one of the three provinces of Bassa Padana, it records much higher nights spent per inhabitant than Mantua and Cremona, but it is the only province that has exhibited a negative growth rate from 2003 to 2019 (-1.1%).

4.5 *Quality of Life*

The concept of quality of life has evolved continuously in social studies, up to recent studies on subjective well-being. In these new strands of research, the aim is to intercept the multidimensionality of tangible and intangible assets of different economic systems [10, 18, 20].

In Italy, a very innovative path coordinated by ISTAT and CNEL has been started: it originates from the international debate on "going beyond GDP", which led to the development of the "BES", which stands for "Benessere Equo e Sostenibile", i.e. "Equitable and Sustainable Well-Being". In 2015 at the global level the 2030 Agenda with the 17 Sustainable Development Goals (SDGs) has been approved: these goals are organized in a system of 169 targets and over 200 indicators, which set the path for sustainable development for the coming years.

To analyze the quality of life, an analysis of the indicators collected annually by the newspaper "Sole 24 Ore" is carried out. These have been recorded since 1988 for all Italian provinces. The set of indicators in its present form includes 90 elementary variables grouped into 6 thematic domains, 15 indicators for each area: standard of living; business and labor; services and environment; public order; population; leisure. In order to analyze the level of well-being of the provinces of Brescia, Verona, Vicenza and Padua, compared with the three provinces of the Bassa Padana area and with Italy, the elementary variables of the "business and labor" domain are considered in detail. In this domain, the provinces crossed by the A4 highway perform significantly better than the provinces in the counterfactual analysis. Compared to Italy, the areas of Alta Padana area show positive differentials in terms of exports on GDP, number of beds in accommodations (especially in Brescia and Verona), propensity to e-commerce, diffusion of home banking, and lower relative number of young people Not in Education, Employment, or Training (NEET).

The economic strength of the Brescia-Padua axis, already highlighted in the previous paragraphs, is thus confirmed. It also emerges that the propensity for entrepreneurship is markedly higher than in neighboring provinces, such as those of the Bassa Padana area, which on the contrary are penalized (even as compared to Italy as a whole) for indicators such as the birth and death rate of enterprises and the quality and quantity of accommodation facilities. The only two indicators that perform well for the counterfactual area are the lower number of accidents at work and the weight of exports, which is very high even as compared to the national average.

4.6 *The Environmental System*

The annual Urban Ecosystem report, produced by Legambiente in collaboration with "Ambiente Italia" and "Il Sole 24 Ore", is analyzed here in order to assess the environmental situation in the four provinces crossed by the A4 highway and the three provinces selected as a counterfactual case. Over the pandemic period, environmental policies and environmental performances have not substantially changed: ecological emergencies 2021 remain similar to pre-pandemic times.

Let's focus the analysis now on the values and relative ranks of the variables measuring air quality and road fatalities. These are negative "environmental externalities," which have different concomitant causes [3, 9]: from agricultural and industrial production to domestic heating, but also private and public transportation.

High levels of fine particulate matter are common to all the provinces analyzed: on the one hand, the situation is worse than the national average; on the other hand, in the last decade the concentration of PM10 decreased by about 20% in both areas (Table 5). One variable potentially related to road and highway infrastructures, namely the number of people killed and injured in road accidents relative to population (Table 6), shows a slight increase over the decade, reaching 5.8‰ in 2021. It should be noted that the provinces in the counterfactual analysis record even worse accident data, despite being less endowed with a highway network, reaching a value of 6.2‰ in 2021 (the national average is 5.6‰).

Table 5 Fine Particulate Matter (Pm10) (average of yearly mean values in µg/mc and ranking 2010–2021)

	2010	2021	% var. 2010–21	rank 2010	rank 2021	rank var. 2010–21
Brescia	41	30,5	−25,6	80	45	+35
Padua	41,4	34,4	−16,9	81	54	+27
Verona	39	31,5	−19,2	78	47	+31
Vicenza	39,3	32,3	−17,8	79	49	+30
Alta Pad	40,3	32,0	−20,4	80	48	+32
Cremona	37,6	34,5	−8,2	73	55	+18
Mantova	43,5	30,3	−30,3	85	44	+41
Rovigo	38,4	32,5	−15,4	75	50	+25
Bassa Pad	40,2	32,3	−19,6	78	49	+29
Italy	31,7	23,8	−24,9			

Table 6 Road fatalities and injuries (number of deaths and injuries in traffic accidents per 1,000 population and 2020–2021 ranking)

	2020	2021	% var. 2020–21	rank 2020	rank 2021	rank var. 2020–21
Brescia	5,0	5,1	1,4	45	44	1
Padua	6,8	7	2,6	83	86	−3
Verona	6,2	6,3	1,4	72	74	−2
Vicenza	4,9	5	1,6	41	42	−1
Alta Pad	5,7	5,8	1,8	59	60	−1
Cremona	7,1	7,3	2,2	91	91	0
Mantova	5,8	5,9	1,3	65	65	0
Rovigo	5,1	5,2	2,1	49	49	0
Bassa Pad	6,1	6,2	2,0	70	70	0
Italy	5,4	5,6	3,7			

5 Conclusions

The role of road infrastructure in territorial economic development processes is the subject of an extensive literature of both theoretical and empirical nature. Statistical evidence confirms the strong correlation between various indices of physical infrastructure - particularly highways - and levels of income per inhabitant, both among States and among European regions. Less significant is the relationship with economic growth, which in Europe, over the past two decades, has been linked more to factors related to foreign direct investment flows to eastern countries and to EU aids to the new member countries to promote their social and economic convergence.

The same evidence is observed in Italy: on the one hand, the correlation between indices of infrastructure endowment and per-capita GDP is confirmed to be consistent at both regional and provincial levels, with many areas in the South suffering from historical lags in terms of public and private investment. On the other hand, the search for the drivers of economic growth highlights the absence of convergence among the poorest areas of the country (beta convergence), the propelling effect of exports and tourist attractiveness, but above all the fundamental role of both social capital - a measure of widespread trust and propensity to cooperate - and physical/infrastructural capital, which is characterized as an indispensable pre-requisite for economic operators and for the quality of life of the inhabitants. Therefore, it is no coincidence that among the strategic policies on a European and national scale, the goal of building and consolidating efficient and sustainable tangible and intangible communication infrastructures is a top priority, as the planned measures in the National Recovery and Resilience Plan demonstrate.

The 70th anniversary of the establishment of the Società Autostrada Brescia Verona Vicenza Padova is a very interesting opportunity to assess how the development of a strategic infrastructure in the economic and social heart of northern Italy has contributed to the growth of these territories. These four provinces, Brescia, Verona, Vicenza and Padua, represent a strategic portion of the Italian system: nearly four million inhabitants in 2021 (6.7% of the country), 121 billion in value added (7.6%), 350 thousand local production units (7.4%), one and a half million employees in local units (8.4%), of which 470 thousand employed in manufacturing activities (12.5%), with a very strong propensity to export, with about 64 billion euros in sales worldwide (12.4% of the national total). In short, a "piece" of the country that has experienced in full the great economic and cultural transformations of Italy since the II World War, with impetuous growth in terms of demographics and production until the beginning of the century, waves of accelerated industrialization and then relocations and then again re-shoring processes in search of production quality and creative innovation. In this story, that is both industrial and social, with very high levels of living standards and quality of life reached—as measured by the annual wellbeing rankings of the Sole 24 Ore—the A4 highway has "accompanied"—more than "caused"—development, by fostering its physical and economic interconnection. The counterfactual analysis, which analyzed the same historical paths of demographic, productive and entrepreneurial development of three provinces that are "neighboring" but not crossed by the A4 highway, further highlighted how the "Serenissima" highway has been a decisive driver of growth in the area. Cremona, Mantua and Rovigo have indeed experienced relative demographic decline, lower economic growth rates, and lower tourist attractiveness, despite starting in the postwar period from comparable socio-economic conditions.

Of course, like all transport and communication road infrastructures, along with the processes of industrialization and urbanization, the A4 highway has also produced negative externalities, which are clearly evident in the environmental indicators that penalize the four provinces of the Serenissima highway in terms of air pollutant emissions, land consumption and accidents. Hence the need to plan investments for the future that will also help the next strategic shift that Italy and the Alta Padana area urgently await: the decarbonization of the economy and the use of sustainable means of transportation toward a green economy that respects ecosystem balances.

References

1. Ciciotti E, Rizzi P (2005) Politiche per lo sviluppo territoriale, Carocci, Roma
2. Dal Bosco U (2001) Autostrada Brescia Verona Vicenza Padua, Grafiche Leardini
3. Dallara A, Rizzi P (2012) A geographic map of sustainability in the Italian local systems. Reg Stud 46(3):321–337
4. Denicol J, Davies A, Krystallis I (2020) What are the causes and cures of poor megaproject performance? A systematic literature review and research agenda. Proj Manag J 51(3):328–345
5. Favari E (2020) Sustainability in (Mega)Project Management—A Business Case for Project Sustainability, Polimi SpringerBrief (2020)
6. Flyvbjerg B (2017) The iron law of megaproject management. In: The oxford handbook of megaproject management. Oxford University Press, Oxford, chap. 1, pp 1–18
7. Gil N, Beckman S (2009) Infrastructure meets business: building new bridges, mending old ones. Calif Manag Rev 51:6–29
8. Gil N (2021) Megaprojects: a meandering journey towards a theory of purpose, value creation and value distribution. Constr Manag Econ 40(7–8):562–584
9. Graziano P, Rizzi P (2016) Vulnerability and resilience in the local systems: the case of Italian Provinces. Sci Total Environ 553:211–222
10. Helliwell JF, Layard R, Sachs J, De Neve J (2020) World Happiness Report 2020. Sustainable Development Solutions Network, New York
11. Huggins R, Izushi H, Thompson P (2013) Regional competitiveness: theories and methodologies for empirical analysis. J CENTRUM Cathedra Bus Econ Res J 6(2):155–172
12. Istat, Censimento della popolazione, Roma, 1961, 1971, 1981, 1991, 2001, 2011
13. Istat, Censimento della Industria e dei Servizi, Roma, 1961, 1971, 1981, 1991, 2001, 2011
14. Kessler AE, Brugnoli G (1972) L'Autostrada Brescia-Verona-Vicenza-Padua La Serenissima dal 1952 al 1972, Verona, Nuova Stei
15. McGahan AM (2020) Where does an organization's responsibility end? Identifying the boundaries on stakeholder claims. Acad Manag Discov 6:8–11
16. Rizzi P (2021) Il territorio nell'anima. Pensiero strategico e politiche territoriali. (a cura di), Vita e Pensiero, Milano
17. Rizzi P, Popara S (2006) Il capitale sociale: un'analisi sulle province italiane. Rivista di Economia e Statistica del territorio 1
18. Rizzi P, Graziano P, Dallara A (2018) A capacity approach to territorial resilience: the case of European regions. Ann Reg Sci 60:285–328

19. Rizzi P, Graziano P, Dallara A (2015) The regional competitiveness: an alternative approach. Rivista Internazionale di Scienze Sociali 3
20. Robinson JA, Acemoglu D (2012) Why nations fail: The origins of power, prosperity and poverty. Profile, London
21. Shapiro A, Lorenz C (2000) Large-scale projects as complex systems: managing "scope creep". Syst Thinker 11(1)
22. Turner JR, Xue Y (2018) On the success of megaprojects. Int J Manag Proj Bus 11(3):783–805
23. Whyte J, Naderpajouh N, Clegg S, Matous P, Pollack J, Crawford L (2022) Project leadership: a research agenda for a changing world. Project Leadersh Soc 3:100044

Implementation of a Project Management Office at the Service of a Mega Mobility Project: The Case of A4 Highway/HSL Corridor Brescia-Padua

Fernando Motter Caregnatto and Pier Mauro Masoli

Abstract The assets and services provided by Autostrada Brescia Verona Vicenza Padova, since 2020, are affected by the construction of the new Brescia-Verona High Speed/High-Capacity railway. It is the first time in its history that the concessionaire has found itself managing a huge "control" operation of a maxi-construction site that deeply concerns it, but which is not within its strict competence. Given the enormity of the task, Autostrada Brescia Verona Vicenza Padova has decided to acquire an external Project Management Office (PMO) focused on the smooth delivery of the project, a pioneering decision for the Italian infrastructure market. In the first phase, PMO was responsible to write the Project Management Plan (PMP), in which the methods of execution, monitoring, control, and closure of project activities were defined to ensure the meet of the Autostrada's needs. The second phase, PMO work to apply the guidelines, procedures and processes defined in the PMP for the governance of the project. In addition to supplying the lack of expertise to deal with an atypical situation to its core business, other aim of the project was the possibility to increase the Project Management culture among the internal structures.

Keywords Project Management · Stakeholder Engagement · Mobility · Infrastructure · Value Creation · New Competencies

F. M. Caregnatto (✉)
Project Manager, Milan, Italy
e-mail: fernando.fmc@hotmail.com

P. M. Masoli
Autostrada Brescia Verona Vicenza Padova, Verona, Italy
e-mail: piermauro.masoli@autobspd.it

E. Favari and F. Cantoni (eds.), *Complexity and Sustainability in Megaprojects*, Lecture Notes in Civil Engineering 342,
https://doi.org/10.1007/978-3-031-30879-6_21

1 Introduction

1.1 The Project Management Office (PMO) approach for a Megaproject

According to the PMBOK [6], the PMO can be described as a management structure that regulates project-related governance processes and promotes the sharing of methodologies, techniques, and resources. The PMO can take multiple forms, one of them, is to offer project support services for planning activities, risk management, project performance tracking, among others.

Considering that megaprojects are strategic for the companies and require a lot of knowledge covering several areas, many resources with different profiles and innovative solutions, it can be assumed as a sophisticated inter-organizational space that also offers a platform for capability development based on rich co-creation with multiple partners. This environment favors the transformation and capability building [10].

One of the main benefits to made use an external group to perform the PMO activities is to acquire a structured and specialized team of professionals with previous experience in this environment to mitigate the lack of knowledge of the company in the management of projects of that nature. Therefore, making use of the expertise of the consultants to ensure that Autostrada's team understand, develop, apply, and value a range of project management skills and competencies, becoming an important added value to the Company.

1.2 The Case of A4 Highway/HSL Corridor Brescia-Padua

The roads administered by the Autostrada Brescia Verona Vicenza Padova, since the year 2020, have been affected by the construction works of a new railway, which are forecasted to be concluded in 2025.

The railway work, for which work has already begun, runs for about 48 km, starting in the municipality of Mazzano, in the province of Brescia, and reaching the municipality of Verona on the western side with the new HS tracks and the new Verona Merci interconnection. The HS/HC route crosses 2 regions (Veneto and Lombardy), 3 provinces (Brescia, Verona and Mantova) and 11 municipalities. The railway route is developed mostly alongside the existing infrastructures in the area, about 30 km parallel to the A4 motorway and about 8 km in alignment with the conventional Milan-Venice railway line. The intervention will make it possible to increase the supply of high-speed, regional and goods transport throughout the territory, ensuring a better separation of traffic flows, with a consequent increase in the capacity and regularity of the service, reduction in travel times and an increase in the frequency of trains. (Cepav Due 2022).

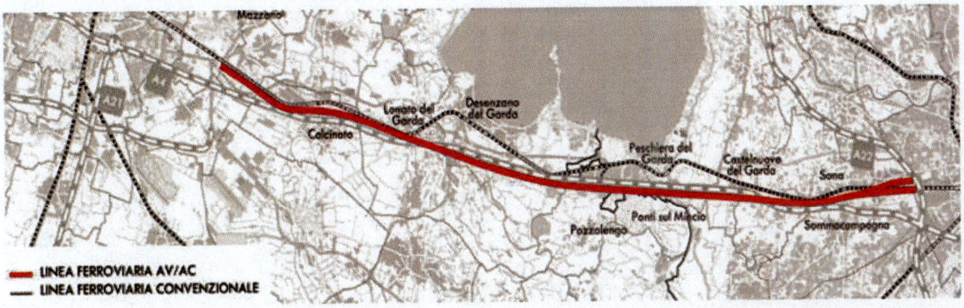

Fig. 1 More than 30 km length of the section highway affected by the interfering works of the construction of the railway line

The responsible for the construction of the Brescia-Verona High Speed/High-Capacity railway line, as a General Contractor, is the Cepav Due Consortium, Eni Consortium for High Speed. The joint venture includes 59.09% Saipem, 27.27% Impresa Pizzarotti and 13.64% ICM Group. (Cepav Due 2022).

Considering the activities develop in a close distance to the highway along 30 km, the numbers are as follows: 11 municipalities affected by the interfering works, quantifiable in the number of 110, to be realized in about 5 years. Going into further detail: 13 tunnels artificial tunnels, 4 natural tunnels, 18 overpasses to be demolished, 12 overpasses to be built, 2 viaducts, 6 underpasses, 3 highway interchanges, and so on. Railway works, in most cases, but also road works, considering the secondary road upgrades or further adaptations. Works, in any case, that in some cases result in the presence of construction sites in the highway platform, with closures of the emergency lanes, modified viability, night construction sites, with related repercussions on traffic safety management, not to mention the innumerable large or small issues concerning the safety of infrastructure, new as well as existing (Fig. 1).

In the first time in its history, Autostrada Brescia-Padova has found itself managing a huge "control" operation of a maxi-construction site that deeply concerns it, just as it deeply concerns its users, but which is not within its strict competence.

Given the enormity of the task, Autostrada Brescia Verona Vicenza Padova has decided to acquire an external Project Management Office focused on the smooth delivery of the project, a pioneering decision for the Italian infrastructure market [1].

1.3 Autostrada Brescia Verona Vicenza Padova Highway–

The A4 Turin-Trieste Motorway is part of the European "Mediterranean" Corridor and therefore of the European Core Network and is one of the busiest arteries in Italy.

The section between Brescia and Padova is about 146 km long, 37 of which are in Lombardy and 109 in Veneto and consists of 3 lanes in each direction (in addition to the emergency lane).

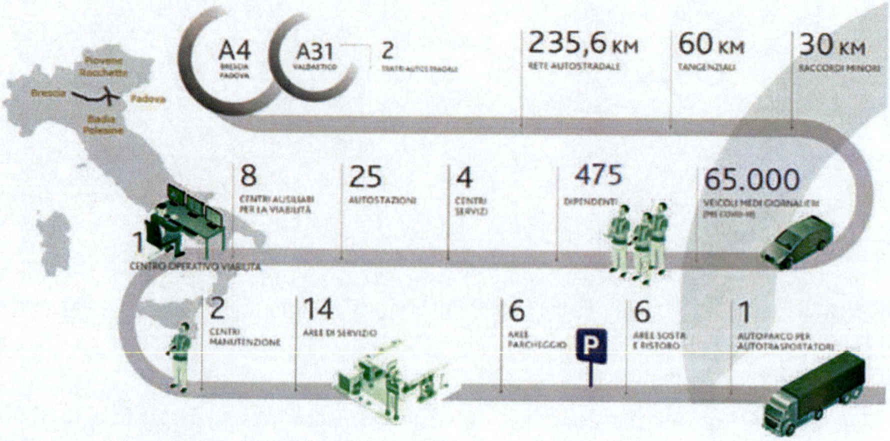

Fig. 2 Autostrada Brescia Verona Vicenza Padova's organization and services

Autostrada Brescia Verona Vicenza Padova was established in June 1952, in the city of Verona, during the economic boom that made to emerge the need to improve the transport's infrastructure of the region. The initiative was taken through a cooperation between public and economic entities from 7 provinces of northern Italy: Milano, Bergamo, Brescia, Verona, Vicenza, Padova and Venezia. The aim of the project was to connect the two already existing motorways, the Milano - Brescia and the Padova - Venezia, to create a direct connection between the main industrialized capitals of Lombardia and Veneto. Since then, it has always been indispensable to the development of the territory.

The infrastructure was built in the 1950s/60 s by Autostrada Brescia Verona Vicenza Padova SpA (still the concessionaire of the work) with 2 lanes in each direction, widened to 3 lanes in the 1970s between Brescia and Verona, and in the 1980s/90 s for the remaining part up to Padova (Fig. 2).

1.4 *The Interferent Project*

The project has a wide range of stakeholders: from public sector players like the Sustainable Infrastructure and Mobility (MIMS), the Italian railway operator Rete Ferroviaria Italiana, their in-house engineering firm Italferr, the concession operator Autostrada Brescia Verona Vicenza Padova, the consortium Cepav Due, as well as a great number of other players including local stakeholders.

The major railway infrastructure work of about 30 km in very close side-by-side with the highway, with as many as two crucial undercrossing nodes, at Lonato (Brescia, from north to south) and then at San Giorgio in Salici (Verona, from south to north again). This is a very large construction site, consisting in works over 48 km in total, 4 new viaducts, 15 overpasses, 4 tunnels natural, 17 artificial tunnels.

The main works are:

- the Lonato Tunnel system (length 7,950 m) in the municipality of Lonato del Garda (BS), the most articulated work in the project.
- the Frassino Tunnel system (3,360 m long) in Peschiera del Garda (VR).
- the San Giorgio Tunnel system (3,708 m long) in the municipality of Sona (VR).
- the viaducts on the Chiese (377 m long) and Mincio (320 m long) rivers.

In the face of this great number of activities, the priority of the "interfered" concessionaire has been to delineate and qualify adequately its own role and prerogatives, through the instrument of an Agreement. A role extremely relevant, and active, guided by what stipulated in ISO 9000. Thus, another milestone consisted in going beyond the "patrimonial principal": that is, could not be considered interferences only those works falling within the highway appurtenances, but also all those new infrastructures that could in some way affect its functionality and efficiency. Here are thus counted the “110 pieces of work” with which the highway would have to deal with. Among them, the new constructions but the most important of all, the related implications: from safety barriers to be uninstalled and re-install in the appropriate class to continue, in some specific cases, with relocations of the telematics backbone or other facilities and sub-services.

Thus, narrowing the field to the new works, it should then be added that in many cases these are infrastructures designed and built by the general railway contractor and destined later to come under the management of the highway. Hence The need to know, in depth, the *modus construendi*, although it was it was “someone else’s” construction site [1].

1.5 The Agreement

The relationships, transitory and permanent, between the two works, as well as between the respective “landlords’ householders” are regulated by a special and articulated convention that has involved the MIMS, in its dual function of road and rail “grantor.“, RFI, Italferr, the general contractor Cepav Due and the highway concessionaire Autostrada- Verona-Vicenza-Padova (headed by A4 Holding, Abertis Group).

The Agreement set out the “principles fundamentals” not to be violated, effectively giving the highway a set of prescriptive powers. That is: safety must be always ensured and the fluidity of traffic, the highway infrastructure must not never suffer compromises of a static, the assets that will be delivered to Autostrada Brescia-Padova must always meet the concessionaire’s own quality standards, finally, the distances between the two works must never fall below 7 m in plan, with francs of 5.50 m always clear. But the principles of this infrastructural cohabitation had to have at their side an organization deputed to manage the interference, with related inspections, to be added to the activity ordinary and proper highway activity. A task force, in fact, two: a PMO and, together, a team of engineers dedicated specifically to inspections [1].

Autostrada faces a twofold challenge within its Project, setting itself two priority objectives:

1. to guarantee the success of the Project, in compliance with the terms stipulated in the Convention, in particular:
   a. art. 3.4, focused on the protection of users and motorway assets, as well as on the fluidity of motorway operations.
   b. art. 25, concerning the amount of the charges that Cepav Due must pay to Autostrada for the activities that it will have to carry out due to the interfering works.
2. favoring towards Cepav Due the construction of the new railway line, in compliance with adequate safety levels of the asset and motorway traffic.

With reference to point 1, Autostrada will have to bear a series of charges in relation to:

A. transfers, concessions of use and temporary occupations of motorway property (Convention, art. 25.2.A).
B. reduction in toll revenues due to the decrease in motorway traffic (Convention, art. 25.2.B).
C. reduction in income earned by Autostrada from royalties on turnover from oil, food and shop services in the service and rest areas (Agreement, art. 25.2.C).
D. post-operam updating of documents and management systems (Convention, art. 25.2.D).
E. publication of announcements for motorway section closures (Convention, Art. 25.2.E).
F. internal structure and external consultants to support the Project (Convention, art. 25.2.F).

In addition, Autostrada will have to bear any additional noise abatement costs due to the concurrence of the existing motorway infrastructure and the new railway (Convention, art. 25.2.A).

The governance model of the Project is based on the processes and rules dictated by the Convention and its annexes. To guarantee the objectives, mentioned before, in fact, the Convention sets out some fundamental gates for the passage to the subsequent phases of the project, placing technical, administrative, and legal constraints on Cepav Due. With reference to the phases envisaged at level 4 of the Project WBS, 3 primary gates are identified.

1. Gate A: authorization of the executive project.
2. Gate B: authorization to start the works with reference to the interfering work.
3. Gate C: check on the conformity of the works carried out.

The gates indicated are to all intents and purposes process control points, aimed at ensuring that the following results have been achieved:

a) the phase has been completed in accordance with Autostrada's constraints and acceptance levels.
b) there are no open points that could prejudice the continuation of the Project in phases after the successfully completed phase
c) for the completed phase, risk factors related to the variables to be monitored (time, cost, quality, communication, stakeholder engagement) have been mitigated or removed.

Formal verification of the overcoming of gates A, B and C takes place through the presentation by Cepav Due of the documentation requested by Autostrada and feedback from the same, as established in the various articles of the Convention (e.g., see articles 3, 5 and 7 for the overcoming of gate A).

2 The Autostrada Brescia Verona Vicenza Padova Approach to the Interfering Project

Following the decision to "buy" an external structure to provide the needed support to carry on the project, a public tender was published in the beginning of 2021 with the following title "Gestione dei servizi di verifica della progettazione esecutiva e della conformità dei lavori (Project Management Office) delle opere interferenti tra la linea ferroviaria AV/AC Brescia-Verona e l'autostrada A4 - seconda fase (attuazione PMP)" [1].

To ensure that the tender could reach its purpose, providing clear information and the main objectives requested by the company, the PMO was divided in two main parts. The first was named PMO 1, which had the duration of a bit more than 4 months, the aim was to draft the Project Management Plan (PMP) in which the methods of execution, monitoring, control, and closure of project activities were defined, with reference to the different knowledge areas (scope, time, cost, quality, risk, resource, communication, stakeholder engagement, procurement, change).

The structure is made up of 2 subgroups of technicians, in relation to the types of activities to be delivered, as required by the mandate given to them by Autostrada:

- "core" team, which deals with the main project management activities, concerning the governance of the different dimensions (e.g. time, costs, risks) of the Project and is therefore made up of various experts for the individual areas. This sub-group is in turn composed of:
  - a transversal co-ordination structure, which is responsible for governing the activities throughout the whole project
  - an operational structure for the day-to-day implementation of the processes defined in the PMP.
- an 'additional' team: this deals with additional activities, which cannot be classified as recurring and ordinary, but rather as specialized and unforeseeable, to be

deployed on specific requirements (e.g., cost–benefit analysis, variant appraisals, support for the implementation of IT systems).

2.1 The Project Management Office – Phase 1

The Project Management Plan takes as a starting point the reference standards, contents and indications provided in the PMBOK (Project Management Body of Knowledge), integrating some aspects in a peculiar and distinctive way, in order to add value to the Project management methods.

A project management plan defines in a clear and detailed manner how the project is to be executed, monitored, controlled, and closed. In this specific case, given the peculiarities and criticalities of the Project, this PMP primarily aims to.

- to clearly and unambiguously define the methods by which the PMO team will have to operate to meet the Project objectives indicated by Autostrada.
- frame all the activities that will be necessary (technical, management, relational, communication, legal) in all phases of the project (pre-construction, construction site, post-construction), specifically defining the roles and responsibilities for their management
- act as a general guideline to inform all stakeholders (both internal and external) on how the Project will be managed and how they will be able to interact within it
- illustrate a structured workflow that can be easily implemented by the subjects in charge of performing PMO activities, in line with international best practices and the guidelines defined in the PMI (Project Management Institute) PMBOK.

The PMP was thought carefully to guarantee that Autostrada's management of the assignment process of the PMO contractor was effective and simple. To reach this purpose, some main points guided the approach of its development process:

- clearly description of project management aspects (e.g., time and cost monitoring, stakeholder involvement), minimizing the potential levels of risk to which Autostrada will be exposed during activities.
- write it in a manner functional to future sharing of project management concepts within the corporate culture, to increase Autostrada's level of maturity on specific issues.
- considering the coexistence of two parallel projects (the Project and the construction of the entire HS/HC Brescia East - Verona line, of which Cepav Due is the General Contractor), with interests that do not always converge.
- maintaining complete consistency with the contractual framework regulating relations between Autostrada and Cepav Due (the Convention), the constraints, rules, and principles of which must be implemented consistently.
- declining the operational aspects for managing the activities that Autostrada will have to put in place in relation to its contract, seeking as far as possible to identify

procedures and tools that can be shared by Cepav Due, in a partnership approach that benefits both parties for the entire duration of the interfering worksites.
- considering that, for some interfering works, the Project has already been started in its pre-site and site phases.

Among the tools and techniques used for mapping the information needed for drafting the individual management plans, can be highlighted the follow:

- internal brainstorming activities within the working group.
- single-issue meetings with the stakeholders active in the project, to assess the individual aspects to be managed within the PMP
- focus groups extended to multiple stakeholders involved in the project, to review in a synergic manner

During the transition phase between PMO 1 and 2, Autostrada was supported by an interim structure, that was appointed to perform the PMO activities for the period that the tender was published until the contract awarding to the selected competitor.

2.2 *The Project Management Office – Phase 2*

The definitive PMO structure started their duties on July 2021, composed by ARUP, PwC and LC&Partners. Finally, they oversaw applying the guidelines, procedures and processes defined in the PMP for the governance of the project, as well as updating the contents of this document to make it adhere to the actual needs of Autostrada.

The approach is dynamic and transversal to ensure the prevention and management of project risks, following the 5 pillars of a PMO structure to support the project.

1) Project control:
   - Monitoring forecasts
   - Quality control
   - Management of changes in progress
2) Interface management:
   - Coordination of figures responsible for interfaces with other management functions of the project, ensuring timeliness and the consistency of information
3) Stakeholder engagement:
   - Management of job communications
   - Involvement of internal and external stakeholders
4) Legal support:
   - Management of legality protocol and financial flows
   - Legal structure support for dispute prevention

- Support for the management of contracts of external structures

5) Digital support:

- Project information and knowledge management
- Implementation of technological tools to support data collection and reporting processes

The PMO, in other words, in collaboration with the concessionaire, which reserves decision-making power and other activities of a strategic and functional nature, oversees "governing" all the processes in place between the motorway and the general contractor within the scope of the contract, as well as the related communication activities, stakeholder engagement and so on, constantly monitoring their efficiency. According to an approach and management workflow consolidated in large international examples.

2.3 *The Engineering Structure*

The structure external to Autostrada, in charge of carrying out project preliminary investigation activities, monitoring of assets and inspection of interfering works under construction. This structure, being made up of a multiplicity of experts from various sectors (geotechnics, structures, plants, routes, hydraulic interferences, safety, construction sites, environment, underground works), is represented by an Engineering Coordinator, who acts as a collector with respect to the work of the individual members of the group. This figure has a direct relationship with the PMO group (with the PMO Team Leader and the "Interface with Engineering" Function) and with PM A4, with respect to particularly significant and/or critical aspects.

To the so-called services of engineering, the technical structure was also divided in two parts. The first, called ING 1, was responsible for the analysis of the definitive project and the impact of the implementation of interfering works on motorway appurtenances, support in the drafting of the Agreement between the parties.

For the following phase, named ING 2, a separate bidding process was developed to cover this specific need, the specification of the work was focused on the activities of inspection of executive project and construction site activities to ensure the compliance with what stated in the Agreement. In this specific case, awarded of the assignment was a grouping composed of Technital, Pro Iter, Studio Bortolami and Politecnica Ingegneria e Architettura.

Carrying out the preliminary analysis of the executive design and site inspections, everything is controlled according to the risk to the users' safety and to the development of the works, as well as the present and future functionality of the infrastructure. One example among the wide range of issues, may concern prescriptions related to the executive activity, remaining always within the perimeter of what is within its competence: one cannot dig under operating traffic, it requires a detailed plan to deviate the section affected by the activities to be held. The structure of inspection takes charge of all the appropriate verifications, analyzing the correctness of the design elements, carrying out compliance checks in the field, including the activity carried out by construction management.

2.4 The Project Development

The project can be summarized in five main processes and for each one, the stakeholders change. The first process is carried out by the general contractor Cepav Due and it consists in the presentation of the executive project.

Posteriorly, the engineering structure starts with the preliminary investigation and verification of project documentation following the coordination of PMO and collaboration of Autostrada. The purpose of this activity, charged to Autostrada and carried out with the support of the engineering group, is to verify from a formal and technical point of view the conformity of the project documents produced by Cepav Due and submitted to Autostrada's analysis, making sure that the General Contractor itself respects all existing constraints and highlighting any modifications that do not comply with the requirements of the stipulated Agreement.

Respecting the time of the Agreement, Autostrada authorizes the final report to be sent to the joint venture, if it is approved, all the documentation must be forwarded also to the MIMS to give its "Nulla Osta". If the result is negative, the report lists the prescriptions and/or observations to be clarified by the responsible for the development of the project. This process repeats until the concessionaire has the necessary information to ensure all the aspects are respecting the previous accorded exigencies and decide if the project is approved.

Prescriptions can be formulated in relation to the following aspects (Convention, Art. 3.4).

1. adequacy of the operating conditions of the A4 during the works (traffic fluidity).
2. safety of motorway users
3. integrity and stability of the motorway infrastructure in operation and its appurtenances.
4. compliance with the geometric constraints necessary to guarantee the possibility of constructing the fourth motorway lane as detailed in Article 4 of the Agreement.

Once the project is approved, it begins the "pre-construction" process, which involves PMO, ING, Autostrada and Cepav Due. An important argument to be discussed at this point, is the on going works of the other internal structures of Autostrada to not be affected by the highspeed rail line works. After a careful analysis, exchange of information and properly coordination between all the interested stakeholders, it is released the official authorization to the construction works to be initiated in the areas owned by the concessionaire.

Once the works initiate, the engineering structure proceeds with on site inspections and the analysis of the procedures and quality documentation shared by Italferr. The frequency of the visits is defined according to the level of criticality of each works, but all the sites are visited at least once a week.

The last process takes place in the end of a work when the structure or the area must be delivered to the Autostrada/MIMS. After a careful evaluation of all the documentation and on-site inspection, the process is ready to be closes.

2.5 Project Outcomes

The introduction of the two task forces, PMO and engineering, also brought a plus in innovation culture. Aiming at quality standards of excellence has already meant reasoning about technological upgrades, for example, in the management of site signs, as well as in sensor technology already known in tunnelling, but less so in other contexts. Furthermore, from a documental point of view, think of the use of software for managing, sharing, certifying, and approving workflows, or dynamic and always up-to-date reporting.

In order to provide a more graphical and clear approach to communicate the data and the status of progress of the project, it was created a dashboard, daily updated. Among the material that stakeholders have access, it is available all the project risks and its assessment, the budget and schedule reports, WBS information, an interactive map in which can be easily found the location of each intervention of the project (Fig. 3).

As mentioned before, the idea of acquiring an external structure as PMO of the project, is an innovation in the Italian market, not having any other references of how to develop this work in the national market, besides being a completely new and

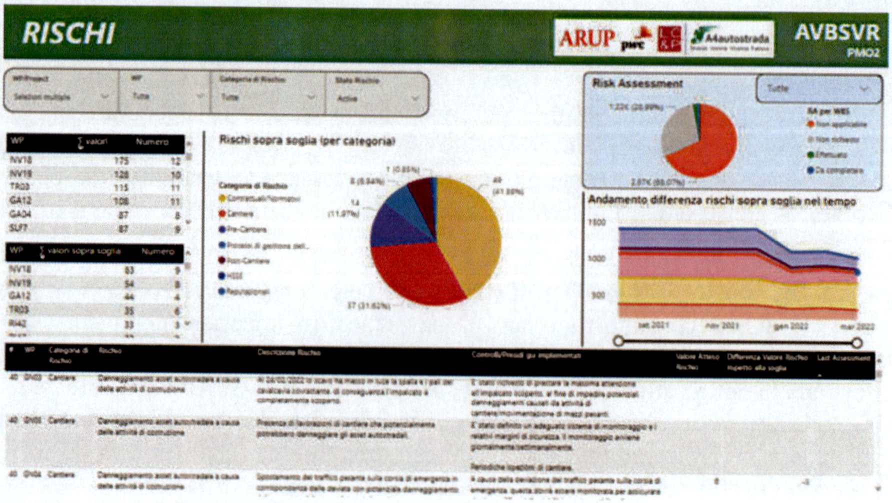

Fig. 3 Dashboard developed by the PMO

unknown situation within the main business of the concessionaire, the organization and configuration of the project were huge challenges, but until now, the decision has achieved its objectives, the process are following the agreed flow and the relationship between the stakeholders is collaborative. Internally the company, the concepts of project management are also more common, and the lessons learned are being replicated in new projects in initiation phase and in others which are already being executed.

References

1. Visionj R (2022) PMO per connettere ferrovia e autostrada. Vision J 84–88
2. Autostrada Brescia Verona Vicenza Padova Homepage. https://www.a4holding.it/en/70-years-of-activity. Accessed 1 Nov 2022
3. Caregnatto FM (2020) Autostrada Brescia Verona Vicenza Padova e la linea ferroviaria AV/AC Brescia Est –Verona: dialogo e collaborazione per ridurre impatto e interferenze. In: Seminario Linea Ferroviaria AV/AC Brescia Est –Verona. Ordine degli Ingegneri di Verona e Provincia
4. 4a Corsia e riqualificazione tecnologica (2020) Autostrada A4Tratta Brescia-Padova
5. Project Management Plan - Project Management Office delle opere interferenti tra la linea ferroviaria AV/AC Brescia – Verona e l'Autostrada A4
6. Convenzioneregolante dei rapporti transitori e permanenti tra Autostrada Brescia Verona Vicenza Padova, RFI, Cepav Due e MIMS, relativamente alla realizzazione delle opere interferenti della tratta AV/AC Brescia Est-Verona
7. PMBOK® guide, 6th edn (2017). Project Management Institute, USA
8. PMBOK® guide 7th edn (2021). Project Management Institute, USA
9. Cepav Due Homepage. https://www.cepavdue.it/progetto/il-progetto/. Accessed 1 Nov 2022
10. Denicol J, Davies A (2022) The megaproject-based firm: building programme management capability to deliver megaprojects. Int J Project Manage 40:505–516

Risk Analysis in Private Building Projects: A Pilot Study in Chile

Gabriel Castelblanco, Harrison Mesa, and Luis Serra

Abstract A proper identification, assessment, and allocation of risks are essential for reducing the likelihood of time and cost overruns. Aiming for lowering the overall cost of claims and disputes, this study aims for the identification, assessment, and allocation of the risks in private building projects in Chile. Based on a thorough examination of the literature, reveals 104 risks that are organized into four categories and eleven subcategories. The allocation, probability, and impact of these risk variables were estimated by experts consulted through questionnaires. Results show that delays in approval/permits, delays in decision-making, poor design, equipment problems, planning deficiencies, poorly trained skilled labor, unclear contract clauses and conditions, late design changes, competition, resource availability, and unrealistic baseline scheduling are the top-ranked risk factors. The study's respondents' recommendations regarding the risk allocation were compared with the contractual risk allocation in the projects analyzed unraveling that over 50% of the most critical risks had disparities between contractual allocation and respondents' recommendations. The study's findings are useful for assisting practitioners in allocating risks to those stakeholders who are better equipped to evaluate, manage, and control those risks. The risks can be prioritized for response planning using the generated risk priority.

Keywords Risk identification · Risk assessment · Risk allocation · Project success factors · From global to local

1 Introduction

The development of architectural and engineering projects generates significant contributions to the economic growth of all the countries around the world. However,

---

H. Mesa · L. Serra
Pontificia Universidad Católica de Chile, Santiago, Chile

G. Castelblanco (✉)
Politecnico di Torino, Turin, Italy
e-mail: gabriel.castelblanco@polito.it

E. Favari and F. Cantoni (eds.), *Complexity and Sustainability in Megaprojects*, Lecture Notes in Civil Engineering 342,
https://doi.org/10.1007/978-3-031-30879-6_22

the complexity of these projects and their increasingly uncertain environments may jeopardize their quality, budget, and time performance [1]. Moreover, the poor performance and recurrent conflict between the contractor and client in these projects frequently are linked to flaws in the risk assessment and allocation [2].

Numerous risks and conflicting relationships exist between contractors and owners throughout construction projects, which typically result in disputes initiated by either party. These conflicts frequently are disruptive, expensive, and time-consuming particularly when they are not solved quickly and escalate to official resolution by a tribunal or a court [3–6].

The construction sector is constantly looking for more effective and efficient dispute resolution mechanisms, such as alternative dispute resolution techniques like adjudication, mediation, conciliation, and other hybrid procedures that are popular and well-liked in the building sector [7]. However, these methods are expensive and, frequently, ineffective in resolving disagreements [8]. Furthermore, the majority of alternative dispute resolution techniques for avoiding litigation and arbitration are not legally binding on the parties until the dispute is finally settled by the end of the project [3].

The allocation of risks involves their transfer via contractual provisions from one party to another [9]. Fair risk allocation is a practical method for preventing expensive disagreements [2, 10]. In effect, unclear risk allocation in building contracts is one of the main causes of disputes [11, 12]. Proper risk management and allocation in building projects require a careful development of contractual governance mechanisms reflected in risk allocation and control [13–16]. For this reason, the identification and understanding of risks should be prioritized in the early phases of the project to limit potential negative consequences of them [17].

Project Management researchers have developed countless papers focused on risk allocation of public work projects [16, 18–25]. Their preference for these projects is due to the availability of public databases with the information required. However, there is a remaining gap in non-public projects because of the lack of transparency on this information, as reported in the literature [26, 27]. Moreover, the construction sector has been the subject of risk assessment research by a number of academics in regions around the world, such as North America, East Asia, and Europe rather than medium size developing countries located in some other regions. To address the aforementioned gap, this study aims for analyzing the risk factors in the Chilean private building construction sector.

2 Background

2.1 Risk Quantification

The Project Management Body of Knowledge (PMBoK) establishes that risks are chance events that may negatively or positively affect the objectives of the project

[28]. Risk is inevitable considering the inherent uncertainty associated with every human action [29]. Considering the frequency and influence of risks on project objectives, decision-makers should conduct risk quantification.

One of the first quantification models proposed was developed by Daniel Bernoulli almost 300 years ago and was focused on the calculation of the utility of weighted averages for the potential outcomes based on an uncertain scenario [1]. This model became the standard reference for two hundred years until the mid-1900s when Neumann and Morgenstern proposed the probability-weighted average of a single utility of combined outcomes but of the utilities of all outcomes [30].

Nowadays, the Expected Utility Theory has become the most widely accepted model to quantify risks. Under this theory, the PI model establishes the estimation of a risk event as the multiplication of its impact (I) and the probability of occurrence (P).

2.2 *Risk Assessment and Identification*

The implementation of risk management is impeded by insufficient resources, insufficient experienced staff, and scarce time for implementation [8]. Cost overruns and time delays are the main reasons for disputes [31, 32]. Moreover, the biggest influence on construction time and cost is financial risk, which is typically correlated with a lack of financial resources, cash flow issues, and claims due to payment delays [33–37]. Construction time and cost also demonstrated being susceptible to specific features such as delayed modular component deliveries, disruptions in the supply chain, capital cost increases, management complexity, environmental sustainability, and stakeholder fragmentation [38–42].

From the perspective of the contractor, construction risk factors include decreased productivity, uncertain ground conditions, ambiguous specifications, and varying construction standards [43, 44]. Additionally, contract-based analyses have identified key risk factors: change in design, design error, unclear clauses and conditions, inflation, Currency exchange rate fluctuations, corruption, and public opposition to projects [8, 45]. However, it is required to complement risk identification with quantification to offer value for project managers and improve risk management.

3 Methodology

3.1 *Risk Identification*

The methodological process started by identifying the construction projects' risk factors. This identification process was focused not only on a list of potential risks

Table 1 Procurement Risks

Category	Subcategory	Risk
Procurement	*Management*	*Planning deficiencies*
		Site acquisition
	Environmental	*Environmental Impacts*
	Contracting	*Tender delay*
		Approvals delay
	Financial	*Funding*
	Institutional	*Environmental approvals*
	Project Site	*Access and rights*
		Unforeseen Underground Structures
		Connections to the site
		Ground conditions
		Easements

but also on a taxonomy to classify them into specific risk categories, as recommended in risk literature [46–48].

Based on a literature review process, 137 risks were identified initially. Risks unrelated to lump-sum contracts were discarded. After this filtering process, a list of 104 risks was reached as shown in Tables 1, 2, 3 and 4, which is consistent with previous risk identification [8, 49].

The risk taxonomy considered three hierarchies. The first is the category that is closely related to the project phase, namely: procurement, design, construction,

Table 2 Design Risks

Category	Subcategory	Risk
Design	*Management*	*Corruption*
		Delays in the approval of suppliers
	Institutional	*Delay in approval/permits*
		Change in law
	Liaison	*Construction Standards*
		Measure Standards
		Information Accessibility
		Client's expectations
	Relational	*Design coordination*
	Technical	*Poor design*
		Changes
		Complexity
		Design delivery delay
		Unexperienced designer

Table 3 Construction Risks (*Short version*)

Category	Subcategory	Risk
Construction	*Management*	*Resources availability*
		Improper risk monitoring
		Delays in decision-making
		Owner's management issues
		Subcontractor
		Suspension of works
		Construction monitoring issues
		Construction initiation delay
		Fast-track costs
	Environmental	*Weather conditions*
		Pollution due to construction works
	Contracting	*Unclear contract clauses*
		Supervision delays
		Unrealistic baseline scheduling
		Contract omissions
		Lack of experience
		Subcontractors issues
		Competitionara>
		Opportunistic behavior
		Litigation
		Unclear contract clauses and conditions

Table 4 Transfer Risks

Category	Subcategory	Risk
Transfer	*Management*	*Conformity inspections delay*

and transfer. Then, the subcategory focused on the risk type such as environmental, management, institutional, liaison, relational, technical, and contracting.

In summary, procurement risks entail six subcategories and twelve risks (Table 1), design risks are made of five subcategories and fourteen risks (Table 2), construction risk is disaggregated into thirteen subcategories and 76 risks (Table 3), and transfer risk related to one single risk and subcategory (Table 4).

3.2 Projects' Data Collection

For this pilot study, Chilean projects were selected. In general, all the projects are lump sum contracts for high-rise residential buildings, 83% of the projects are Design-Bid-Build, and 17% Design-Build.

3.3 Risk Quantification

For this pilot study, an exploratory analysis was conducted based on questionnaires with various experts. The questionnaire's respondents included representatives from the owner (37%) and the contractor (63%). The respondents were filtered to exclude practitioners with less than three years of working experience and avoid the least experienced and qualified practitioners.

The questionnaire started by asking for respondents' general information. Then, the severity (impact) and probability of occurrence of each of the risk factors were requested using a Likert scale, where five (5) and one (1) mean very high and very low, respectively. Finally, the allocation of each risk factor was asked, giving the options of the contractor, owner, or shared, as recommended by risk literature [8].

The risk factor was computed based on the associated risk impact and risk probability as follows:

$$\text{Risk Factor} = \text{Risk Impact} * \text{Risk Probability} \tag{1}$$

Consequently, for a respondent that assesses a low (=2) probability and high (=4) impact of a certain risk, the Risk Factor is eight (8).

3.4 Risk Allocation

The contractual documents of each project were analyzed through a content analysis for identifying the risk allocation. This analysis technique allows for replicability in the inferences through the coding and interpretation of text [46, 50–53].

A risk allocation rubric was developed for the analysis of the contractual documents, as recommended in risk allocation studies [54]. Then, the risk allocation was filled in the form based on the content analysis of the contractual documents.

The respondents were also asked for their recommendations regarding the risk allocation for each of the risks. This allows for contrasting the real risk allocation with the experts' perspectives.

3.5 Validation

To evaluate the consistency and reliability of the questionnaire, Cronbach's Alpha Coefficient was calculated as follows:

$$\alpha = \frac{K}{K-1} \times \left(1 - \frac{\sum Vi}{Vi}\right) \tag{2}$$

where α is Cronbach's Alpha Coefficient, *K* is the number of risk factors in the questionnaire, *Vi* is the variance of risk *i*, and *Vt* is the total variance. According to the extant literature, when this coefficient exceeds 70%, the questionnaire may be considered reliable [8].

4 Findings and Discussion

The reliability and consistency of the questionnaires were confirmed based on Cronbach's Alpha values over 0.95. Based on the data gathered in the questionnaires, the Risk Impact was calculated. The most critical risks according to the Risk Impact (>20) are shown in Table 5.

Although design risks only represent around 10% of the risks analyzed in this study, 20% of the most critical risks were related to design issues. The first and third risks are within the design category, namely: "delay in approval/permits and "poor design." In effect, a significant proportion of the risks may not necessarily affect the critical path of the project and, consequently, will not affect the project duration; "delay in approval/permits" undoubtedly affect the critical path of the project and, consequently, the project duration. Additionally, previous literature has

Table 5 Most critical risks

Rank	Risk	Category	Subcategory
1	*Delay in approval/permits*	*Design*	*Institutional*
2	*Delays in decision-making*	*Construction*	*Management*
3	*Poor design*	*Design*	*Technical*
4	*Equipment problems*	*Construction*	*Execution*
5	*Planning deficiencies*	*Procurement*	*Management*
6	*Poorly trained skilled labor*	*Construction*	*Execution*
7	*Unclear contract clauses and conditions*	*Construction*	*Contracting*
8	*Late design changes*	*Construction*	*Technical*
9	*Competition*	*Construction*	*Contracting*
10	*Resources availability*	*Construction*	*Management*
11	*Unrealistic baseline scheduling*	*Construction*	*Contracting*

emphasized the common practice of owners to impose tight schedules for the delivery of the drawing and design that could be in some cases unrealistic [8]. Moreover, "poor design" may also have significant detrimental effects on the quality, cost, and time performance of the project. Additionally, there is one single critical risk within the procurement category, namely, "planning deficiencies." In effect, project management literature has emphasized the relevance of planning processes in order to establish a proper scope and accurate budget and schedule [55].

Conversely, eight out of the eleven critical risks within the construction category, namely: "delays in decision-making", "equipment problems", "poorly trained skilled labor", "unclear contract clauses and conditions", "late design changes", "competition", "resources availability", and "unrealistic baseline scheduling." Overall, the insufficient time and untimely efforts devoted by the owner in the planning along with organizational issues may result in delays triggered by the owner affecting the time for key decisions, lack of clarity in the contract clauses, late design changes (and their consequent cost and time overruns), and unrealistic schedule.

4.1 Risk Allocation

An essential aspect of risk management is risk allocation, which is the practice of allocating risks to the most appropriate contracting party. The importance of assigning risk to the party that can manage it the best is that a suboptimal risk allocation is likely to result in controversies, drive up costs, and significant delays.

The allocation of the risks was analyzed in both the contractual documents and according to the experts' criteria. For the analysis of questionnaires, this study adopted the literature's recommendation to assign each risk to the party when there is more than 50% of respondents agree to allocate that risk to it [8, 56]. Conversely, risks that have a threshold of less than 50% are classified as "non-decided."

The allocation recommended for each critical risk in this study is shown in Table 6. There are only four risk factors listed that should ideally be allocated to the owner: "Delays in decision-making"; "Poor design"; "Unclear contract clauses and conditions"; and "Late design changes". Most of these risks (60%) belong to the construction category and the remaining 40% to the design. Evidently, these risks entail untimely decisions of the owner that will impact the project outcomes so as poor contractual documents and design (for DBB projects).

Conversely, Table 6 shows that five risk factors are assigned to contractors. These risks are "Equipment Problems", "Planning deficiencies", "Poorly trained skilled labor", "Competition", and "Resources availability". Most of these risks (80%) belong to the construction category and the remaining 20% to the procurement category. This pattern is closely related to general practice for allocating construction risks to the contractor.

The remaining two critical risks ("Delay in approval/permits" and "Unrealistic baseline schedule") were recommended to be shared.

Table 6 Experts' recommendation for risk allocation for the most critical risks

Rank	Risk	Contractor (%)	Owner (%)	Shared (%)	Allocation
1	*Delay in approval/permits*	*0*	*50*	*50*	*Shared*
2	*Delays in decision-making*	*12.5*	*50*	*37.5*	*Owner*
3	*Poor design*	*0*	*87.5*	*12.5*	*Owner*
4	*Equipment problems*	*62.5*	*12.5*	*25*	*Contractor*
5	*Planning deficiencies*	*62.5*	*12.5*	*25*	*Contractor*
6	*Poorly trained skilled labor*	*100*	*0*	*0*	*Contractor*
7	*Unclear contract clauses and conditions*	*12.5*	*87.5*	*0*	*Owner*
8	*Late design changes*	*12.5*	*87.5*	*0*	*Owner*
9	*Competition*	*62.5*	*25*	*12.5*	*Contractor*
10	*Resources availability*	*75*	*12.5*	*12.5*	*Contractor*
11	*Unrealistic baseline scheduling*	*25*	*25*	*50*	*Shared*

A comparison between the contractual risk allocation in the Chilean contracts analyzed and the recommendations made by experts is shown in Table 7. Interestingly, almost 50% of the most critical risks presented discrepancies between contractual allocation and what respondents have recommended, as seen in the table. These differences and ambiguities in risk allocation could lead to recurrent claims and conflicts.

Table 7 Risk allocation in the contractual documents vs. experts' recommendation

Rank	Risk	Contractual Allocation	Allocation Suggested
1	*Delay in approval/permits*	*Contractor*	*Shared*
2	*Delays in decision-making*	*Owner*	*Owner*
3	*Poor design*	*Shared*	*Owner*
4	*Equipment problems*	*Contractor*	*Contractor*
5	*Planning deficiencies*	*Contractor*	*Contractor*
6	*Poorly trained skilled labor*	*Contractor*	*Contractor*
7	*Unclear contract clauses and conditions*	*Contractor*	*Owner*
8	*Late design changes*	*Shared*	*Owner*
9	*Competition*	*Contractor*	*Contractor*
10	*Resources availability*	*Contractor*	*Contractor*
11	*Unrealistic baseline scheduling*	*Contractor*	*Shared*

5 Conclusions

The findings show that the assessed risks' impact and probability on the project's goal were moderate on average. In effect, only 11% (11 out of the 104) risks analyzed were critical and 17% of the risks were significant. Conversely, 36% of the risks were moderated, and 30% minor.

Although only around 10% of the risks examined in this study were design risks, 20% of the most critical risks included design-related difficulties. In practice, design risks undoubtedly affect the critical path of the project and, as a result, the project duration. Within the procurement category, there is only one single major risk, which indicates the need for planning procedures in order to establish an appropriate scope, precise budget, and reliable timetable.

Contrarily, eight out of the eleven significant risks in the construction category fall into the category of critical risks, It illustrates how organizational problems, combined with the owner's inadequate planning efforts and timing, can lead to mistakes in the scope, budget, and schedule definition.

Both the contractual papers and the recommendations established by the experts were used to examine how the risks were allocated. There are only four significant risks that pertain to the construction and design categories and should ideally be assigned to the owner.

Contrarily, five critical risks are allocated to contractors in the construction category, while the remaining are allocated to the procurement category, This is very similar to the custom of transferring construction risk to the contractor. It was suggested that only two critical risks be disclosed.

It was found that over 50% of the most critical risks had disparities between contractual allocation and what respondents suggested. These discrepancies and ambiguities in risk allocation may give rise to repeated disputes and claims.

References

1. Farooq MU, Thaheem MJ, Arshad H (2018) Improving the risk quantification under behavioural tendencies: a tale of construction projects. Int J Proj Manag 36(3):414–428. https://doi.org/10.1016/j.ijproman.2017.12.004
2. Castelblanco G, Guevara J, Mesa H, Flores D (2020) Risk allocation in unsolicited and solicited road public-private partnerships: sustainability and management implications. Sustain 12(11):1–28. https://doi.org/10.3390/su12114478
3. Marques RC (2018) Is Arbitration the right way to settle conflicts in PPP arrangements? J Manag Eng 34(1):1–7. https://doi.org/10.1061/(ASCE)ME.1943-5479.0000564
4. Osei-Kyei R, Chan APC, Yu Y, Chen C, Dansoh A (2019) Root causes of conflict and conflict resolution mechanisms in public-private partnerships: comparative study between Ghana and China. Cities 87(Aug):185–195. https://doi.org/10.1016/j.cities.2018.10.001
5. Cruz CO, Marques RC, Cardoso P (2015) Empirical evidence for renegotiation of PPP contracts in the road sector. J Leg Aff Disput Resolut Eng Constr 7(2):05014003. https://doi.org/10.1061/(asce)la.1943-4170.0000151

6. Harisankar KS (2013) Rethinking Dispute Resolution in Public – Private Partnerships for Infrastructure Development in India. https://doi.org/10.1177/0974930613488292
7. Levitt RE, Henisz W, Scott WR, Settel D (2010) Governance Challenges of Infrastructure Delivery: The Case for Socio-Economic Governance Approaches, pp 757–767
8. Hiyassat MA, Alkasagi F, El-Mashaleh M, Sweis GJ (2020) Risk allocation in public construction projects: the case of Jordan. Int J Constr Manag 22:1–11. https://doi.org/10.1080/15623599.2020.1728605
9. Zhang S, Zhang S, Gao Y, Ding X (2016) Contractual governance: effects of risk allocation on contractors' cooperative behavior in construction projects. J Constr Eng Manag 142(6):04016005. https://doi.org/10.1061/(asce)co.1943-7862.0001111
10. Castelblanco G, Guevara J (2020) Risk allocation in PPP unsolicited and solicited proposals in latin america: pilot study in Colombia. Constr Res Congr 2020:1321–1329. https://doi.org/10.1061/9780784482889.141
11. Marcellino M, Castelblanco G, De Marco A (2022) Contract renegotiation in PPPs: evidence from Italy. In: IOP Conference Series: Materials Science and Engineering
12. Castelblanco G, Guevara J, Mesa H, Hartmann A (2022) Social legitimacy challenges in toll road PPP programs: analysis of the Colombian and Chilean cases. J Manag Eng 38(3):1–15. https://doi.org/10.1061/(ASCE)ME.1943-5479.0001010
13. Hodge G, Greve C, Boardman A (2017) Public-private partnerships: the way they were and what they can become. Aust J Public Adm 76(3):273–282. https://doi.org/10.1111/1467-8500.12260
14. Yan L, Zhang L (2020) Interplay of contractual governance and trust in improving construction project performance: dynamic perspective. J Manag Eng 36(4):04020029. https://doi.org/10.1061/(asce)me.1943-5479.0000791
15. Dewulf G, Garvin MJ (2020) Responsive governance in PPP projects to manage uncertainty. Constr Manag Econ 38(4):383–397. https://doi.org/10.1080/01446193.2019.1618478
16. Guo F, Chang-Richards Y, Wilkinson S, Li TC (2014) Effects of project governance structures on the management of risks in major infrastructure projects: a comparative analysis. Int J Proj Manag 32(5):815–826. https://doi.org/10.1016/j.ijproman.2013.10.001
17. Loosemore M, Raftery J, Reilly C, Higgon D (2006) Risk Management in Projects. https://doi.org/10.1142/9781908979513_0009
18. Meng X (2012) The effect of relationship management on project performance in construction. Int J Proj Manag 30(2):188–198. https://doi.org/10.1016/j.ijproman.2011.04.002
19. Bing L, Akintoye A, Edwards PJ, Hardcastle C (2005) The allocation of risk in PPP/PFI construction projects in the UK. Int J Proj Manag 23(1):25–35. https://doi.org/10.1016/j.ijproman.2004.04.006
20. Wibowo A, Mohamed S (2010) Risk criticality and allocation in privatised water supply projects in Indonesia. Int J Proj Manag 28(5):504–513. https://doi.org/10.1016/j.ijproman.2009.08.003
21. Carbonara N, Costantino N, Pellegrino R (2014) Concession period for PPPs: a win-win model for a fair risk sharing. Int J Proj Manag 32(7):1223–1232. https://doi.org/10.1016/j.ijproman.2014.01.007
22. Nasirzadeh F, Khanzadi M, Rezaie M (2013) Dynamic modeling of the quantitative risk allocation in construction projects. Int J Proj Manag 32(3):442–451. https://doi.org/10.1016/j.ijproman.2013.06.002
23. Wang H, Liu Y, Xiong W, Song J (2019) The moderating role of governance environment on the relationship between risk allocation and private investment in PPP markets: evidence from developing countries. Int J Proj Manag 37(1):117–130. https://doi.org/10.1016/j.ijproman.2018.10.008
24. Lam KC, Wang D, Lee PTK, Tsang YT (2007) Modelling risk allocation decision in construction contracts. Int J Proj Manag 25(5):485–493. https://doi.org/10.1016/j.ijproman.2006.11.005
25. Jin XH, Zhang G (2011) Modelling optimal risk allocation in PPP projects using artificial neural networks. Int J Proj Manag 29(5):591–603. https://doi.org/10.1016/j.ijproman.2010.07.011

26. Saeed AM, Duffield C, Hui FKP (2018) An enhanced framework for assessing the operational performance of public-private partnership school projects. Built Environ Proj Asset Manag 8(2):194–214. https://doi.org/10.1108/BEPAM-07-2017-0041
27. Agarchand N, Laishram B (2017) Sustainable infrastructure development challenges through PPP procurement process: Indian perspective. Int J Manag Proj Bus 10(3):642–662. https://doi.org/10.1108/IJMPB-10-2016-0078
28. Project Management Institute (2013) A guide to the project management body of knowledge (PMBOK® guide), 5th Edition. Project Management Institute, Pennsylvania
29. Loosemore M, McCarthy CS (2008) Perceptions of contractual risk allocation in construction supply chains. J Prof Issues Eng Educ Pract 134(1):95–105. https://doi.org/10.1061/(asce)1052-3928(2008)134:1(95)
30. Sánchez-Silva M, Klutke GA (2016) Reliability and Life-Cycle Analysis of Deteriorating Systems. Springer, Piscataway. https://doi.org/10.1080/13854049708407039
31. Oduro Asamoah R, Offei-Nyako K, Ampofo-Twumasi K (2021) Relative importance of triggers influencing cost contingency determination for building contracts - the perspective of quantity surveyors. Int J Constr Manag. https://doi.org/10.1080/15623599.2021.1930638
32. Moazzami M, Dehghan R, Ruwanpura JY (2011) Contractual risks in fast-track projects. Procedia Eng 14:2552–2557. https://doi.org/10.1016/j.proeng.2011.07.321
33. Flyvbjerg B et al (2018) Five things you should know about cost overrun. Transp Res Part A Policy Pract 118(July):174–190. https://doi.org/10.1016/j.tra.2018.07.013
34. Flyvbjerg B, Bester DW (2021) The cost-benefit fallacy: why cost-benefit analysis is broken and how to fix it. J Benefit-Cost Anal 12(3):395–419. https://doi.org/10.1017/bca.2021.9
35. Priemus H, Flyvbjerg B, van Wee B (2008) Decision-Making on Mega-Projects: Cost-Benefit Analysis, Planning and Innovation. Edward Elgar Publishing Limited, Cheltenham
36. Castelblanco G, Guevara J, Mendez-Gonzalez P (2022) PPP Renegotiation Flight Simulator: A System Dynamics Model for Renegotiating PPPs after Pandemic Crisis. https://doi.org/10.1061/9780784483978.011
37. Castelblanco G, Guevara J, Mendez-Gonzalez P (2022) In the Name of the Pandemic: A Case Study of Contractual Modifications in PPP Solicited and Unsolicited Proposals in COVID-19 Times. https://doi.org/10.1061/9780784483978.006
38. Castelblanco G, Guevara J, Salazar J (2022) Remedies to the PPP crisis in the Covid-19 pandemic: lessons from the 2008 global financial crisis. J Manag Eng 38(3):1–18. https://doi.org/10.1061/(ASCE)ME.1943-5479.0001036
39. Castelblanco G, Guevara J (2022) Crisis driven literature in PPPs: a network analysis. In: IOP Conference Series: Earth and Environmental Science. IOP Publishing, Melbourne - Australia. https://doi.org/10.1088/1755-1315/1101/5/052002. https://oxford-abstracts.s3.amazonaws.com/bbb900d1-74cf-4dbb-b573-81550cf6ef90.pdf
40. Mesa HA, Molenaar KR, Alarcón LF (2016) Exploring performance of the integrated project delivery process on complex building projects. Int J Proj Manag 34(7):1089–1101. https://doi.org/10.1016/j.ijproman.2016.05.007
41. Castelblanco G, Guevara J (2022) Building bridges: unraveling the missing links between public-private partnerships and sustainable development. Proj Leadersh Soc 3(100059):1–10. https://doi.org/10.1016/j.plas.2022.100059
42. Castelblanco G, Guevara J, Rojas D, Correa J, Verhoest K (2023) Environmental impact assessment effectiveness in public-private partnerships: study on the colombian road program. J Manag Eng. https://doi.org/10.1061/JMENEA/MEENG-5015
43. Bloomfield P, Westerling D, Carey R (1998) Innovation and risk in a public-private partnership: financing and construction of a capital project in massachusetts. Public Product Manag Rev 21(4):460–471
44. Ogunlana SO (2010) Beyond the 'iron triangle': stakeholder perception of key performance indicators (KPIs) for large-scale public sector development projects. Int J Proj Manag 28(3):228–236. https://doi.org/10.1016/j.ijproman.2009.05.005
45. Nguyen D, Garvin MJ (2016) Risk allocation and management practices in highway PPP contracts: pilot study of Virginia, pp 2861–2871

46. Siraj NB, Fayek AR (2019) Risk identification and common risks in construction: literature review and content analysis. J Constr Eng Manag 145(9):03119004. https://doi.org/10.1061/(ASCE)CO.1943-7862.0001685
47. Yescombe ER, Farquharson E (2018) Public-Private Partnerships for infrastructure: Principles of policy and finance, 2nd edn Butterworth-Heinemann. https://doi.org/10.1016/b978-0-08-100766-2.00001-2
48. Almarri K, Alzahrani S, Boussabaine H (2019) An evaluation of the impact of risk cost on risk allocation in public private partnership projects. Eng Constr Archit Manag 26(8):1696–1711. https://doi.org/10.1108/ECAM-04-2018-0177
49. Tembo Silungwe CK, Khatleli N (2020) An analysis of the allocation of pertinent risks in the Zambian building sector using Pareto analysis. Int J Constr Manag 20(4):321–334. https://doi.org/10.1080/15623599.2018.1484853
50. Krippendorff K (2004) Content Analysis: An Introduction to Its Metholody, 2nd edn, vol 31, no 6. SAGE Publications, Thousand Oaks. https://doi.org/10.1103/PhysRevB.31.3460
51. Elo S, Kyngäs H (2007) The qualitative content analysis process. J Adv Nurs no. Cole 1988, 107–115. https://doi.org/10.1111/j.1365-2648.2007.04569.x
52. Kolbe RH, Burnett MS (1991) Content analysis research: an examination of applications with directives for improving research reliability and objectivity. J Consum Res 18(2):243–250
53. Afjei SM (2015) A Content Analysis of Sustainability Dimensions inAnnual Reports. Florida International University. https://doi.org/10.25148/etd.FI15032103
54. Nguyen DA, Garvin MJ, Gonzalez EE (2018) Risk allocation in U.S. public-private partnership highway project contracts. J Constr Eng Manag 144(5):04018017. https://doi.org/10.1061/(asce)co.1943-7862.0001465
55. Wang L, Kunc M, Bai S (2017) Realizing value from project implementation under uncertainty: an exploratory study using system dynamics. Int J Proj Manag 35(3):341–352. https://doi.org/10.1016/j.ijproman.2017.01.009
56. El-Sayegh SM (2008) Risk assessment and allocation in the UAE construction industry. Int J Proj Manag 26(4):431–438. https://doi.org/10.1016/j.ijproman.2007.07.004

Case Study: The Olona Valley Masterplan. Soft Mobility Infrastructures as Innovative Strategies for the Sustainable and Adaptive Regeneration of a Territory

Maria Cristina Fregni, Enea Sermasi, Sara Angelini, Alessio Valmori, and Anna Giusti

Abstract Could something apparently "light" like a **soft mobility infrastructure** become the **backbone** of a territorial megaproject? Yes, if it represents the **trigger action** of a strategy for the regeneration of a wide area from Swiss to Milan, involving **multiple public and private stakeholders** and investors and supporting **long-lasting requalification** of a territorial system in its economic, social and ecological components. The Olona Valley Masterplan is the expression of this strategy, based on a new bicycle lane that crosses the territory of Varese alongside the Olona river and, by touching **dismissed industrial areas, cultural sites and precious landscapes**, brings new life to them, reinterpreting their identity and roles, moving people and goods, giving back significant sites to local communities. The masterplan brings together economic subjects, representatives of the public administration, researchers and innovative companies and local communities towards a vision of the future based on resilience, adaptability and the rediscovery of the intrinsic potential of the territories.

Keywords Change of approach · Stakeholder engagement · Project success factors

M. C. Fregni (✉)
Politecnica Ingegneria ed Architettura, 220 Via Galilei, Modena, Italy
e-mail: mcfregni@politecnica.it

E. Sermasi
Politecnica Ingegneria ed Architettura, 220 Via Galilei, Modena, Italy

S. Angelini · A. Valmori
de Gayardon Bureau, Via mura F., Comandini, 2, 47521 Cesena, Italy

A. Giusti
IUAV Venezia, Venezia, Italy

E. Favari and F. Cantoni (eds.), *Complexity and Sustainability in Megaprojects*, Lecture Notes in Civil Engineering 342,
https://doi.org/10.1007/978-3-031-30879-6_23

1 Purpose

The context in which we have to intervene is the "Valle Olona" which takes its name from the river that flows through it and which, in correspondence with the territory of Varese, has a length of about 10 km and runs its course with a predominantly North–South direction. From an environmental, settlement and infrastructural point of view, this Valley constitutes an external axis with respect to the municipal territory of Varese. There are artisanal and industrial settlements, public services and a large shopping center. The set of activities present appears to be devoid of an organic distribution of open spaces that often follow one another in a residual manner with respect to the volumes creat-ed and without reciprocal relationships between them and without direct connection with the watercourse. The Olona river measures 71 km in length and has a hydro-graphic basin that extends over part of the provinces of Varese, Milan and, minimal-ly, Como, also affecting a part of Swiss territory as a small part of the Gaggiolo basin, its tributary, it belongs to the Canton of Ticino. With its 911 km^2, the Olona catchment area occupies 5% of the surface of Lombardy and hosts approximately 1,000,000 inhabitants.

According to those premises, Varese and the Olona Valley can become a reference pole of a "wide area" which includes the neighboring provinces of Milan, Monza and Brianza, Como and Lecco in Lombardy and of Verbania-Cusio-Ossola and Novara in Piedmont, going beyond the catchment area of the river. It is in fact **a strategic location in the heart of the North West of Italy** connected with the main European infrastructural lines (France, Switzerland and Germany).

The area is also located just 6 km from the border with **Switzerland** and, thanks to the presence of the state road SS 342, which forms the backbone of the Varese ring road system, is directly connected to the motorway network.

The recent completion of the TILO Lugano-Malpensa railway connection with the Varese-Lugano-Como section, moreover, it has consolidated the territorial system on a regional scale of the three cities of Varese, Lugano and Como, united by a pub-lic transport system that connects them with a travel time between one and the other of less than 30 min.

This infrastructure marks the birth of a **new territorial figure** along the transalpine route that connects Milan with Zurich and Bern in the north of the Alps.

Along the same route, other urban events of regional importance are changing the scenario within which all agglomerations will have to redefine their location: the new research and training center of Rho Fiera, the design of the new railway stations in Milan, the upgrading of the ports of Genoa and Rotterdam, the new production areas along the Franch-German border, the City-Ticino project, the improvement of modal accessibility to airports, to name the most emblematic.

Valle Olona is also a territory with a strong identity for its inhabitants.

In the second decade of the nineteenth century in the province of Varese began to develop what soon became a **thriving industrial activity** and which today, due to long periods of economic crisis often accentuated by strong international competition, leaves a great heritage of industrial archeology on the territory of the Province.

In most cases, these new production activities were born side by side (or represented their evolution) with the many and more ancient water mills already present in the area and evolved by transforming the previous activities of grinding seeds and flour, husking of rice and cutting of wood into boards, towards more modern work-ing structures.

Around 1820 and in the midst of the **industrial revolution**, these once artisanal re-alities were transformed into organized processes based on a large production that, specializing in tanning leather processing, weaving and its derivatives, fulling of fabrics or paper and also for the firing of bricks or limes soon became a reality des-tined to grow further.

In the period between 1870 and 1900 their development underwent such an increase that it often required the construction of **collective connection structures** (rail transport with derivation tracks to companies) and, thanks to this advantage, many industries in the Province of Varese came to be among the first and most important Italian production companies.

At the beginning of the twentieth century the paper and textile sectors prevailed, but there was no lack of chemical industries or tanneries and mechanical industries, furnaces and companies where animal bones were processed also played their part. The enlargement and construction of new industrial structures continued until the 1930s and, after the interruption due to the Second World War, they maintained a constant level of production until the end of the 1950s.

In the sixties the entire production system suffered a crisis which, becoming increasingly hard, led some sectors towards closure and others towards a strong downsizing: from the seventies many of the textile, chemical and tanning industries closed, in the eighties there was the decline of the entire paper production, while from the nineties the downward phase of the mechanical sector began (Fig. 1).

Today, the territory of the valley is **very fragmented**, with many small urban villages all interconnected in a whole urban **disfunctional system**, historical and non-historical production areas, still active or abandoned, residual areas that are not built and often inter-closed, a confused and impactful infrastructural system, with recent roads made to cross quickly the territory, **indifferent** to its morphology, its history, and often even hostile to displacements on the local scale.

At the same time, it's clear that the area very rich in potential, due to its cultural and environmental nature, attractive for visitors and tourists both from the local district and from abroad.

Moreover, there are many potentialities for socio-economic development not linked to tourism, there are thriving productive realities, as said the position is strategic in relation to Milan, Switzerland, Malpensa airport.

In this context, the local Municipality assumed an original and brave choice: try to exploit all these potentials, to generate a **territorial system** based on a new concept of green production, in favor of tourism but also for the quality of daily life of the local population. And try to reach this goal through a **cycle path**! A long,

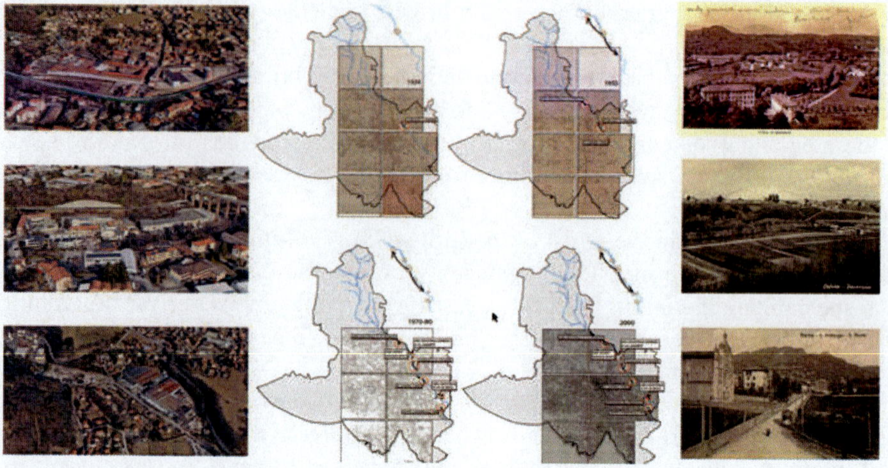

Fig. 1 Historical evolution of the territory and current pictures of the dismissed industrial areas

structured and innovative cycle path, which crosses the whole valley in a north–south direction, connects to the already existing cycle paths and the railway system towards Switzerland and Milan, allowing to mend the tears of the territory, to make the existing and potential attractors accessible in a sustainable way. A cycle path that, assimilating the American lesson of cultural linear corridors, structures a re-interpretation of the identity of the Olona valley territory and acts as **the backbone of a new integrated development system.**

In this sense, the Olona River Masterplan represents an innovative example of **mega-project**. In fact, you want to induce and support a significant territorial transformation on a supra-local and cross-border scale, with the reactivation of settlements and communities, not through the construction of an imposing building or infrastructural work, but through an element of sustainable mobility, which generates networks, activates synergies, it connects opportunities and involves various actors.

2 Design/Methodology/Appoach

The tool selected by the Municipality in order to implement this vision is a Masterplan, intended, in a very international and challenging perspective for the size and structure of Varese Municipality, as a **management tool**, able of prefiguring a physical transformation that was however the **form of a strategy**, therefore also made up of environmental and socio-economic assessments, action cards, discussion tables with investors, citizens and institutions, in order to address issues of **scale, flexibility and opportunities.**

The tender published by the Municipality for the engagement of an engineering firm to be in charge of the development of the masterplan reads:

The Municipality of Varese intends to draw up the Valle Olona Masterplan … as a territorial project that prefigures, in ***spatial and relational terms****, the desired city including both public transformations (land project) and private ones for public use (unitary project).*

This document must make possible to decline in terms of land consumption, density, endowment of services and functional structure the entire sector and this declination must take place through the preparation of a series of documents (reports, volumetric plans, sheets and a three-dimensional model) that explain the new urban form in terms of planivolumetric structure and relations with the rest of the city.

Based on the aforementioned strategic lines, the professional, starting from the context analysis, will have to identify the points of strengths and weaknesses of the starting situation and define the elements on which you intend to intervene, identify possible project scenarios by appropriately developing all the themes that represent priorities for the Administration.

In this phase, the successful tenderer will have to proceed with the ***identification of a perimeter*** *of the masterplan that the same is not already defined.*

It was immediately clear to us that, to respond to this ambitious goal, our work obviously had to move at different scales and with the involvement of different disciplines. On the one hand we had to investigate the technical feasibility of a cycle path that crosses a busy territory, morphologically designed on several levels, cut by highways and rivers, with the aim of making a cycle path that is not simply functional, but also able of being **perceptually and strategically connected** to the "pearls of the necklace" located along its line.

The selected approach was therefore the one based on the experience of the American National Park Service and the Landscape Preservation Group of Politecnico di Milano related to the **Linear Cultural Landscape**.

The historical landscape constitutes, in its role of visible projection of the territory, a document of great interest, as a palimpsest, a system of signs and meanings, through which to proceed to the unveiling of the identity of the places, as well as to the identification of resources to build the future. In the territory, therefore, identity becomes objective and becomes visible, it becomes landscape. It, as a cultural heritage resulting from the prolonged interaction in different societies between man, nature and the physical environment, testifies to the evolutionary relationship of society and individuals with their environment.

In this sense, we can speak about "A landscape systems": that is a territorial object that works at different relational scales and which presupposes the enhancement of the traces present in the places through a process of synthesis of the specific characteristics of the landscape.

And, in this sense, we can approach a design activity focused on a territorial scale object, with a main element that plays the role of the "cathalist" or the "generator"of the landscape.

In the case of the landscape where the main element is a linear element, **the linear infrastructure** assumes the value of tracing the shape of the territory often in the horizon of the dispersed or abandoned landscape, and this landscape has to be to be investigated, recognized, interpreted through the analysis and the project. The field of investigation belongs to a dimension of **intersection between territory and infrastructures**, The general idea is that an infrastructure does not simply generate a path or a functional sign, but is able to originate a **linear system of open and built spaces**, characterized by strong territorial connections with other landscape, architectural or environmental elements. In this sense, when we work with the design of an infrastructure, that does not simply concern the definition of a formal structure, but it is morphogenetic, that is, it has within itself the **ability to induce transformations** on the settlement and landscape structure.

So, when we work with existing linear elements (like the Olona River, its valley and a new cycle path along them) and their related landscapes, we can talk about **Linear Cultural Landscape**: a path of historical interest, along which people and / or goods have moved, in which it is evident that the natural environment has been transformed by man over the centuries. It is therefore **a linear cultural landscape**, i.e. a linear constructed (or cultural) landscape, in which nature and anthropogenic transformations are closely connected by an element with a prevailing direction.

The enhancement of this linear cultural landscape seemed immediately a good strategy to take up the challenge of the tender and support the Municipality's strategy in an effective way.

Therefore, the concept of the new cycle path was based on some principles able to assure the mix between physical, social and economic elements and the let interact the linear dimension with a wider territorial context. This approach could be called "the 3D strategy", as follows:

D as DENSITY: opportunity to find more than something in a defined area. Not only one monument or one service but many, not only ancient traces but also innovation's components, not only one level of meaning but a multilayered experience. The result is the compound could be vital, the territorial density allows people to move on foot or by bicycle and to use and enjoy the local offer;

D as DIVERSITY means offering more opportunities to the local communities as well as visitors. Not only one type of tourism, not only one target of local users, not only cultural value, not only services but an interconnected system. And diversity of choice means also offering the opportunity to decide during the experience of the place, that is create a system of signage and wayfinding able to let decide the user in realtime;

D as DESIGN means good design, that is a design able to support all these goals, a design that doesn't aim at its own visibility but rather at the enhancement of the local values, the offer of different opportunities, the easy interconnection (visual, functional, physical etc.) between the elements. Finally, a design at the service of the territory and the communities.

3 Findings

The masterplan responds to the need for a new sustainable, ecological and inclusive development through the development of the following **main strategical lines**:

- Urban regeneration of former industrial abandoned areas with the right to demolish the volumes pending the full implementation of the planned interventions;
- Analysis of the relationship between full and empty space in the compartment and in the individual lots;
- Analysis of the quantitative and qualitative relationship between public space (green, squares, streets, ...) and private space (appurtenant areas of the building);
- Analysis of the functions that can be established with particular attention to innovative uses related to scientific research, social welfare services, technologically advanced economic activities;
- Analysis of the need for the establishment of infrastructures linked to public services of various kinds (transport, energy, social, educational, etc.);
- Study on the ideal orientation of buildings also in consideration of the settling functions;
- Environmental redevelopment of the **Olona river** through the organization and restoration of green areas placed astride it;
- Enhancement and safeguarding of the historical-cultural and historical-monumental aspects that characterize the River Valley auction;
- Mitigation of the impact of the road bypass parallel to the watercourse;

For the implementation of the redevelopment's strategic vision and according to the methodology described above, the masterplan defines **three main thematic lines**, called "layers", which, intertwining and interacting in synergy, have the strength to reactivate the territory:

- Layer 1: **the cycle path** - sustainable mobility infrastructure and spine of the valley
- Layer 2: **the transformation's areas** - new architectures and re-functionalization of interstitial public spaces
- Layer 3: **the Olona river park** - a system of urban and naturalistic parks which densify the course of the Olona river with values and opportunities and make it permeable and usable at the same time by the flows of users.

3.1 Layer 1: The Cycle Path

In order to reach all these goals, the masterplan identifies as a connecting element of the various interventions, as well as as a structural element of the new 2.0 productive and tourist identity of the valley, the realization of a cycling route for use by both citizens and tourists. This cycle path, keeping together all the souls, places and objectives of the masterplan, allows an organized redesign and an organic management of the interventions. Moreover, the bicycle line should support the enhancement of the intermodality of the railway line connecting Switzerland - Canton Ticino, in relation to the mobility of cross-border workers.

This new cycle line acts like a backbone and the corridor made by all areas around it move like a ribe cage, widening and narrowing around the spine according to the "breath" that the territory wants to express in that point of the valley.

The cycle path will have to inspire and enhance the places it crosses as well as connect points, from A to B. The cycle path will be the bearer of redevelopment and will become a reference point in the Valley. In its present condition, the main landmark is the elevated ring road that crosses the stretch between the Poretti Brewery and the Iper Shopping Center with a succession of viaducts and roundabouts. The Olona river remains embedded in the valley, it is no longer visible nor usable. The existing bypass road has taken all the spaces and tracing a cycle path that underpasses or overpasses both this infrastructure and the Olona river has repercussions both on an economic level and on the effective usability of the route itself.

The masterplan proposes to think of the cycle path as the natural lateral expansion of the bypass. The cycle path of the Olona valley will run contiguously with the it, following its path. The Olona valley cycle path will reconcile two important functional and aesthetic needs:

1. The absolute comfort and clarity of the route (it will be the shortest and safest route to go from A to B)
2. It will be a bicycle architecture with materials, colors and lighting integrated into the iconic and recognizable landscape

The cycle path will therefore be a 3-m wide lane reserved for cyclists in both directions. The cycle path and the ring road will be separated by barriers and these will be able to support the lighting system of the cycle path to make it accessible and safe even at night. The route will be at high altitude when the ring road will be on a viaduct and ramps - real installations in the landscape - will allow cyclists to descend at the notable points of the route to continue on other itineraries. The track will therefore have nodes where it will "touch the ground" and when this happens there will be opportunities for redevelopment both for the green areas and for the nearby urban fabric. In the attached thematic table it is possible to see some examples of implementation of these concepts in Europe and of how the investment in a visually striking but functional cycle path can radically change both the urban/natural landscape and the perception and the use of those who live in the area.

The interventions in the transformation areas and public parks will contribute to strengthening the culture of cycling through functions dedicated to bicycles (bicycle parking/exchange parking/shops and resales/cycle-workshops/parks and paths dedicated to bicycles/possibility to connect to existing itineraries/etc. ...).

Once the main cycle path has been defined, the masterplan suggests widespread cycle connections both on the road and in public spaces and squares in the transformation areas and in public parks. Like the meanders of a river, the cycle paths multiply making the territory porous and crossable not only longitudinally (from north to south) but also transversely (from east to west) by opening gaps and overcoming obstacles, the one that at present they are the enclosures of disused or disused activities, the infrastructures and the terrain-vague along the banks of the river (Fig. 2).

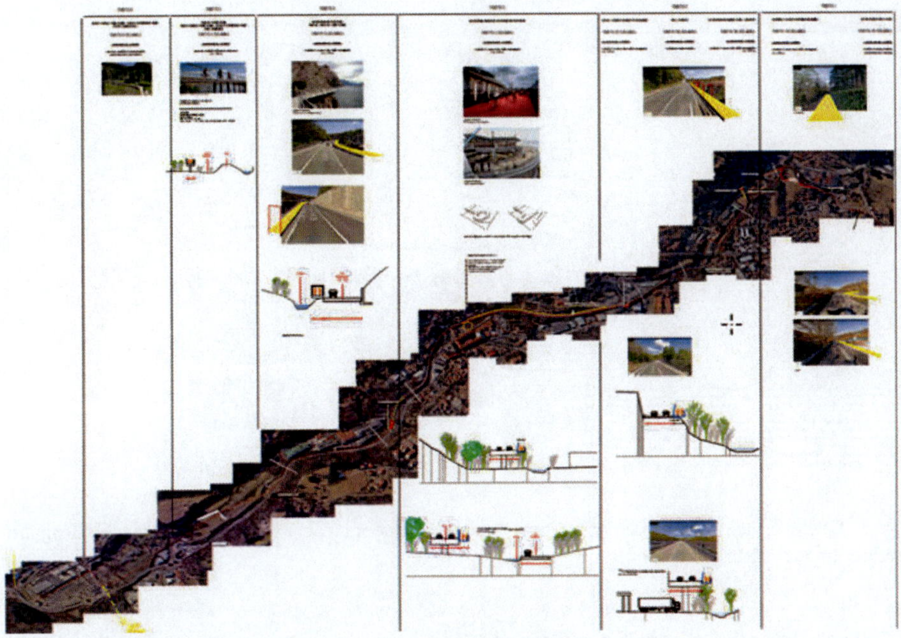

Fig. 2 Path, sections and main intersection of the proposed cycle path

3.2 *Layer 2: The Areas of Transformation*

For each area, the masterplan has determined a main vocation and has studied - through the analysis of similar case studies - possible scenarios of transformation and redevelopment. The transformation areas will be the bearers of new urbanity and will also be attractors towards the Olona Valley.

Through a selected functional mix it will be possible to graft new forms of living and producing more sustainable and innovative forms into the valley that renew the image of Valle Olona and the Varese area.

Each area is linked to the Olona river and interacts with it in a unique and peculiar way. Each area is still very strongly linked to history and its territory; the master-plan tries to update this link, update it to the present condition and, why not, to the future vision of the next few years (Fig. 3).

3.3 *Layer 3: The Olona River Park*

A river park is established along the Olona river. This park will be a continuous succession of (existing) natural areas and urban parks also with an educational vocation capable of reactivating the banks of the river in a fruitive and leisure key. The river landscape is currently dormant waiting to be enhanced and made accessible.

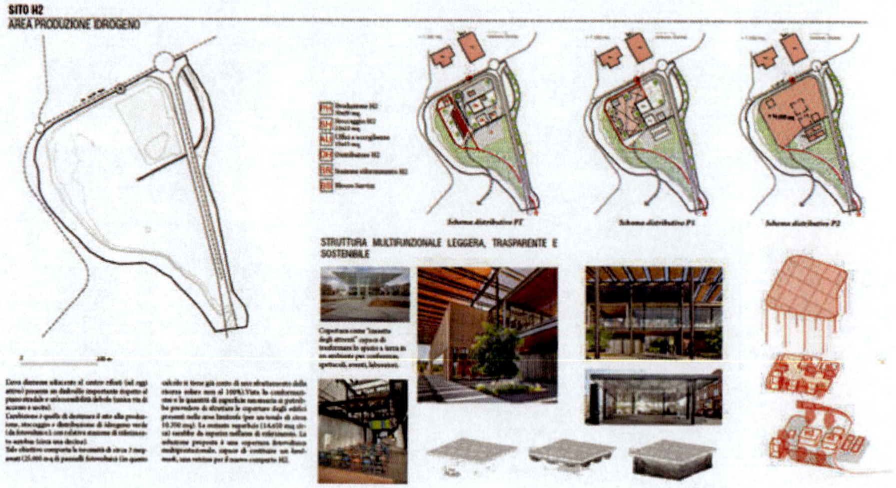

Fig. 3 One of the transformation's areas and the concept for its reactivation as Hydrogene production and distribution points

The cycle path will cross parks and green areas as its main spine, connecting them seamlessly and making them accessible with a recognizable and easily accessible route.

The urban parks will be close to the main centers and will complete the new nodal spaces of the epicenters that will reactivate the building and former industrial heritage.

Through the synergy of infrastructure - nature and architecture it will be possible to create a virtuous and positive system that will equip the city of Varese with new urban spaces suited to biodiversity and ecology. A green district that derives from the blue infrastructure of the Olona River and the production of solar energy.

The masterplan identifies **three park areas** to be designed and made accessible again and which are:

- URBAN PARK - OLONA VALLEY
- EDUCATIONAL PARK - RIVER EXPANSION CASSA
- SPORTS PARK - IPER

And it identifies three urban areas to be regenerated to make the programmed functions more integrated with the context in which they are inserted. These new public spaces are (Fig. 4):

- PIAZZA AND GREEN SPACES SYSTEM - CONCIARIA VALLE OLONA
- PIAZZA DELLA COGANA - MULTIFUNCTIONAL PUBLIC SPACE
- GARDENS AND COURTS CARTIERA VARESE

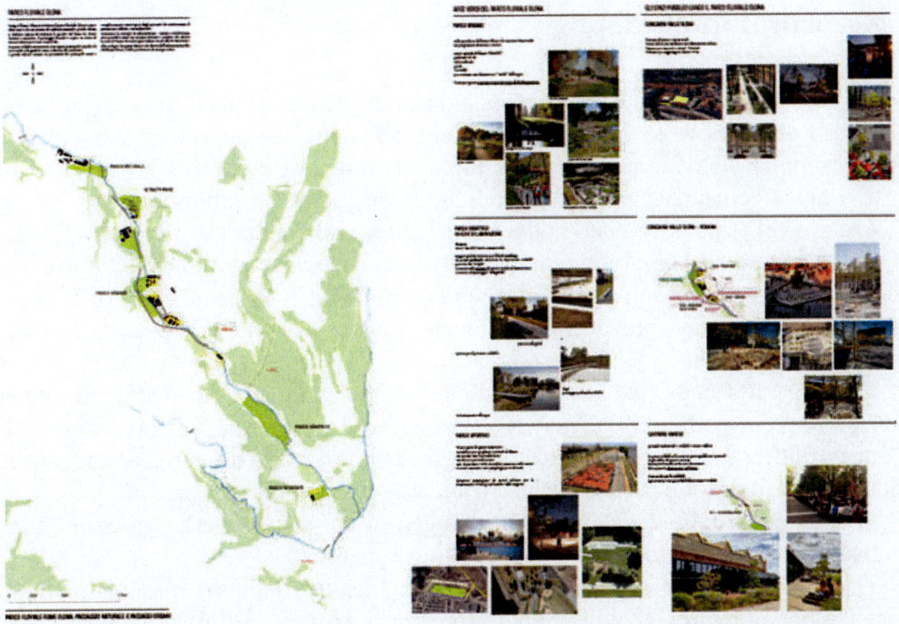

Fig. 4 The selected abandoned green areas along the river, reconnected to create a system of parks

4 Originality/Value

According to the complexity of the issues involved in the masterplan, the local communities will decide how wide the range of action of the masterplan must be, not only through their administrators, but also directly, thanks to **stakeholder engagement activities** for the identification of subjects that can be involved in the financing of the interventions, in the activation of public–private partnerships, in the co-planning and co-management of spaces. A WBS of presentations, listening and discussion meetings is therefore underway, in physical and/or digital mode, which will bring together the representatives of the associative system, industry and services connected to the automotive, mechanics, electronics, ICT, energy and the environment. Stakeholder engagement activities, in addition to the need to gather interest and financial support for the project, also aim to increase stakeholder participation by limiting and anticipating, thanks to the development of analytical arguments, possible side effects by groups of the local community that can cause a slowdown in the process.

Furthermore, in the locality of Olona, are located two important and active production's companies (Lindt and Carlsberg/Poretti), for which the masterplan will be able to evaluate any development opportunities, in mutual relationship with the territory.

5 Practical Implications

During the development of the strategical lines of the masterplan, the stakeholders discussed about how to improve the ecological value of the valley, beyond the preservation of the river and the support to the sustainable mobility.

The final decision has been to prompt the **hydrogen production and use**.

The project to be developed in the Olona valley is linked to the high-level vision of the Italian government on the role that hydrogen can occupy the national path of decarbonization. This is in accordance with the Integrated National Plan for Energy and Climate, with the broadest environmental agenda of the European Union by 2050.

For the next decade, the government expects the application of hydrogen in the transport sector, particularly heavy (for example long-haul trucks) in railways and industry, with specific reference to those segments in which hydrogen is already used as a raw material, for example in the chemical sector and oil refining.

In addition to this, the mixing of hydrogen in the gas network can be used to an-ticipate and stimulate growth of the hydrogen market.

The "hydrogen valleys", ecosystems that include both the production and consumption of hydrogen, will be able to provide areas for the diffusion of hydrogen by 2030 leading to its possible application in other sectors.

The Masterplan, based on the objectives of the regeneration of the Olona Valley, will therefore be oriented to support the creation of a research and development/experimentation hub on the production and use of hydrogen through the implementation of **pilot projects and prototypes** capable of attracting investments.

Consistent with this approach, the masterplan proposes to valorize another resource that lies dormant along the river: the **roofs of industrial buildings** (paper mills or tanneries) that have stratified over time along the valley. These are several square meters that could generate many kilowatts if properly redeveloped. These large surfaces today lie inert, bringing no benefit to their immobility.

It is necessary to exploit this ready-made resource to make the valley a large photovoltaic park that could be used to produce green hydrogen, but also to supply energy to production activities and settlements, contributing to the ecological transition and at the same time implementing the requalification of the existing real estate assets.

The engagement for the masterplan's development forsees also the reuse of a dismissed building located in the middle point of the valley, which will act as a **catalyst** for the transformation, hosting a **research and training hub on sustainable mobility.**

This renovated building, with its innovative function, should become the emblem of the regeneration of the territory, point of reference and activator of energies and collaborations.

6 Research Limitations/Implications

According to what said until now, it is clear that the effectiveness of the masterplan is strictly related to the ability to engage institutional and not-institutional actors.

The main toll for the implementation of the masterplan's strategy is the Public–private partnerships (PPPs), that have become an increasingly popular way to get major infrastructure projects built. Compared with traditional procurement solutions, these arrangements show a significantly increased level of private-sector participation, with the goal of boosting the efficiency and effectiveness of the project through its entire life cycle, from development to the end of the operating phase. However, PPPs should not be seen as an instrument to **solve public-sector budget constraints** or financing gaps, but rather a tool to deliver effective, cost-efficient projects and associated services. That means, in our case, that the implementation of the masterplan requires, besides the funds for the realization of the cycle path, additional public money, at least to support activation's actions as well as stakeholders engagement activities.

Moreover, a strong recourse to PPPs involves great management skills and an effective control room, which coordinates the various players, puts activities in sequence, defines priorities and interactions. The great challenge will therefore be the ability for the local authorities involved to cover these roles and act proactively and dynamically, which are not usually the main feature of Italian institutional bodies.

The awareness of this challenge has meant that the Municipality has already decided to equip itself not only with a staff of technical consultants linked to the physical planning of the territory, but also with economists and sociologists, while the political component has taken steps to personally mediation with institutions, citizens and businesses in the area.

Sustainability Challenges of High-Speed Railway Megaprojects from a Systems Thinking Lens

Zahra Shams Esfandabadi, Dario Cottafava, Laura Corazza, and Simone Domenico Scagnelli

Abstract Megaprojects are intended to generate positive impacts for society at regional, national, or international levels. However, the potential unintended consequences associated with the activities in various stages of the megaprojects' lifecycle, such as air pollution, waste generation, and social challenges, can adversely affect the sustainability of these projects. In this article, based on a systems thinking approach, a causal-loop diagram is developed to address the sustainability challenges of high-speed railway megaprojects. The two stages of construction and operation of the project have been considered and the effects of relevant variables on the three pillars of sustainability are discussed. The analysis and provided insights shed light on the significance of considering both short-term and long-term effects of the activities directly or indirectly linked with megaprojects and confirm systems thinking as an appropriate approach to analyze the sustainability issues of these projects.

Keywords System Dynamics · Systems thinking · Turin-Lyon high-speed railway · Complexity · Triple bottom lines

---

Z. Shams Esfandabadi (✉) · D. Cottafava · L. Corazza · S. D. Scagnelli
Department of Management, University of Turin, Corso Unione Sovietica, 218 Bis, 10134 Torino, Italy
e-mail: zahra.shamsesfandabadi@unito.it

D. Cottafava
e-mail: dario.cottafava@unito.it

L. Corazza
e-mail: laura.corazza@unito.it

S. D. Scagnelli
School of Business and Law, Edith Cowan University, 270 Joondalup Dr, Joondalup 6027, Australia
e-mail: s.scagnelli@ecu.edu.au

E. Favari and F. Cantoni (eds.), *Complexity and Sustainability in Megaprojects*, Lecture Notes in Civil Engineering 342,
https://doi.org/10.1007/978-3-031-30879-6_24

1 Introduction

Megaprojects provide the required crucial infrastructure for a country's development, hence, are recognized as the backbone of societies. These large-scale projects generally cost over US$1 billion [1, 2] and differ from the traditional infrastructures in terms of cost and the number of stakeholders involved, amplitude, lifespan and size of territory impacted, and underlying complexity [3, 4]. The high intrinsic complexity of the megaprojects, which lies in every single feature from the technical side to the social and environmental counterparts, has been highlighted in several pieces of research in the literature. The megaprojects' complexity [5–7] and the increasingly conflictual situations among stakeholders and local communities [8] have become one of the greatest challenges. In fact, while megaprojects are intended to generate positive impacts and help improve society, they are usually largely contested by civil society [9–11] due to their potential negative unintended consequences (i.e., rebound effects [12, 13]) and the direct imposition of international, national and local governments.

In this regard, Turin-Lyon high-speed railway can be mentioned as an example of a contested global megaproject that has been used to elaborate the model herein presented [14]. This project is a part of the Trans-European Network-Transport (TEN-T) policy and is a complex megaproject promoted by the European Union (EU) and the Italian and French governments. The construction of the high-speed railway line between Turin (in Italy) and Lyon (in France), which is designed to cross the Susa Valley (at the Italian-French border) and the Alps has been discussed since the 1990s. However, it has faced severe opposition (at least on the Italian side) by local communities, civil society organizations, and even governments over the years [1]. The overt contest of the local communities of the Susa Valley (near Turin) can be considered a key reason for delays in the completion of this project [14]. Therefore, the challenges in terms of socio-economic [14], and environmental aspects [3, 9] of such projects deserve to be studied from a holistic point of view to highlight the existing and potential sources of challenge.

In this article, an overview of the sustainability challenges in megaprojects, with a specific focus on high-speed railway projects, from the lens of systems thinking is provided. The remainder of the article is organized as follows. Section 2 presents a general introduction to the applied methodology based on systems thinking. Section 3 is devoted to the developed causal-loop diagram (CLD) to illustrate the causalities among the identified variables affecting the sustainability challenges in high-speed railway megaprojects and the relevant discussions on the main causal loops identified. Finally, Sect. 4 concludes the paper.

2 Methodology

Based on the systems thinking approach and qualitative System Dynamics modeling [15], a general CLD has been developed in this research to illustrate the interconnections among the variables causing sustainability challenges in high-speed railway megaprojects.

CLDs are often used to facilitate systems thinking [15] and provide visual representations of interrelated complex systems [16]. They have been used in standard quantitative System Dynamics practices both to articulate the dynamic hypothesis and to summarize and communicate feedback insights based on the simulation model [17]. However, with the advent of qualitative analysis in the 1980s, CLDs started to be used for detailed system descriptions and also stand-alone policy analysis [17]. CLDs explicitly present the structural and agent system elements that may endogenously generate the dynamics in the behavior of the system or organization being studied [16]. Therefore, they have been widely used to analyze complex systems and their challenges, such as in the analysis of the sustainability transitions [18, 19] and their associated rebound effects [12, 20], technology diffusion [21], supply chain management [22, 23], knowledge transfer in megaprojects innovation [24], and the imbalance of markets [25].

The key variables used in order to build the CLD in this research were extracted from previous research conducted by Norese et al. [3] and Corazza, et al. [1, 4, 26] on the case of the high-speed railway line between Turin and Lyon. Further analysis was conducted by the authors based on the literature to identify and illustrate the causalities and interconnections among the variables.

3 The Developed CLD and Key Loops

The developed CLD to address the sustainability challenges of the high-speed railway megaprojects in terms of the three social, economic, and environmental pillars is illustrated in Fig. 1 and the discussion of its main identified loops is presented in this section. Based on the diagramming convention, the arrows linking every two variables in the CLD presented in this figure show the causality among them (i.e. the arrow starts from the cause and points at the effect). Where there is a delay between the cause and effect, the sign "||" has been placed on the arrow. Positive (+) and negative (-) signs on the arrows indicate their polarity (i.e. (A↑ → B↑) and (A↑ → B↓), respectively). The indicated polarities form reinforcing and balancing loops, which cause dynamics in the system being studied.

In Fig. 1, the triple bottom lines have been highlighted at the top of the CLD in green and the causalities leading to changes in each of these three pillars of sustainability are shown. While variables and their shadows (i.e. copies of the variables) are shown in black and gray, respectively, the exogenous variables known as control variables are presented in orange. Control variables in the model are variables outside

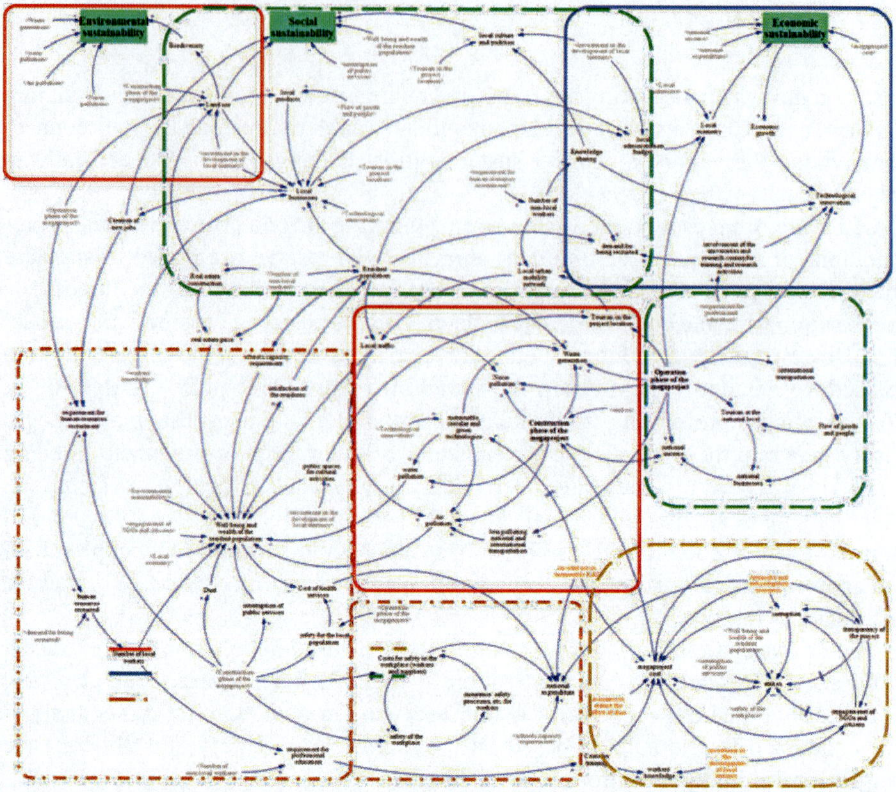

Fig. 1 The developed CLD to illustrate the sustainability challenges in high-speed railway megaprojects and the identified main causalities leading to macro dynamics in the system

the model boundary that are not affected by variables within the model boundary but affect them and therefore, can change the system behavior over time [15, 16]. Changing control variables in the model and monitoring their resulting behavior in the system can help decision-makers have a wider understanding of the potential behavior of the system in terms of changes.

In the developed CLD, the construction and operation stages of the high-speed railway project have been separated due to their different effects on the sustainability pillars. For instance, while railway transportation does not directly release CO_2 during its operation phase (if electricity is used for transportation, and we ignore the CO_2 emission resulting from the electricity generation), a huge amount of CO_2 is emitted at the construction stage [27]. Although end-of-life management of the project can be of high importance in the realm of sustainability [28, 29], this aspect has not been considered in the model, assuming too-far end-of-life for this type of megaprojects in terms of time.

The behavior of a system arises from its underlying causal feedback structure [15]. Figure 1 consists of several reinforcing and balancing loops, leading to the dynamics of the whole presented system over time and changes in the triple bottom lines with regard to the considered type of megaproject. As discussing all these loops is not possible in this paper, a selection of the most relevant causalities is discussed in Sect. 3.1. These causalities refer to (i) the dynamics in the air pollution over time (increasing air pollution during the construction stage and its improvement during the operation stage), (ii) strikes and protests due to factors such as negative environmental impacts of the project construction, lack of engagement of the workers and the community, and land use change in the project location, which can cause delays and other issues in the construction phase, (iii) dynamics of the technological innovation due to factors such as collaboration with universities and knowledge sharing with non-local workers, (iv) positive and negative impacts of tourism on the culture, tradition, and businesses in the project location and other territories, and (v) the dynamics in the well-being and satisfaction of the residents and workers linked with the safety of workers and the resident population in the project location. Besides, in order to provide insight regarding the managerial implications of the model, in Sect. 3.2, the effect of control variables on the general dynamics of the system is discussed.

3.1 Main Causalities Leading to the System's Dynamics

Due to the high complexity of the CLD developed for the considered type of megaproject, to obtain useful insights, Fig. 1 is read and analyzed by looking at single branches separately. As briefly introduced in the previous subsection, when looking at the causal dynamics, among different aspects and effects for megaprojects, there are a few emerging relevant implications.

First, regarding air emissions and dust, contrasting dynamics between short and long-term impacts emerge from the construction and operation phases of a railway transportation megaproject. First, as shown in Fig. 1 (central part of the model), the construction of a megaproject may increase air emissions and dust in the short term due to an increase in local traffic, as a consequence of the increase in the resident population and the transportation of materials (trucks, extraction of raw materials, …). On the contrary, during the operation phase, the transportation infrastructure's long-term effect results in a reduction of air pollution emissions and dust, thanks to the usage of electricity to power the trains. This is true for any transportation megaproject/infrastructure such as a railway line or a new harbor that uses less polluting sources of energy. In the case of a polluting transportation infrastructure (e.g. a new airport), the long-term environmental effect is reversed (new flights will increase air pollutant emissions). However, the presented analysis ignores the air pollution emission at the electricity generation stage, which depends on the energy mix used to generate the required electricity to power the considered transport means. Taking into account the energy mix used to generate electricity can affect the level of air pollution at the operation stage of the transportation megaproject. The greener and less-pollutant sources of energy used in the energy mix, the more environmental-friendly the operation of the transportation megaproject. On this basis, following different loops and causalities, the consequences of a new megaproject on environmental and social sustainability are mainly negative in the short-term (due to an increase in pollution, more local traffic, increase in waste generation) and potentially positive in the long-term, as the negative consequences may be canceled out in the operation phase of the megaproject.

Second, as widely reported in the literature about megaprojects, during the construction phase, strikes and protests from the local population may arise for several reasons [1, 26, 27, 30], including inter alia environmental concerns (typically in the short-term), change in land use and land expropriations, public services interruptions, and corruption (see the bottom-right part of Fig. 1). Basically, the increase in strikes and protests strongly affects the cost of the construction due to induced delays, political debates, or the need to manage external stakeholders [4]. Conflicts with the local population, as well as political debates, should (and could) be properly managed with specific actions and measures, such as an increase in transparency about the megaproject, the engagement of local NGOs and citizens, anti-mafia, and anti-corruption measures. Although there are several stakeholders often with conflicting requirements to manage in a megaproject that makes their engagement challenging [31, 32], stakeholder engagement is a direct approach to improve their satisfaction and resolve potential conflicts [33, 34]. In this vein, stakeholder engagement in the initial stages of the project can reduce the conflict among them in the later stages. In general, the short-term effect of conflicts is a deterioration in social sustainability, which, if protracted over time, can be a reason for hindering or slowing down even other projects that could bring added value and development. At this point then, comes the crucial piece of stakeholder involvement within the decision-making stages of any megaproject.

Third, with regard to innovation and R&D activities and also the local (and national) effect on the development of new innovative technologies, it can be highlighted that the construction of a megaproject positively impacts local R&D in different ways (see top-right part of Fig. 1). The technical requirements for a megaproject construction attract non-local workers with high-level skills, since generally, megaprojects are at the front-end of technological innovation and local workers have not adequate skills and competencies. This human capital attraction should improve both local workers and local businesses through a knowledge-sharing mechanism. Similarly, the requirement for specific and innovative skills and technologies affects local professional training and the involvement of universities in R&D activities. This aspect, typically positively affects local and national economic sustainability both in the short and long term.

Fourth, the construction of new transportation infrastructure can have an impact, positive and negative, on the local culture and tourism both in the short and long term (see the top part of Fig. 1). The short-term effect on the residents may be summarized as an effect of an increase in non-local workers migration[1] and of the increase in tourism. In rural territories, the migration of hundreds, or thousands, of non-local workers may disrupt local traditions and culture, and consequently, affect social sustainability and the conservation of local culture, languages, and/or activities. Similarly, both the construction and the operation phase may increase local tourism, and even international tourism if the operation phase of a large transportation infrastructure similar to the Turin-Lyon high-speed railway between two countries is considered. The outcome in the long term is twofold. On the one hand, the increase in tourism may positively affect local businesses and consequently, social and economic sustainability; on the other hand, mass tourism (as an effect of a large transportation megaproject) may impact local culture and tradition by weakening their conservation and consequently, negatively impact social sustainability.

Finally, in terms of the safety of the workers and local population (see the bottom part of Fig. 1), the construction phase of a megaproject negatively affects the safety and well-being of people. This aspect, if not properly managed through adequate investments or policies, may have a negative impact both on the well-being and wealth of people and on the rise of conflicts, protests, and strikes by the local population and workers. Generally, a lack of safety measures may negatively affect both social sustainability (by reducing well-being and wealth) and economic sustainability (due to an increase in demand in the local and national health system or due to an increase in strikes that have an impact on the cost of the projects).

3.2 Managerial Implications Based on the Control Variables

The general dynamics described in the previous subsection has not to be intended as unavoidable and certain effects. The discussed causalities and their possible impacts on environmental, social, and economic pillars of sustainability have to be properly managed during all life cycle phases of a megaproject by mitigating and reducing the negative effects and improving the positive ones through virtuous cycles and loops. Such virtuous management can be controlled (up to a certain degree) and led, thanks to ad-hoc policies and investments (made by local or national governments and authorities as well as by the companies in charge of the megaproject). In this regard, there are several managerial actions and investments that can be performed. In Fig. 1, a few "control variables" are shown in orange. The control variables represent possible investments or policies to mitigate environmental or social negative impacts.

For instance, by increasing "investments in sustainable R&D", new sustainable, circular, and green technologies may be developed and adopted with a corresponding reduction of the environmental negative impacts (e.g. waste generation, air emissions, water pollution, ...). Such investments may, hence, improve environmental sustainability in the long-term (see delay sign between investment and new technologies

[1] This effect is noteworthy especially for small towns (e.g. in mountain or rural territories). The same is not necessarily true by considering large cities or metropolitan areas.

development), while in the short-term, these investments increase national expenditure. Thus, adopting a long-term or short-term approach to analyze the effect of such changes can affect the decisions made by the decision-makers and consequently, impact the megaproject.

Similarly, strikes and protests of the local population may be reduced by taking ad-hoc measures such as “Anti-mafia and anticorruption measures”, an increase in the transparency of the project, and the engagement of NGOs and citizens. All these control variables have a short-term effect on national expenditure and the costs of the megaproject (e.g. more transparency needs more investments, more public engagement initiatives need more investments, etc.). However, they can reduce the likelihood of future protests and strikes with a corresponding reduction in project costs.

Finally, “investment in the development of local territory” may mitigate the negative effect on local culture and tradition due to mass tourism or the migration of a large number of non-local workers. In this case, the long-term effects pass through the creation of new aggregative and cultural spaces[2] to preserve the local culture and traditions. Such investments may also activate virtuous and positive feedback loops by further stimulating cultural and ethnographic tourism, which consequently supports local businesses and the local economy.

4 Concluding Remarks and Future Research Agenda

Although megaprojects are basically intended to generate positive impacts and improve economic, social, and environmental aspects of human lives toward higher sustainability impacts, the activities conducted at different lifecycle stages of these projects may result in unintended negative effects. Besides, investments or activities at any stage and any part of the project may have diverse impacts on the system in the short and long term. Therefore, a holistic approach is required to analyze the sustainability challenges of these projects during various stages of their lifecycle.

In this article, following the systems thinking approach, a general CLD was developed to address the potential dynamics in the movement toward sustainability in high-speed railway megaprojects. Among the several causalities presented in the model, selected causalities were analyzed and discussed in the article. These causalities highlight the challenges regarding (i) the dynamics in air pollution over time, (ii) strikes and protests causing delays and other issues in the construction phase, (iii) dynamics of the technological innovation due to factors, such as collaboration with universities and knowledge sharing with non-local workers, (iv) positive and negative impacts of tourism on the culture, tradition, and local businesses, and (v) the dynamics in the well-being and satisfaction of the residents and workers linked with the safety of workers and the resident population. The provided analyses targeted both the construction and operation phases of the megaprojects lifecycle and shed light on the significance of considering both the short-term and long-term effects

[2] This can also refer to temporary events such as popular festivals.

of megaprojects activities. Systems thinking is highlighted as a proper approach for analyzing sustainability challenges in this regard.

Based on the limitations and the insight gained in this research, directions for improving the presented model and also potential further investigations into the sustainability of megaprojects can be suggested as follows. First, although the relevant dimensions and indicators to develop the CLD in this research have been derived from previous research [1, 4, 26], the built model can benefit from the opinion of experts and stakeholders to be approved in terms of validity and inclusiveness. Hence, involving experts, both scholars and practitioners, in ad-hoc focus groups or semi-structured interviews [35] to strengthen the causalities among the different indicators could be considered in future research. Focus groups represent a proper approach for validating qualitative CLD as they allow exploring deeper levels of meaning and emphasizing different views and contradictions. Second, the model considers only the construction and operation phases, while not considering the design and planning and the dismantling/end-of-life phases can affect the overall assessment of the activities and hence, impact the decisions made by the authorities and decision-makers. These phases are suggested to be considered in future research to provide a broader view of the effects. Third, the control variables introduced in the model can be further analyzed to study the dynamics they can impose in the system. Fourth, the presented model refers to a subset of megaprojects, i.e. railway line megaprojects (under the transport sector) in a mountain environment. Thus, a few considerations, causalities, and feedback loops may apply only to this specific case. For instance, the impact of mass tourism on small towns cannot be modeled in the same way for a transport megaproject within a large city and urban environment. Similarly, not all megaprojects generate an innovation virtuous loop (e.g. highways that do not need large investments in innovation). It is recommended that a similar approach be adopted for other types of megaprojects to help better and more effective decision-making. Fifth, beyond the triple bottom lines, the model can be enriched with more details to target different capitals, including inter alia social, cultural, and environmental capitals. And finally, the CLD model is suggested to be transformed into a quantitative simulation model using System Dynamics tools to quantitatively depict the behavior and dynamics of the variables over time for specific case studies. In this regard, the model can be used in a few case studies, such as the Turin-Lyon high-speed railway line [14] or the Sagrera Train Station in Barcelona [36, 37].

References

1. Corazza L,Torchia D, Cottafava D, Tipaldo G (2022) Considering the social and economic implications of infrastructure megaprojects: theoretical contributions, practical challenges and managerial implications. In: The Impact of Corporate Social Responsibility, Routledge, pp 134–160. ISBN 9781003182276
2. Flyvbjerg B (2014) What you should know about megaprojects and why: an overview. Proj Manag J 45:6–19. https://doi.org/10.1002/pmj.21409

3. Norese, M.F.; Corazza, L.; Cottafava, D (2021) Environmental Monitoring of the Socio-economic Components of the Impact of a Mega Project. In: Doumpos, M, Ferreira FAF, Zopounidis C, (eds) Multiple Criteria Decision Making for Sustainable Development, pp 131–149. ISBN 9783030892777.
4. Corazza L, Cottafava D, Torchia D (2022) Stakeholder identification and mapping in highly contested megaprojects: interpreting stakeholders ecosystems through relational stakeholder theory. SSRN Electron J1913:0–2. https://doi.org/10.2139/ssrn.4165936
5. Nyarirangwe M, Babatunde OK (2021) Megaproject complexity attributes and competences: lessons from IT and construction projects. Int J Inf Syst Proj Manag 7:77–99. https://doi.org/10.12821/ijispm070404
6. Giezen M (2013) Adaptive and strategic capacity: navigating megaprojects through uncertainty and complexity. Environ Plan B Plan Des 40:723–741. https://doi.org/10.1068/b38184
7. Chapman RJ (2016) A framework for examining the dimensions and characteristics of complexity inherent within rail megaprojects. Int J Proj Manag 34:937–956. https://doi.org/10.1016/j.ijproman.2016.05.001
8. Di Maddaloni F, Davis K (2017) The influence of local community stakeholders in megaprojects: rethinking their inclusiveness to improve project performance. Int J Proj Manag 35:1537–1556. https://doi.org/10.1016/j.ijproman.2017.08.011
9. Greyl, L., Healy, H., Leonardi, E., Temper, L (2012). Stop that train! Ideological conflict and the TAV. Econ. POLICY ENERGY Environ. 193–218. https://doi.org/10.3280/EFE2012-002009
10. Adityanandana M, Gerber J-FJ-FF (2019) Post-growth in the tropics? Contestations over tri hita karana and a tourism megaproject in bali. J. Sustain. Tour. 27:1839–1856 https://doi.org/10.1080/09669582.2019.1666857
11. Esposito G, Terlizzi A, Crutzen N (2022) Policy narratives and megaprojects: the case of the lyon-turin high-speed railway. Public Manag Rev 24:55–79. https://doi.org/10.1080/14719037.2020.1795230
12. Ranjbari M, Morales-Alonso G, Shams Esfandabadi Z, Carrasco-Gallego R (2019) Sustainability and the sharing economy: modelling the interconnections. Dir y Organ 68:33–40. https://doi.org/10.37610/dyo.v0i68.549
13. Castro CG, Trevisan AH, Pigosso DCA, Mascarenhas J (2022) The rebound effect of circular economy: definitions, mechanisms and a research agenda. J Clean Prod 345:131136. https://doi.org/10.1016/j.jclepro.2022.131136
14. Esposito G, Nelson T, Ferlie E, Crutzen N (2021) The institutional shaping of global megaprojects: the case of the lyon-turin high-speed railway. Int J Proj Manag 39:658–671. https://doi.org/10.1016/j.ijproman.2021.06.001
15. Sterman JD (2000) Business dynamics: systems thinking and modeling for a complex world. Isenberg S, Shelstad JJ (ed) ISBN 0072311355.
16. Papachristos G, Struben J (2020) System dynamics methodology and research: opportunities for transitions research. In: Moallemi EA, de Haan FJ, (eds) Modelling Transitions: Virtues, Vices, Visions of the Future, Routledge: New York, pp 119–138. ISBN 978-0-429-05657-4
17. Homer J, Oliva R (2001) Maps and models in system dynamics: a response to coyle. Syst Dyn Rev 17:347–355. https://doi.org/10.1002/sdr.224
18. Struben, J, Chan D, Dubé L (2018) Dynamics of market transformations for sustainability : the challenge of healthy and equitable food. early Draft (Please do not Circ. or cite) 1–50
19. Kump B, Fikar C (2021) Challenges of maintaining and diffusing grassroots innovations in alternative food networks: a systems thinking approach. J Clean Prod 317:128407. https://doi.org/10.1016/j.jclepro.2021.128407
20. Shams Esfandabadi Z, Ravina M, Diana M, Zanetti MC (2020) Conceptualizing environmental effects of carsharing services: a system thinking approach. Sci Total Environ 745:141169. https://doi.org/10.1016/j.scitotenv.2020.141169
21. Struben J, Sterman JD (2008) Transition challenges for alternative fuel vehicle and transportation systems. Environ Plan B Plan Des 35:1070–1097. https://doi.org/10.1068/b33022t

22. MacAskill S, Mostafa S, Stewart RA, Sahin O, Suprun E (2021) Offsite construction supply chain strategies for matching affordable rental housing demand: a system dynamics approach. Sustain Cities Soc 73:103093. https://doi.org/10.1016/j.scs.2021.103093
23. Falagara Sigala I, Sirenko M, Comes T, Kovács G (2022) Mitigating personal protective equipment (PPE) supply chain disruptions in pandemics – a system dynamics approach. Int J Oper Prod Manag 42:128–154. https://doi.org/10.1108/IJOPM-09-2021-0608
24. Liu H, Yu Y, Sun Y, Yan X (2020) A system dynamic approach for simulation of a knowledge transfer model of heterogeneous senders in mega project innovation. Eng Constr Archit Manag 28:681–705. https://doi.org/10.1108/ECAM-01-2020-0077
25. Shams Esfandabadi Z, Ranjbari M, Scagnelli SD (2022) The imbalance of food and biofuel markets amid Ukraine-Russia crisis: a systems thinking perspective. Biofuel Res J 9:1640–1647. https://doi.org/10.18331/BRJ2022.9.2.5
26. Corazza L, Cottafava D, Torchia D (2020) Toward sustainable infrastructural megaprojects. In: Leal Filho W, et al (eds) SDGs in the European Region, Implementing the UN Sustainable Development Goals – Regional Perspectives, Springer, Switzerland (2022). ISBN 9783030912611. https://doi.org/10.1007/978-3-030-91261-1_14-1
27. Zucchetti M et al (2017) The turin-lyon high-speed rail: a technical assessment. Int J Ecosyst Ecol Sci7:141–148
28. Wang G, Wu P, Wu X, Zhang H, Guo Q, Cai Y (2020) Mapping global research on sustainability of megaproject management: a scientometric review. J Clean Prod 259:120831. https://doi.org/10.1016/j.jclepro.2020.120831
29. Invernizzi DC, Locatelli G, Velenturf A, Love PE, Purnell P, Brookes NJ (2020) Developing policies for the end-of-life of energy infrastructure: coming to terms with the challenges of decommissioning. Energy Policy 144:111677. https://doi.org/10.1016/j.enpol.2020.111677
30. Bearfield DA, Dubnick MJ (2009) All mega-projects are local? citizen participation lessons from the big dig. J Public Budgeting Account Financ Manag 21:392–425. https://doi.org/10.1108/JPBAFM-21-03-2009-B003
31. Leung M, Yu J, Liang Q (2013) Improving public engagement in construction development projects from a stakeholder's perspective. J Constr Eng Manag 139. https://doi.org/10.1061/(ASCE)CO.1943-7862.0000754.
32. Lee C, Won JW, Jang W, Jung W, Han SH, Kwak YH (2017) Social conflict management framework for project viability: case studies from Korean megaprojects. Int J Proj Manag 35:1683–1696. https://doi.org/10.1016/j.ijproman.2017.07.011
33. Rowe G, Frewer LJ (2005) A typology of public engagement mechanisms. Sci Technol Hum Values 30:251–290. https://doi.org/10.1177/0162243904271724
34. Ninan J, Phillips I, Sankaran S, Natarajan S (2019) Systems thinking using SSM and TRIZ for stakeholder engagement in infrastructure megaprojects. Systems 2019(7):48. https://doi.org/10.3390/systems7040048.
35. Stewart DW, Shamdasani PN, Rook DW (2015) Group dynamics and focus group research. In: Focus Groups: Theory and Practice, vol 3 (2015)
36. Peters D, Novy J (2012) Train station area development mega-projects in europe: towards a typology. Built Environ 38:12–30. https://doi.org/10.2148/benv.38.1.12
37. Baron N (2019) Who killed barcelona's high speed station promises ? Infrastructure political capture, scaling narratives and the urban fabric. Eur J Geogr 10:6–22